Neurobiology
of Feeding and Nutrition

Neurobiology of Feeding and Nutrition

Jacques Le Magnen

Collège de France
Paris, France

Academic Press, Inc.

Harcourt Brace Jovanovich, Publishers

San Diego   New York   Boston

London   Sydney   Tokyo   Toronto

Academic Press, Inc.
San Diego, California 92101

United Kingdom Edition published by
Academic Press Limited
24–28 Oval Road, London NW1 7DX

Library of Congress Cataloging-in-Publication Data

Le Magnen, Jacques.
    Neurobiology of feeding and nutrition / Jacques Le Magnen.
        p.   cm.
    Includes index.
    ISBN 0-12-443340-5
    1. Hunger--Physiological aspects.   2. Appetite.   3. Neurobiology.
I. Title.
    [DNLM/ 1. Brain--physiology.   2. Feeding Behavior--physiology.
3. Neurobiology.   4. Nutrition.   WL 300 L545n]
QP147.L4   1991
612.3;91--dc20
DNLM/DLC
for Library of Congress   91-18268
                                                                      CIP

PRINTED IN THE UNITED STATES OF AMERICA
91 92 93 94   9 8 7 6 5 4 3 2 1

Contents

Preface

Feeding is one of the most common daily behaviors from birth to death. This behavior is of primary survival value because its ruling mechanisms are incorporated into the overall mechanisms involved in the most important physiological macro-regulation: the regulation of body energy and composition.

In this book (the first to deal with such a broad aspect of problems), brain, sensory, and endocrine involvement in both the behavioral and nutritional regulatory processes are studied in terms of neurobiology.

An initial survey of works on the normal feeding of an animal model, the rat, and of humans, emphasizes the basic periodicity of the behavior and the significance of this periodicity. Two parameters of the cumulative intake, (1) the stimulation to eat and (2) the size of meals and the relationships of these parameters, are identified in three different conditions of feeding, *ad libitum,* induced food deprivation, and scheduled feeding.

The stimulation to eat, or meal initiation, results from a combination of two stimuli: a systemic or metabolic stimulus of hunger arousal and sensory stimulation by the food or "palatability" of that food. This book describes works by the author and his school that led to the discovery of the systemic stimulus of hunger. The brain mechanisms involved in the detection of this stimulus and in the behavioral output are discussed. The sensory stimulation to eat and its state and sensory specificities are examined. Palatabilities, taste preferences, or aversions as unconditioned or conditioned responses are demonstrated. The respective role of oral, gastrointestinal, and systemic factors in determining the size of meal and, therefore, the induction of satiety or satiation, are discussed along with the brain mechanisms involved.

The second half of this book is devoted to the overall regulatory processes and the role of the behavior in these processes. The regulation of body energy balance, body fat mass, and body weight, and the requirements of specific nutrients are successively described in their neural and peripheral mechanisms. Finally, some aspects of the relations between feeding and obesity and the ontogeny of the behavior are delineated.

A new understanding of this major physiological regulation emerges from this analysis of an enormous body of connected works. Throughout this analysis many classical concepts are questioned and revised, and new concepts are proposed.

Because studies on feeding in its regulatory aspect are interdisciplinary in nature, a wide range of various specialized scientists and technologists are interested in this topic. Readers of this book will be distributed among these varied specialists. For some of them, the interest will likely be limited to some chapters.

One difficulty in writing this book was the enormous and widely distributed literature on the subject. About 15,000 articles have been published in the last 40 years on feeding and relevant topics. Of course, a selection of the citations and an analysis of this body of work was needed. This was easy for many redundant works or for others that, after 10 or 20 years and sometimes sooner, were dated.

The writing of this book has been made possible thanks to recent and decisive advances that clarified the complex network of involved mechanisms. A number of these advances come from works of this author and his co-workers and pupils. They will be described in some detail, partly because some of them, initially published in French, are largely unknown. However, complete agreement among scientists has not been reached on many points; many discrepancies remain. Some of these points are very controversial. In these cases, alternative and opposed data and theories will be reviewed and discussed. Finally, gaps in our knowledge—often surprising—do exist. Further investigations are necessary to solve these problems and will be suggested.

Eating is a constraint. In humans, it is, alternatively, a pleasure and an aesthetic activity. The characteristic behavior of humans is to promote all their sensory modalities and all their behaviors up to an esthetic level. Preparing foods—the high cuisine (a specific human behavior)—is a cultural activity. Tasting and eating preferred foods become aesthetics. This aspect will not be forgotten in this book, written in France.

I dedicate this book to my wife, Régine, who as ever, has been my supporter and was helpful in the achievement of this work.

Special thanks are addressed to France Bellisle for her efficient assistance and to others among my ex-co-workers: Paul Laffort, Michel Devos, Christiane Larue-Achagiotis, Arthur Campfield, and Solange de Saint-Font, for their help in preparing the manuscript.

Introduction

Eating is an everyday behavior from birth to death. In animal species and humans, seeking or producing and eating foods to survive is a constraining and highly time-consuming activity. Animals must actively seek and select their foods, an activity that distinguishes the animal kingdom from the vegetal one. In wild conditions, animals spend most of their active time hunting, foraging, and feeding their young and themselves. For human beings, thanks to the division of work in evolved populations, time devoted to food and feeding is fortunately limited; however, a large population around the world works full time in food production, manufacturing, transportation, trade, etc. Nevertheless, many people still spend 2 or 3 hours per day buying, preparing, and eating meals. In general, food and feeding represent at least half of human activities.

Because it is the requisite for individual survival, eating is the most vital need in all living organisms. It also represents a strong selective pressure in the evolution of animal species and differentiations of organs. Before reaching the age of reproduction, animals must survive by, first, being nursed and, later, seeking and finding their foods, while fighting against their predators. Only the best in these performances will survive. Only the best can, through natural selection, transmit and fix the anatomical and functional bases for these successful performances. Presumably, humans emerged from animal species because, originally, they have the best brain to master feeding and to ensure their own safety. It is impossible not to mention here that unfortunately this cruel law of natural selection still operates after millions of years of evolution in populations suffering starvation.

We eat to be fed. Feeding is a "regulatory" behavior. Its mechanisms are closely incorporated in the macroregulatory process of body nutrition.

All living organisms expend (i.e., degrade) energy. These energy expenditures must be matched by acquiring energy from the environment and/or from internal sources of fuel. It is not necessary to detail here the origin and various compartments of these energy requirements: tissue metabolic turnover, heart, muscular and kidney functions, homeothermia, etc. This energy is supplied by cell oxidation and the mitochondrial synthesis of adenosine triphosphate from both oxygen and energy metabolite uptakes. In a steady-state condition, the basal rate of energy output, called (as is well known) the "basal metabolic rate," is assessed by measuring either the oxygen consumption or, directly, the heat production, in the resting and short-term fasting conditions. This basal metabolic rate has been shown to be proportional to the 0.75 power of body weight (I; Chapter 1). Around this mean, cell oxidation varies widely from tissue to tissue. As recalled later, the energy substrates used in various tissues also vary. In the brain, the energy substrate is only glucose. In humans, the brain (2% of the body weight) consumes 25% of the total body oxygen. Added to these basal expenditures, muscular activity, feeding, and cold-induced thermogenesis constitute the total metabolic rate. It may be considerably higher than the basal metabolism and it fluctuates extremely over time. Still, in addition to the basal metabolism, part of the energy requirement is due to energy retentions mainly represented by the constitution or restoration of body energy stores, essentially as body fats. These stores will eventually be used.

To match these energy expenditures and, thus, to achieve a body energy balance, the body cannot use any source of energy. Only materials that provide a metabolizable energy must be selected in the environment. As is well known, those materials, called "foods," are carbohydrates, fats, and proteins. These three macronutrients are interchangeable energy substrates. In addition, their supply corresponds to specific metabolic requirements. In a steady-state condition, a proportion of the supply of each of them is needed. However, this requirement of the distribution of energy intake among the three macronutrients is stamped by some degrees of internal conversions: fat synthesis from glucose, fatty acids substituted for glucose oxidation, and neoglucogenesis from amino acids. A limit in the interchangeability and interconversion is a specific need for essential amino acids and nitrogen and for some other molecules not synthesized by organisms: essential free fatty acids, vitamins, and minerals. Thus, the specific metabolic turnover of these compounds in addition to the energy source must be balanced by a specific amount of their intake.

Humans and, most of all, animal species do not wait for the knowledge of what foods and what amount of them they must eat to achieve their body energy balance and their overall body nutrition. Spontaneously, they eat and do not eat something at any time and in any amount. Physiological mechanisms govern feeding behavior. In studying this behavior, the main question is how and how much these physiological mechanisms participate in the general process of nutritional homeostasis. In the past, and up until now, many studies have neglected this regulatory aspect of the behavior.

Comprehensive studies on feeding in its regulatory function require a measurement both of short- and long-term feeding responses and of parallel metabolic and neural events such as body weight, blood parameters, and neuronal activities in the brain. Multidisciplinary physiological approaches have allowed decisive advances during the last decades toward understanding feeding mechanisms. They will be extensively reviewed in this book.

Most experimental investigations on feeding have been carried out in animal models in laboratory settings, essentially in the rat. The ecological aspect of feeding and, thus, food seeking and foraging of animal species in wildlife conditions will not be examined, nor will these aspects in various vertebrate and invertebrate species be compared. The comparative viewpoint will be limited to some mammal species. The main comparison will be between rats and humans. Although humans are not identical to rats, many basic mechanisms, experimentally identified in the animal model, are found unaltered in humans; however, all striking differences between rat and human feeding behavior will stress the characteristics of the human dimension.

Chapter One

Normal Feeding

In every science, the first step in research is observing and measuring phenomena. The second step is to look for their causal relationships, i.e., determine their mechanisms. Before examining the neurohumoral mechanisms that account for eating behavior, observing and measuring normal spontaneous behavior is required to dissociate its various parameters and to elucidate their relationships.

I. *Basic Concepts: Methods of Observation and of Measurement*

A. Oxygen and Food Intake

Although widely fluctuating in amplitude over time, body heat production is continuous. At the cellular level, particularly in brain cells, this vital continuity immediately depends on a continuous supply of both oxygen and fuel to cells, via the blood dispenser. A sudden failure of the fuel supply to the brain (induced, for example, by an overdose of insulin) is as lethal as anoxia. However, the supply of continuous and required amounts of oxygen and fuel to the blood is very different. In terrestrial animals, the oxygen load of the blood, in the absence of substantial stores, depends on respiratory movements and on their regulation by internal sensors on a short-term basis. On the other hand, the oral intake of food is essentially periodic in nature in all animal species including humans. Continuous body consumption and the need for energy and other metabolites are matched by a discontinuous intake. The causes of this difference between oxygen and food intake in terrestrial animals are self-evident. Air is permanently and immediately available in the environment. Respiration being reflexly organized and an active behavior, it is compatible with sleeping and with all active behaviors in the awake animal. Unfortunately, foods are not so gratuitously available. They must be sought, selected, prepared, and then eaten by humans. This cannot be done of course during sleep, and eating foods and other active behaviors must be organized.

The continuous energy outputs and internal inputs to tissues, opposed to a discontinuous intake, lead to three preliminary conclusions. The first is that, on a short-term basis, the body is in a permanent state of imbalance between output and external input and that the body energy balance can only be achieved over time on a medium- or long-term basis. The second is that a measure of the temporal patterning of feeding is an essential requisite of a comprehensive analysis of the behavior. The third is that the current input of fuel into the blood and then to tissues does not come directly from the intermittent intake but, rather, from internal stores filled by previous or concurrent intakes.

The role of these various stores and the interaction of their respective control mechanisms with the control of intake will be extensively examined later; at this point, however, it will be convenient to recall that three different, either small or large capacity, stores exist. The first one (rarely considered as such) is the stomach. The stomach is a tank exactly comparable to the gasoline tank of a car, filled temporarily and rapidly by the oral intake and slowly emptied via the intestinal absorption to feed the systemic compartment. A role for the filling and emptying of the gastrointestinal store in generating the signals governing initiation and end of intake will be a decisive one. The second and smaller store is the liver and muscle glycogen. However, the liver glycogen content at its highest level is by far lower than the energy content of the stomach loaded by a normal meal. The third and most important reserve of energy metabolites is the body fat mass. This store of energy metabolites (alternately repleted and depleted) plays a major role in the overall body energy balance. It is a true energy ballast by which the short-term energy imbalance is buffered and by which food deprivation and fastings are overcome by the use of the internal fuel. It is of interest to note that triglyceride is the most energy-dense molecule of the body (8 Kcal/g) and permits the highest storage in a minimal weight or load of the body. With 20% of weight as fats, a human of 70 kg bears the equivalent of 50 days of their mean daily energy expenditures as stored energy.

B. Three Conditions of Normal Feeding

Three different conditions of feeding can normally occur. They must be dissociated and compared in feeding studies because their underlying mechanisms differ widely. An animal model, such as the rat, eating freely foods permanently available at all times and in excess is in the so-called "*ad libitum* condition of feeding*." In these conditions of free access, a dual (prandial and diurnal) periodicity of feeding is exhibited (Fig. 1.1). The observation and measure of this feeding pattern may be done either in a short period of time or permanently during 24 hr or more.

Another condition of feeding is the response to food deprivation, fasting, or restriction. The response is recorded in either a short or a longer period at the restitution of free access. The previous food deprivation may be a mild or short-term deprivation, induced by the removal of food for some hours. It may be a 12-hr night or day deprivation. It may be a long-term fasting or starvation, as distinguished from a long-term restriction of food availability. In the short-term deprivation, the time of day is an important parameter. In all conditions of deprivation, the time of the record response also is an important variable. For many reasons (which will be detailed later), the underlying mechanisms of this food deprivation-induced feeding are very different from those of the *ad libitum* feeding condition. The result of any manipulation in a mild or overnight food deprivation cannot be generalized to an effect of the same manipulation performed prior to or after a spontaneous meal in the free-fed animal.

The third condition of experimental observation of a normal feeding is "the feeding schedule." Every day, rats and dogs are offered an access to food during a programmed and limited time or until spontaneous and persistent satiety. These scheduled meals are offered at either fixed or irregular hours, during different times of the day. In such an imposed feeding pattern, a freely determined meal-

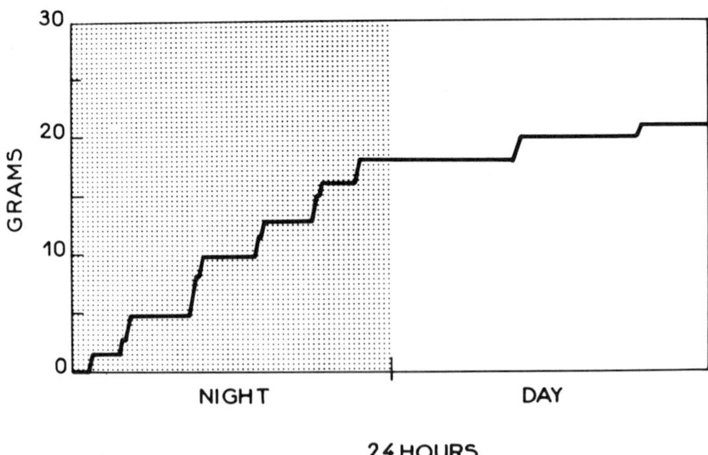

Figure 1.1 Twenty-four-hour free-feeding meal pattern in a normal rat.

to-meal interval is prevented, and only sizes of meals determine the daily cumulative intake. In humans, a feeding schedule of three to four meals at fixed hours during the day is normal, i.e., the more common feeding condition. Its ontogeny will be discussed later, but already it may be noted that such a feeding condition in humans (in which only sizes of the meal can be eventually adjusted to energy expenditures) creates a specific condition of the achievement of body energy balance.

C. The Offered Foods

If all conditions regarding the internal stimulation to eat are equal, eating is a current response to the food. It is a combined response to a series of food properties and to their respective activities at oral, gastrointestinal, and systemic levels before, during, and after eating. These various properties should be, if possible—and it is not always possible—separately controlled and eventually manipulated.

A food possesses various oropharyngeal sensory properties; i.e., it stimulates a series of oropharyngeal sensory receptors. In each sensory modality, intensity and quality of the stimulation by foods are discriminated. These various sensory activities are independent of each other. An olfactory stimulus is generally not a taste stimulus and vice versa. The mechanical properties (liquid or solid) and texture of the latter are important variables because they determine the motor performance of eating: licking, chewing, and swallowing.

Another series of food parameters is weight, volume, and weight : volume ratio, i.e., food density. They are crucial factors and are taken into consideration at the gastrointestinal level as parameters of the stomach content. The amount eaten is classically expressed in terms of weight of food eaten. But inasmuch as this food supplies calories and, for the purpose of feeding studies, is used to assess a caloric control of food intake, the caloric content and caloric density of the food are the major parameters to be controlled or manipulated. Then, the amount eaten or stomach content is expressed in terms of amount of calories derived theoretically from the various energy metabolites present in the food. These energetic properties of foods are, or are not, correlated to their sensory properties. Sweetness is or is not correlated to calories according to the sweetener. Sweet intensity is never correlated to caloric density. The smell of food is not related to its caloric property.

The third series of food properties to be controlled and possibly manipulated is, of course, its composition in the three macronutrients—carbohydrates, fats, and proteins—and also in other metabolites, vitamins, and minerals. The food is either a low- or high-carbohydrate diet, a low- or high-fat diet, and a low- or high-protein diet, with numerous possibilities of mixtures. However, these possibilities of varying nutritive properties are limited in that, for example, a high-fat diet is necessarily a low-carbohydrate diet. The relation between the nutrient content and the caloric density and, through this relation, the possibility to study the respective roles of calories and of the specific supplier of these calories are difficult to manage. High- and low-fat diets may or may not be made isocaloric. Another difficulty is that changing the caloric and nutrient content of the food also changes its sensory properties. The high-fat diet is also highly palatable. Caloric dilution by adding inert material to the food alters its sensory activity. To discriminate between the roles of sensory and nutritive postingestive properties in the control of intakes, a method of choice is the use of a flavored, more exactly "odorized" diet, the same food being labeled in various versions by traces of odorants or noncaloric sweeteners.

Finally, in models of normal feeding and experimental feeding behavior, an important difference exists between the single and multiple presentations of foods. In the first condition, a nutritionally complete or imbalanced diet, called "stock" or "chow diet," familiar or not, is offered. Then, the spontaneous selection of foods by the animal or human subjects is excluded. This single feeding is mainly an experimental condition of feeding. It is not realized by animal species in their natural environment. It is the opposite of normal human food habits, in which variety and choice within meal and from meal-to-meal are the rule. The multiple-food presentation is of different modalities. It may be the successive presentation within a feeding episode of a series of foods differing only by their sensory properties (flavored foods) or differing by both sensory and nutritive properties. These two methods demonstrate the phenomenon of "sensory-specific satiation" (discussed later). A similar condition is called "the cafeteria regimen." High-palatability foods of varied nutritive properties are permanently and simultaneously presented. It is similar to the preceding model in that the rats are shown to sample successively the different food items in huge meals. This regimen, which induces hyperphagia in rats, is a good animal model of human meal eating.

Another and extensively used method of observing feeding behavior is the choice between two items: Two solutions or two flavored foods are offered and generally intended to test food preferences or aversions. The very short-term response, often called "the brief exposure technique," is distinguished from the respective intake on the two items until satiety. Finally, free access to a simultaneous and permanent presentation of the three macronutrients is a classic method fruitfully compared to the free access to a single complete diet.

D. Observations and Measurements of Feeding in the Steady State of Energy Demand

Significant information is drawn from human and animal subjects through procedures in which the feeding response is observed when an energy supply is added or substituted for the oral intake via gastric, intestinal, or intravenous routes. Gastric loads preceding a spontaneous meal or prefeeding, which interferes with the satiating process or satiation, must be distinguished from the same acute loads given long before or after a meal, which interferes with "satiety" and the initiation of a subsequent meal. Via the intraportal and the intravenous routes, the significance of results also is different according to the time of infusions relative to the meal. The procedures of enteral and parenteral feeding in animal models and in human patients are highly informative about the relationships between energy demands, the substrates of energy supply, and the control of feeding. In the following chapters, mention will be made of other procedures in which an artificially modified condition of the behavior is observed. It is the case of sham-feeding compared to real feeding, of self-intragastric or intravenous feeding. It is also the case of the old and now-abandoned procedure of operant eating of food by lever-pressing.

E. Measurements of Food Intake

The various available techniques presently used to measure feeding responses in the preceding conditions of feeding will be detailed later simultaneously with experiments and results. The essential measure is, of course, that of food intake and particularly of the feeding pattern. Surprisingly enough, this self-evident requisite is often overlooked. Very rough techniques of measuring feeding responses are still used, in contrast to the sophisticated techniques of brain microinjections of immunocytochemical determinations of neuronal pathways, etc. One such technique is the measure of the cumulative intake during some hours (12 or 24 hr) by the manual weighing of rests in the food cup. To understand the respective mechanisms involved at peripheral and brain levels (1) in meal initiation or stimulation to eat, (2) in satiety or no stimulation to eat, and (3) in the satiating effect of eating a food that determines meal size, recording the meal pattern is absolutely necessary. A recording of cumulative intake provides almost no insight on these mechanisms, regardless of the experimental manipulation. Moreover, taking into account the total contrast of the feeding pattern and of its metabolic background during nocturnal and diurnal periods, a manipulation and a measure performed during the day (as is generally the case) is not interpretable if not compared with the same measure performed at night. As early as the 1960s, Le Magnen and Tallon (2) developed techniques for an automatic recording of the free-feeding pattern of rats. Surprisingly, the use of such a technique has not yet been generalized after more than 30 yr.

Measures of blood and metabolic parameters and of brain events associated with the measures of normal or altered feeding patterns will be described in relevant chapters.

II. Ad Libitum *Feeding Patterns in an Animal Model: The Rat*

A quantitative recording of the temporal pattern of eating a permanently available and familiar food by rats exhibits the essential feature of feeding behavior—its periodicity. The rat, like other animals and humans, does not eat continuously; it is not a nibbler. It comes to the food cup periodically. It initiates an episode of eating defined as "a meal." After eating various amounts, it terminates the meal and remains, without eating the available food during a given time before initiating a new meal. This sequence of meal initiations and terminations and of meal-to-meal intervals (MMI) is called "the meal pattern" of the rat. The first examination of this pattern reveals the two basic parameters of feeding: the size of meals and their frequency. The cumulative intake within a period longer than the occurrence of these short-term events, i.e., in a succession of meals, is, of course, the product of the mean meal size (MS) by meal number. Recording such meal patterns and taking into consideration the two parameters of MS and MMI were a major step in the progress of feeding studies for the last three decades. This allowed experimenters to raise various questions and to investigate their answers. What event causes the rat to initiate a meal and, in so doing, to end the no-eating interval since the end of the preceding meal? What causes the rat to eat a particular amount of food up to a persistent stop? Are MMI and MS related to each other? If these two parameters and their relationships vary over time or in various circumstances, what is the origin and significance of such variations?

Historically, the first recordings of *ad libitum* meal patterns in rats by the achievement of appropriate techniques and analysis of these patterns were undertaken in the early 1960s (2). Since then, these pioneering recording techniques have been improved and their analysis refined.

A. *Ad Libitum* Meal Pattern in Steady-State Conditions

As already mentioned, a dual periodicity of feeding is apparent in the free-feeding condition. The rat takes meals separated by intervals of no eating throughout the day—the prandial periodicity (Fig. 1.1; Section I). It takes large meals separated by short intervals during the night and small meals separated by long intervals during the day. This is the diurnal, or circadian, periodicity superimposed on the prandial periodicity. As a result of large meals and high-meal frequency at night and small meals at low frequency during the day, 70–90% of the 24-hr intake occurs at night. The night : day intake ratio in rats maintained in a 12-hr night : 12-hr day cycle expresses the diurnal periodicity. Inasmuch as this diurnal periodicity results from a change of the prandial one during the two periods, their respective characteristics will be examined here simultaneously.

The following description and analysis are based on an extensive work of 200 daily recordings of meal pattern (20 consecutive days in 10 rats), which completed and, in some cases, revised many of the earlier works by Le Magnen and Devos (3).

During the 20 consecutive days, rats ate a very constant interindividually and almost identical daily amount of food. No day-to-day periodicity was observed. A highly significant negative correlation was observed between the nocturnal and subsequent diurnal intakes as well as between diurnal and subsequent nocturnal intakes (Fig. 1.2). The high significance of these compensations between subsequent nocturnal and diurnal periods will be discussed later in terms of underlying mechanisms.

Before an analysis of meal pattern, a study of the successive 60-min

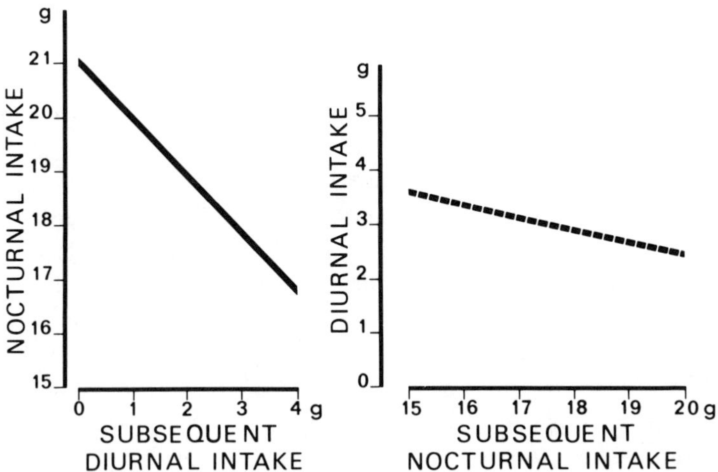

Figure 1.2 Correlations of food consumption between night and subsequent day and between day and subsequent night. Left, r = −0.51 (N = 200); right, r = −0.47 (N = 190).

cumulative intake throughout the day reveals typical features. At night, an initial and moderate peak during the second and third hours was followed by a plateau during the following 4 hr. Then, the hourly intake progressively increased until it reached a peak at the eleventh hour. At this time, the intake reached about twice the intake of the first hour. This intake peak at the end of the night, achieved by the largest meal occurring at this time, was studied by Strubbe *et al.* (4), who showed that the removal of the food during the last 2 hr of the night does not induce a peak during the hours preceding the removal. Instead, the lacking intake is exactly compensated for by an increase of subsequent diurnal intake. This demonstrates that the intake at the end of the night is an anticipatory response and contributes to the normal low intake during the day hours. Consistent with these observations and interpretations, the gastrointestinal content of food was shown to be, in rat, at the highest level at the end of the night (5). The low diurnal intake and the prolonged MMIs by which it is achieved are entirely explained by a storage of energy metabolites during the preceding night, as will be seen later. When, as in the Strubbe experiments (4), the night is curtailed to 10 hr and the day is correspondingly prolonged, the anticipatory peak of intake reappears at the end of the curtailed night. It is one among many other anticipatory responses that will be found, particularly in the food deprivation-induced feeding.

During the day, the rat is almost aphagic during the first 4–6 hr. Almost all the diurnal intake occurs during the last 6 hr.

B. What Is a Meal?

The study of feeding in terms of meal pattern depended on the definition of a meal. The food taking of rats within an episode of eating is not continuous. Graphic recordings show pauses between continuous bouts. It was decisive to know what interruption could be considered as the end of the meal. The choice of criteria of MS and MMI was made somewhat arbitrarily by investigators, who used anywhere from 5 to 40 (or more) min of interruption to indicate when a meal is terminated and another meal initiated after this time. We used the criterion of 40 min and validated this time by a statistical analysis of durations of all interruptions of eating >2 min throughout the day in 200 recordings. The percentage of eating interruptions lasting 10–19, 20–29, 30–39, 40–49, 50–59, and >60 min is illustrated in Fig. 1.3. Apparently, the probability that rats resume eating within 10 min after stopping for 2 min is maximal, decreasing from 10 to 40 min, being minimal at this time, and increasing again for longer times. In other words, when a rat stops for 2 min, the probability is maximal for starting eating again by far less or by far more than 40 min. Clear evidence

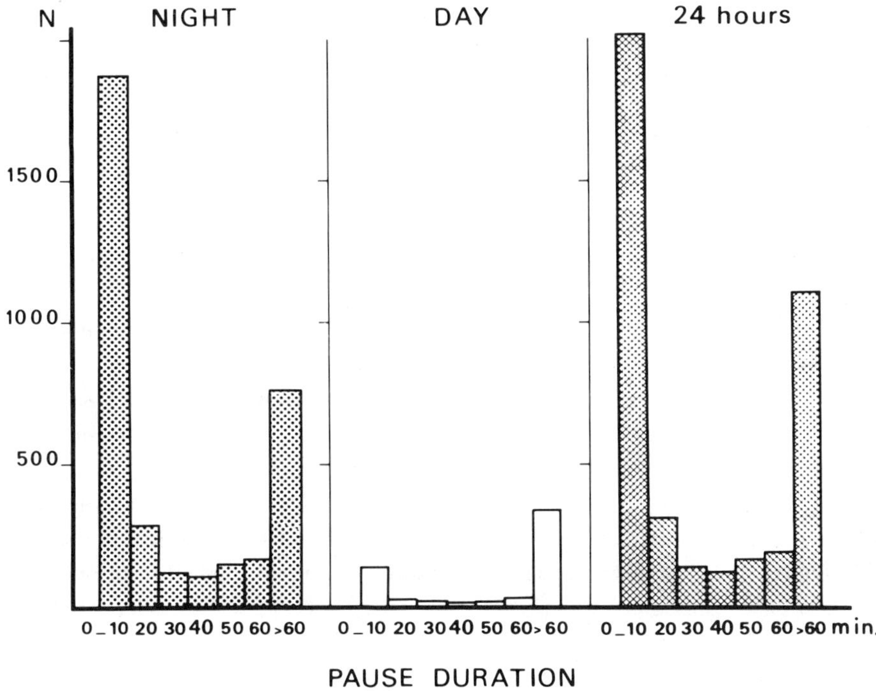

Figure 1.3 Frequency distribution of pauses between bouts of eating >2 min.

indicates that 40 min is a cutting point separating pauses <41 min within meals and MMIs lasting >41 min. A meal is thus defined as an eating episode initiated after at least 40 min, elapsed since the end of the preceding one and separated from the subsequent meal initiation by at least 40 min.

Almost identical results were found by Kissileff (6) using the slightly different eatometer technique. In this analysis, the percentage of pauses <10 min was also higher than that of 10–19, 20–29, and 30–39 min. A low 5-min criterion combined in one meal 6.5% of successive eating bouts. A 40-min criterion combined in one meal 21.5% of eating bouts. Other results validated the 30–41-min interval as a definition of meals. Signs of behavioral satiety (rest and sleep) appear 41 min after the end of a continuous or shortly interrupted feeding episode (7).

MS and MMI being so defined, meal pattern analysis revealed that, at night, rats spend 26.4% of the time eating the successive meals and 73.6% of the time not eating. By contrast, during day, only 3.6% of the time was spent eating meals and 96.4% not eating. The mean MS is 22 gr at night and 1.04 gr during the day. As indicated by the hourly pattern, the last nocturnal meal is the highest one: twice as large as the first one (Fig. 1.4).

C. Intake Rates

The amount of food eaten from the beginning to the end of a meal, plotted against meal duration, indicates an almost identical intake rate of 0.05 g/min during the night and 0.11 g/min during the day (Fig. 1.5). When intrameal pauses are subtracted, the actual eating speed irrespective of meal sizes was, in average, 0.2 g/min at night and 0.18 g/min during the day. In another study, this intake rate was shown to be perfectly constant from the beginning to the end of meals, except for a slight acceleration in the first 2–3 min. Therefore, in every meal, 50% of the food eaten throughout the meal is eaten in 50% of meal duration. This monotonic intake rate is an important characteristic of the *ad libitum* intake. As described below, an initial high intake rate, increasing with previous food deprivation, will be, by contrast, one of the main features of the deprivation-induced meal pattern. In rats offered a liquid food, licking rate also increases as a function of both previous food deprivation and the sensory stimula-

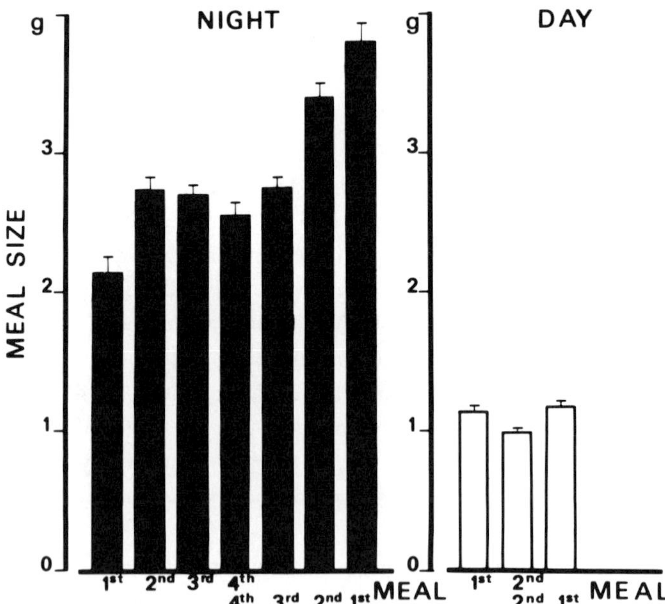

Figure 1.4 Average sizes of meals in their temporal order during the nocturnal and diurnal periods.

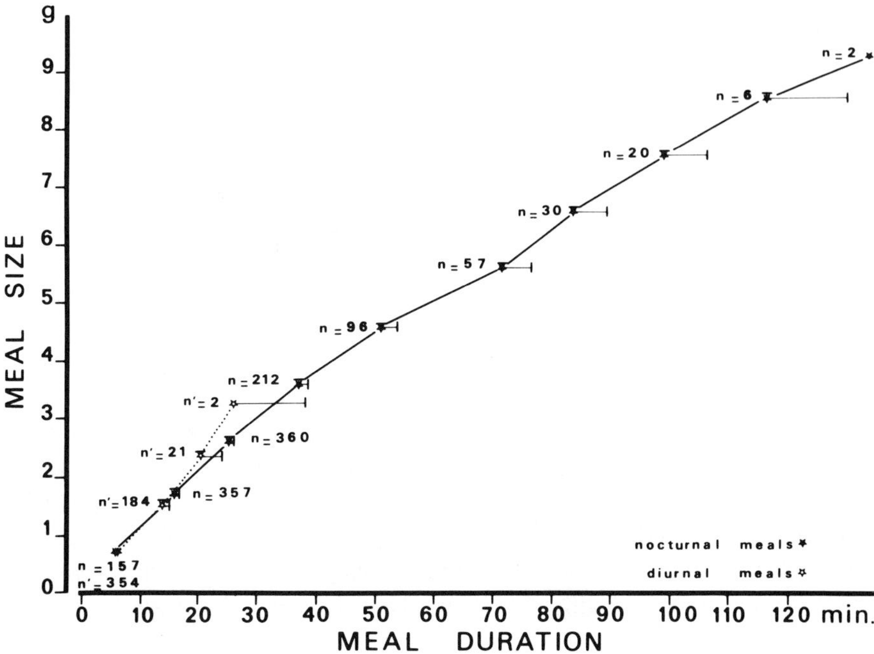

Figure 1.5 Meal size–meal duration relationship: intake rate.

tion to eat (or palatability) of the offered solution (8, 9). This initial rate of eating a solid or liquid food will be taken, among other characteristics of the micro-structure of the meal, as a measure of the strength of stimulation to eat at the beginning of the feeding episode. The fact that this rate is not affected by premeal intervals and their differences during night and day, and does not vary with MS in the *ad libitum* condition, is the first indication that in this condition the strength of the stimulation to eat at meal onset is independent of previous feeding and metabolic events.

D. Meal Size and Meal-to-Meal Interval Relationships

As soon as a recorded meal pattern was available, researchers were interested in examining a possible relationship between MSs and the time elapsed before their onset since the preceding meal and after its end until the onset of a new meal. As early as 1963, it was reported that no correlation existed between the MS and the duration of premeal interval, during either night or day. Later, this lack of a "preprandial correlation" was universally confirmed by other investigators in all animal species studied. This indicates that in *ad libitum* conditions the amount eaten from the start to the spontaneous end of a meal is independent of the time elapsed since the preceding meal. By contrast, during the night—and only during this period—a high correlation was shown to exist between the MSs and the time elapsed from their onset to the onset of the subsequent meal. The time of meal onset, which determines the end of the premeal interval duration, was thus found to depend on the size of the preceding meal. Consequently, the size of a meal, independent of the premeal interval, is also independent of the preceding meal intake. During the day, this postprandial correlation is either absent or of low statistical significance. Both the MSs and the time of their onset are therefore independent of both preceding and subsequent events during this period.

Studies of this postprandial correlation at night have been widely confirmed since its earlier (and somewhat rough) statistical analysis (2) by its detailed reexamination in the 200 recordings of meal patterns. At night, the postprandial positive correlation is highly significant at the 0.01 level in all rats. Regression lines of these correlations during the two periods are illustrated in Fig. 1.6. The different slopes of the regression lines in night and day provide another indication of the different relationships between MS and postmeal inter-

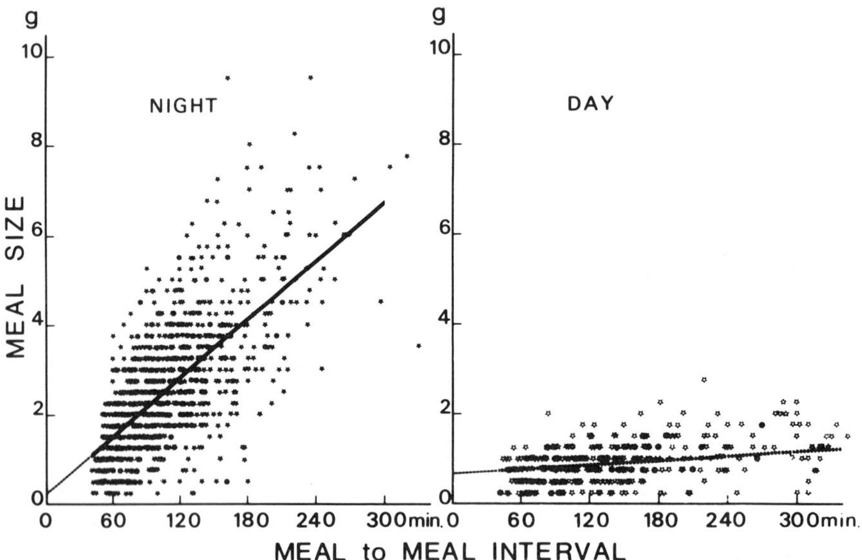

Figure 1.6 Postprandial correlation in normal rat. Night, r = 0.67 (N = 1097); day, r = 0.38 (N = 383).

vals for the two parts of the diurnal cycle. While the slope of the regression line at night is lower than unity, the MS:MMI ratio (later designated the onset ratio) increases with increasing MS. When MSs are plotted against this ratio of the two parameters, some interesting facts appear. At night, this ratio (expressed in calories eaten in a meal to the postprandial interval in minutes) reaches a value of 50 or more for meals >1 g. It increases to a very high value for meals >5 g. Thus, a relative curtailing of the time elapsed after a meal and, thus, a precipitated onset of the subsequent meal occurs after big meals. Meals, $\geqq2$ g being considered separately, it appears that the postprandial correlation was more significant with larger meals. This law of null postprandial correlation after small meals was also found by Levitsky (10).

No difference was found between the level of postprandial correlations and the place of the meal within 12 nocturnal hr. However, when the correlation was separately computed for the succession of meals during the first and last 6 hr of the period, the duration of the postmeal interval (also called postprandial satiety) was shown to be shorter relative to the MS than at the beginning of the night (Fig. 1.7). Later, this fact will be interpreted as an effect of the high rate of fat synthesis and of food utilization from meal to meal at the beginning of the night. The first and last 6 hr of the night also differ when rats are offered a choice of the three macronutrients. During the first 6 hr, they select mainly carbohydrates (70% of the intake in the first meal) and during the end of the night fats and proteins (11).

Postprandial correlations have been compared when calculated from meals separated by criterion of either 10, 30, 40, or 60 min. In all rats, the magnitude of the correlation increases strongly with 10–30-min criterion and slightly with 30–40-min criterion. Furthermore, after a pause of <30–40 min, a meal is not terminated; this validates again the meal definition by a criterion of 40 min.

The presence of this relationship between the amount eaten in the meal and the subsequent time elapsed before the initiation of a new meal in rats was confirmed by many investigators. It was a basic phenomenon in investigating the mechanisms of stimulation to eat or not to eat, i.e., of hunger and satiety (12–19).

However, some authors failed to find the correlation (20). In all cases, this failure was due either to the computation from all nocturnal and diurnal meals or to the use of a low criterion in the definition of meals and MMIs.

A same correlation between MSs and postprandial satiety was found in various other animal species: hummingbirds (21), fowl (22), geese (23), dogs (24), rabbits (25, 26), and monkeys (27). It was apparently absent in cats (28), pigs (29), and guinea pigs (30). We will see later that it is present in human when placed in a free-running condition out of a habitual feeding schedule.

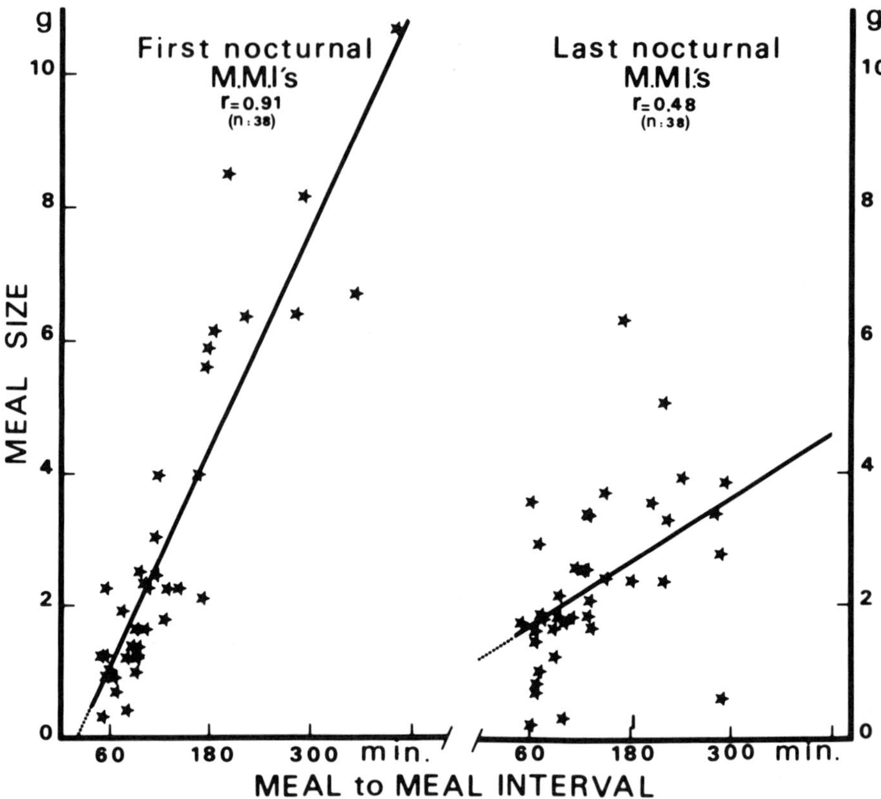

Figure 1.7 Difference of the postprandial correlation at the beginning and end of the nocturnal periods.

E. Onset and Satiety Ratios

While the dark : light phase ratio expresses the diurnal periodicity, the MS : postmeal interval ratio, as a consequence of the presence of a postprandial correlation, expresses the prandial periodicity of free feeding. We proposed earlier to express this index in calories/minute, i.e., the number of food calories eaten in the meal divided by the time separating the onset of the meal from the onset of the subsequent one in minutes. This index allows a direct comparison between the meal-to-meal rate of the caloric intake and the metabolic rate. It was proposed to call this index the onset ratio, indicating that this ratio provides a measure of the time at which the stimulation to start a meal appears following a meal of a given size. According to our earlier suggestion, another index of the

MS–postmeal interval relationship was used by many investigators under the name of satiety ratio. Unfortunately, they have markedly altered the initial notion by using the ratio of the time from the end of a meal to the start of the following one divided by the size in grams of the first of these two meals.

Mean onset ratios were found to be 74.5 ± 0.96 at night and 15.51 ± 0.77 during the day. The marked difference of these mean indices between night and day clearly confirmed the finding that during the day rats remained satiated on average three times as long as at night after an identical caloric intake.

Important conclusions may be drawn from the observation of MS–MMI relationships.

1. Two distinct mechanisms are in action in determining the free-feeding pattern. One of them is involved in meal initiation and its timing and, therefore, in the satiety duration that precedes the meal. Another mechanism determines MS and is not affected by either the premeal or postmeal satiety durations.

2. This dual mechanism, clearly in action at night, is not evident during the day. The absence of a preprandial correlation and the presence of a poor postprandial correlation in this period suggests that meal initiation and its timing do not solely depend on the previous intake nor on the same factors involved at night. However, as during the night, the MS is independent of both premeal and postmeal events. These suggestions were fully confirmed by the finding of the neurohumoral stimuli involved, respectively, in the prandial and the diurnal periodicities of feeding (see Chapter 2).

F. Chewing–Swallowing Patterns in Animals

Recordings of the above-mentioned food intake are those of rats taking food from a food cup. Attempts at recording the actual eating patterns of chewing and swallowing in animal models were generally unsuccessful. However, we must not forget or ignore that eating is taking a food in the mouth, lapping or sucking fluids and soft foods, and chewing hard foods before swallowing.

These motor patterns of mandibular and tongue movements were studied in great detail in various animal species using conventional or X-ray cinematography and other techniques. Their neurophysiological mechanisms, i.e., the motor aspect of the overall neural mechanism of feeding, will be examined in Chapter 7, devoted to "the meal." In animals such as rats and cats, after the initial biting of a solid food by incisors, food particles are transported by tongue movements in two phases. In the first one, food pieces are transported from the anterior part of the mouth to the postcanine teeth and posterior surface of the tongue, where hard foods are chewed at a rate of 4–6 cycles/sec (31). The second phase is the transport from the postcanine teeth to the back of the tongue, where the salivated and masticated bolus is ready for swallowing. The condition of this bolus to reach the threshold of swallowing was also studied in detail, and those interested in this specific literature should consult the extensive reviews and articles by Hiiemae *et al.* (32).

G. Intragastric and Intravenous Self-Feeding

Rats trained to press a lever to get a liquid food in the stomach slowly learned to press to obtain the same amount of calories taken via the mouth. When the liquid diet was diluted, rats adjusted their lever-pressing but failed to perform this adjustment with accuracy (33). At first the authors concluded that rats could self-feed intragastrically without oropharyngeal stimulation. However, they showed that oral stimulation by saccharin, concurrent with the gastric self-feeding, energized the lever-pressing and that indeed oral cues were a requisite of the learned self-administration. On the other hand, it was reported that the nasopharyngeal catheter, through which the liquid food was self-injected, in fact provided contingent oral cues. These oral cues being eliminated, rats no longer learned to press for intragastric feeding (34).

Rats were shown also to be able to learn intravenous self-feeding. However, they succeeded to do so only after a considerable weight loss (-40%) (35). At this extreme level of leanness, rats succeeded in taking a small amount of calories intravenously, sufficient to maintain their weight at this low level, but they failed to recover their initial weight.

These two experiments of intragastric and intravenous feeding argue in favor of the notion that the orosensory control, on the one hand, and the filling of the gastrointestinal tract by eating a food, on the other, are critical determinants of a normal regulatory feeding pattern.

III. *Food Deprivation-Induced Feeding*

A. Food Deprivation and Hunger

Extrapolating from humans to rats, many investigators in the past thought that food deprivation in the rat makes it "hungry." Consequently, they thought that various behavioral changes during food deprivation, other than the response to foods, could be taken as symptoms or measures of what they called "motivation to eat" or "hunger" and of its intensity.

Varner (36) used the obstruction method. Rats were obliged to get their food by passing an electrical grill after increasing time of food deprivation. The number of passages increased progressively up to 4 days of fasting in males and decreased afterward; this fall occurs sooner in females (27 hr). Another obstruction procedure was used by Miller (37). The increasing concentration of quinine needed to block eating in increasing states of deprivation was proposed as a measure of hunger and of the time course of its evolution. Heron and Skinner (38) took the bar-pressing for foods as an indicator of the so-called motivation to eat. Rats deprived of foods were submitted daily to a 1-hr session of bar-pressing for food in a variable schedule (one small pellet every 2 min). The rate of bar-pressing increased day-to-day, parallel with an increase of general activity, and then suddenly fell. The authors concluded that under food deprivation, hunger in rats increased over time according to a characteristic curve exhibited by the evolution of their bar-pressing for foods. Such a conclusion was based on the postulate that the instrumental response rewarded by the delivery of food pellets was a more reliable measure than that of the free intake itself.

Beyond these pioneering works, an ambiguity remains about the notion that either a time of food deprivation or the amount eaten in response to these deprivations is a reliable measure of the overall systemic and sensory stimulation to eat. As described in the following section, the response to an overnight fasting is quite different from the response to a daytime 12-hr fast. A strong increase of the subsequent intake is experienced after 3 hr of food removal at night; it is null at the beginning of the day. In a defined deprivation regarding the time of its occurrence, the recorded feeding response may vary widely. It differs according to the offered food. It also differs when recorded either during the first meal only or during some hours at the restoration of food access. In fact, neither the time of food deprivation nor the responses *per se* can be taken as measures of the stimulation to eat. Only the measurement of identified sources of this stimulation will be a true measure of what is called "hunger" and "appetites."

B. Short-Term Responses to Food Deprivation

The response to short-term food deprivation up to 12 hr strikingly differs from responses to 24 or more hr of fasting. Many conclusions in the literature result from a neglect of these differences.

First, what is a food deprivation condition compared to an *ad libitum* condition? It seems that the augmentation of the spontaneous premeal interval by half of its duration through food removal is sufficient to change the shape and size of the first subsequent meal typical of the food deprivation-induced intake (39). Thus, taking into account that the MMI is two to three times longer during the day than at night, the required food removal is also two to three times longer in the former diurnal condition. However, experimenting with responses to short-term or mild food deprivation always requires a comparison with the *ad libitum* pattern of controls at the same time of day.

Contrasting with earlier works limited to studies of responses to 24-hr (or longer) fasting periods, Bare and Hunger (39) and Bare and Cicala (40) were the first to study the basic phenomenon of feeding after some hours of deprivation. In six groups of rats, responses to 2, 4, 6, 12, 18, and 24 hr of food removal beginning at 6 A.M. were recorded. Lever-pressing for food, not free-food intake, was used. Results gave evidence of the role of the diurnal cycle. However, the fact that the time of both deprivations and testing overlapped or not according to the group, the dark or light period, obscured the result. In subsequent experiments, the authors attempted to dissociate the time of deprivation and of phases of the cycle in which it occurs from the timing of refeeding. Groups of rats were deprived from 0 to 24 hr and then tested during the 24 hr beginning either at night or during the day. Results again gave evidence of an interaction among these three factors. Bellinger and Mendel (41) observed that the 2-hr responses to deprivation were higher at night than during the day, regardless of the duration and locations of the deprivation in the cycle. The nocturnal response was 30–60% at night in all conditions regarding the deprivation. However, the overlap of food deprivation time in the two parts of the diurnal cycle again somewhat obscures the results and their interpretation in terms of mechanisms. Tagliaferro and Levitsky (42) tested during a 24-hr period the responses to 1.5, 2, 4, 6, and 8-hr food deprivations only during the night. At the restoration to food access, they showed an increase of the first meal as a function of deprivation duration. After the longest deprivation, rats continued to overeat during the subsequent day, and in the 24 hr, the cumulative intake increased by 21–50% compared to that of *ad libitum* rats.

Clearly, it was necessary to compare responses to <12 hr of deprivation occurring either at night or during the day and responses tested during the same period. In a series of experiments, we used such procedures to study the short-term food deprivation-induced feeding. The main purpose of these experiments was to identify the metabolic correlates of recorded responses. Here we will examine only the shape and temporal patterning of the responses.

In the first experiment (43), feeding responses to 4–10 hr of deprivation starting at the beginning of the period was tested at night in one group and during the day in another group. A food cup weighing device recorded the meal pattern at the restoration of food access. Closing and opening access to the food cup were automatically monitored. Each deprivation time was repeated three times in random order. In addition, the *ad libitum* meal patterns during two subsequent 12-hr periods were also recorded.

At night, the size of the first meal after the restoration of food access increased as a function of the time elapsed since the last diurnal meal. A significant linear correlation exists between the size of these first meals and the premeal fast duration (Fig. 1.8). Each hour of fasting added 0.85 g to the first meal. During the day, the same increase of the first meal as a function of fasting duration was also observed. However, the slope of the regression line was markedly lower than at night, indicating a lower efficacy of the same deprivation for increasing the size of the first meal. Each hour of previous fasting added only 0.33 g to the first meal. During the dark period beyond the high first meal, the

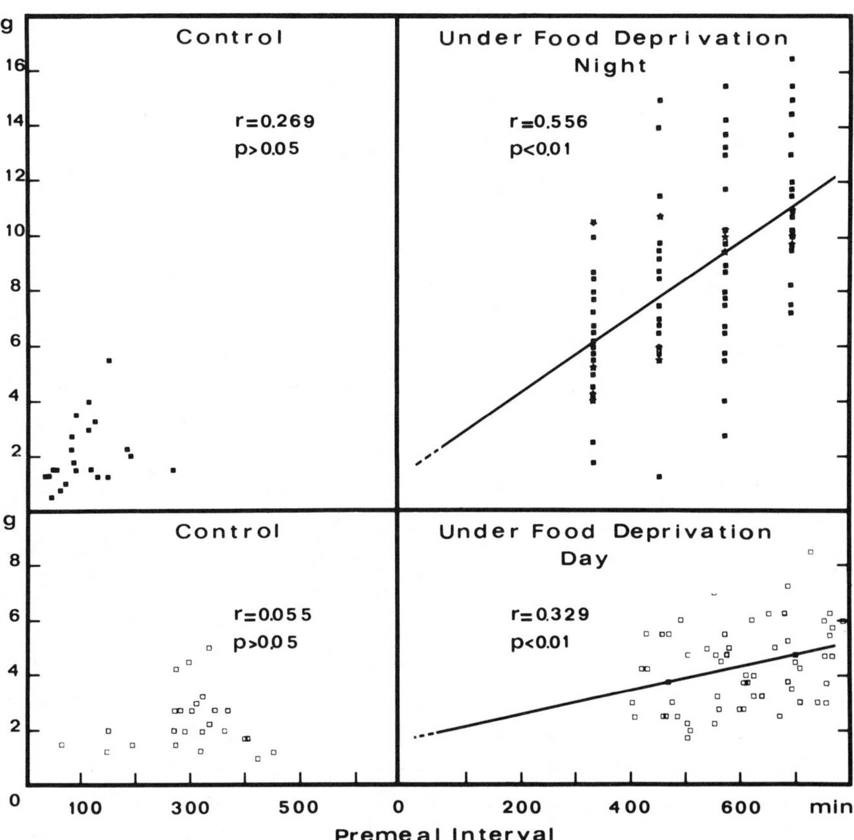

Figure 1.8 Correlation between durations of previous short-term food deprivation and the size of the first meal at the restoration of food access.

previous deprivation induced a typical change in the subsequent meal pattern until the end of the period. The successive meals as well as the postmeal durations were decreasing but, as soon as the first meal, the duration of postmeal interval relative to the size of the preceding meal was strongly shortened (Fig. 1.9). Consequently the OR of the second meal was by far higher than the mean OR at night for such big meals in the free-feeding condition. In other words, food deprivation induced not only an augmented first meal but also a subsequent precipitated temporal patterning of meal initiations. It will be seen further that both the size of the first meal and the shortening of postmeal satiety will be correlated to and predictable from blood glucose and free fatty acid levels at the time of restoration of food access after the various deprivations. Due to the precipitated meal patterning, the nocturnal *ad libitum* cumulative intake was restored during the remaining hours of refeeding during the period. During the light period, the *ad libitum* intake was also restored after 4–10 hr of deprivation (Fig. 1.8 and 1.9). In other experiments, the striking differences between the response of deprivation at the beginning of the night and of the day was emphasized. At night, 2–6 hr of deprivation had a strong promoting effect on the subsequent 6-hr intake at the end of the period. This effect was almost null during the day (44).

Implicitly or explicitly, these different responses to food deprivation during the two parts of the diurnal cycle were attributed to darkness or lightness *per se*, excluding a role for metabolic differences and their impact on feeding mechanisms. This suggestion was ruled out by the following experiment. Rats were tested for their responses to 6 hr of deprivation during the night and the day. Their responses were compared with the effect of the same time of deprivation occurring during darkness prolonged during the previous light period and during lightness prolonged on the previous nocturnal period. In *ad libitum*-fed rats, the noctural and diurnal intakes were not significantly changed in a 24-hr prolonged night or day. The 6-hr intake after 6 hr of deprivation during the prolonged light period (preceding the night) was three times higher than the intake after the same deprivation during the prolonged night (preceding the day). Both responses were not significantly different from those observed during a normal night or day.

As mentioned earlier, not only the size but also the shape of meals is altered after food deprivation. The initial rate of eating in the first meal was shown to increase as a function of the length of the previous deprivation (45). (We will discuss this finding later.) This effect was maximally observed in the day response to 12-hr nocturnal deprivation (Fig. 1.10). Acutely, this response begins with a huge meal of 6–8 g in which a high initial rate followed by a decreasing rate causes 65% or more of the total intake consumed during the first

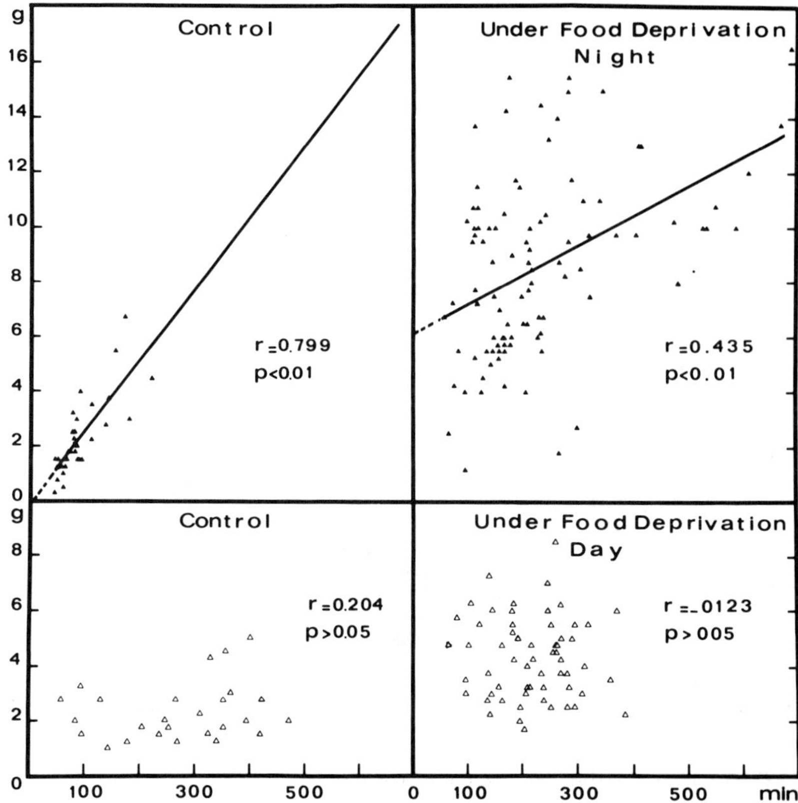

Figure 1.9 Postprandial correlations following short-term food deprivation.

50% of the eating time. When this condition of day feeding was chronically maintained, a normal meal pattern identical to that present at night was progressively established (46).

The response to 12 hr of food deprivation either at night or during the day provides decisive information on the metabolic and neuroendocrine factors responsible for the diurnal cycle of intakes. A nocturnal 12-hr fast immediately induces a four- to fivefold increase of the subsequent intake. By contrast, a diurnal fast only induces a modest increase of the intake at the beginning of the subsequent nocturnal period (10–20%). This fact and the study of metabolic backgrounds in these two fasting conditions argues strongly in favor of the notion (today, well demonstrated) that the high intake at night causes the low intake during the subsequent day.

Thus, short-term food deprivation-induced feeding differs strikingly from

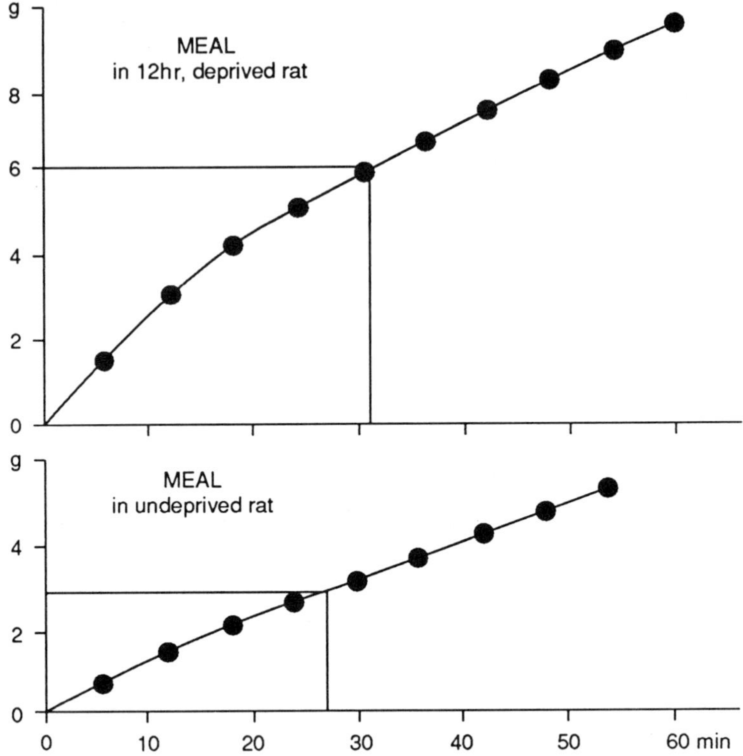

Figure 1.10 Daytime cumulative intake following a 12-hr nocturnal fast.

the *ad libitum* feeding. In the latter condition, the size, not the shape, of the meal is affected by the length of the spontaneous premeal satiety duration. By contrast, both the size and the shape of meals vary proportionally to the duration of a premeal deprivation. This strongly suggests that an "all or none" signal, which triggers the start of meals in the *ad libitum* condition and does not affect its size, is graded in its amplitude by food deprivation and affects both the size and the shape of meals at the restoration of food access. In addition, the relative shortening of postmeal satiety during refeeding suggests that the conditions of successive meal initiations after deprivation also are modified and contribute to the compensatory response.

C. Feeding Responses to 24–48-hr Food Deprivation

Convergent results of many experiments demonstrated poor responses to 1- or 2-day fasts. By contrast to the regulatory response to short-term deprivation at night, rats refed after 24 or 48 hr of fasting increase their food intake but fail to recuperate the prevented intake during 2 or 3 days of refeeding. However, a slow recovery of weight loss after 24 hr of fasting occurs within 4 days of refeeding. A transitory and moderate hyperphagia did not account for this recovery (47). According to Levitsky *et al.* (48), this postfast hyperphagia or excess of intake over control is the same after 24–72 hr of fasting and is equal to 24 hr of the *ad libitum* intake. It is independent of the weight loss. Refed rats restricted on their daily previous intake nevertheless regained their weight. In another study (49), a 24-hr fast was shown to produce only a 10% increase in the free intake of the subsequent day. A 48-hr fast induced 20% overeating on the first day, which is maintained at a lower level during the 2 following days. On these 3 days, this overeating represents only 20% of the deficit of 48-hr deprivation. However, after the 24- and 48-hr fasts, more than half of the daily increase in intake occurred during the first 12 hr followed by a fall of intake for the subsequent 12 hr below the control level. The enhancement in the first 12 hr was due entirely to augmented MS associated with a maintained meal frequency and, therefore, to an increase of the MS : postmeal interval ratio. On the contrary, the reduction in intake during the subsequent 12 hr was due to prolonged meal-to-meal times.

In the absence of adequate postfasting hyperphagia, the weight recovery can only be due to a sparing of energy output. This saving of energy is partly the effect of a reduction of activity. After up to 72 hours of fast, the refeeding of rats is associated with a sudden and profound hypoactivity, called the "starvation syndrome" (50). Later it will be suggested that a reduction of heat production by some tissues may also participate in this energy saving after fasting (Chapter 2, Section II).

D. Anticipatory Response to Food Deprivation

In earlier experiments, it was suggested that the response to food deprivation could be improved by a repetition of the same deprivation and that the appropriate response could be learned. For example, Ghent (51) submitted rats to a daily feeding schedule of a 23-hr fast and a 1-hr refeeding. He showed that the 1-hr intake increased progressively from day-to-day. The following experiment demonstrated an anticipatory capacity of rats. Rats received daily their familiar food in three 1-hr presentations at 4 A.M., 12 P.M., and 8 P.M. After 18 days of this feeding schedule of three meals separated from each other by 7 hr, one of them was suddenly omitted. A gap of 15 hr was thus created between the two remaining meals. To eliminate the effect of the diurnal cycle, the omitted meal was a different one in three groups. The evolution of consumptions in the two remaining meals was followed during 18 days. During the initial control, intakes in the meal before the 15-hr gap, the meal later to be omitted and the meal following the 15-hr gap were 6.04 ± 0.38, 7.57 ± 0.27, and 7.54 ± 0.29, respectively. Initially, the 15-hr deprivation elevated the intake in the subsequent meal, but after some days the intake in the meal preceding the permanent gap of 15 hr was observed to be increased three times more than the post deprivation meal. At this time, 6.90 g previously consumed in the omitted meal were reported to the meal before deprivation versus 0.67 g in the postdeprivation meal (Fig. 1.11). This phenomenon was designated "a provisional appetite" (52). (We will discuss this far-reaching finding later.)

The result imposes the distinction between the acute and chronic responses to food deprivation. The fact that, after some delay, the MS becomes proportional to the length of deprivation following the meal (i.e., a postprandial correlation is established) provides evidence for a learned anticipation by the rat of the subsequent fast and contemporary metabolic expenditures. The same learning process by which the rat anticipated its further expenditures will be confirmed in responses of rats submitted to a feeding schedule of a single daily meal.

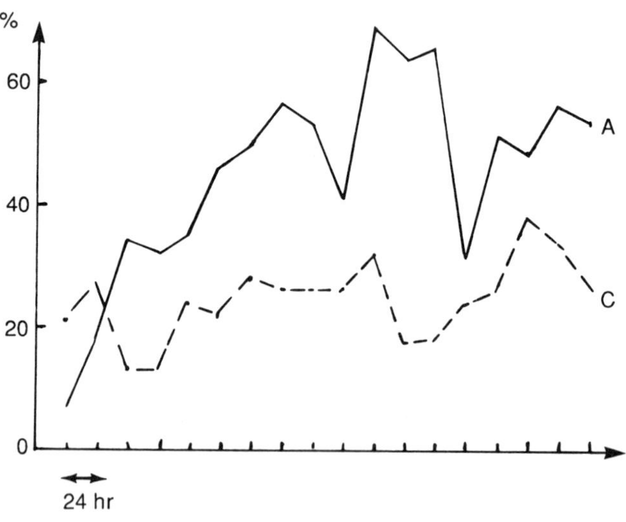

Figure 1.11 Acquired anticipatory response to food deprivation.

E. Long-Term Starvation and Subsequent Recovery of Weight Loss

The rehabilitation of the initial body weight after weight loss induced by starvation or restriction is a typical feature of the regulation of body energy balance, which will be commented on later. Is an hyperphagia involved in this recovery? And, because it is indeed involved, how can the difference be explained with the poor response to a medium-term fast?

Evidence was presented indicating that weight loss and time of deprivation were related according to a power law. Food intake would increase as a function of a percentage of the induced weight loss, itself related to a percentage of the time of starvation (53). In a detailed study (54), two groups of rats were submitted either to starvation or to restriction until a 25% weight loss and then refed *ad libitum*. In the two groups, the initial weight was recovered after a mean of 14 days. A linear negative correlation in log–log coordinates was present between the percentage of weight gain during the first week of refeeding and the total time of body weight recovery. Hyperphagia was exhibited by rats in the two groups, higher in rats leaned by restriction than in starved rats. In the latter, no hyperphagia was observed during the first 3 days of refeeding. In restrained rats, the hyperphagia was maximal as soon as the first day. The food efficiency for weight gain was calculated to be 6.1 kcal/g dry tissue in starved and >10 kcal/g dry tissue in restrained rats during the first week. Meal pattern recording showed a huge meal at the beginning of the night the first day of refeeding and an increased intake higher during the day than during the night. Thus, after such a weight loss, the induced increase in food intake seems to account for the recovery of body weight, at least in restrained rats. A complementary experiment showed that the relative deficiency of the regulation in starved rats was due to starvation diabetes: disturbance of carbohydrate metabolism, which affects the first day of refeeding (54).

The food composition appears to be an important factor in this starvation-induced hyperphagia. After a 6-day fast, the caloric intake was shown to be augmented on a high-fat diet only. In a choice among the three macronutrients, the increase of caloric intake was achieved from the augmented fat intake only (55).

IV. *Feeding Schedules*

A feeding schedule in which rats are offered their food during one or several daily presentations at regular or irregular times is the third condition of feeding to be observed in rats. In this condition, MSs and their variations can only adjust the daily intake to daily expenditures. The question is how does this adjustment occur without the regulation of intermeal intervals in action in free-feeding. In addition, the interest is to mimic the most common condition of feeding in human and animal models. Finally, the interest is still in the fact that such a forced-feeding schedule is used to feed domestic and reared animal species. In this managed feeding, the question is what is the best frequency pattern and meal times for weight maintenance or weight gain.

A. Adjustment in the Shift from *ad Libitum* to Scheduled Meals

Rats that were offered food once a day during 1–4 hr initially took less food in this time than in the 24-hr *ad libitum* control. Then, they progressively increased MS until a plateau of intake that was generally lower than the previous free daily intake. However, at this reduced level, they exhibited weight maintenance or a slight weight gain. In a 4-hr daily meal, a plateau of 70% of the *ad libitum* intake was reached within 7–10 days. During the 4-hr, the rat initially took four distinct meals. The first one was limited to 20 kcal (5 g). After 3 or 4 days, this initial meal reached 28 kcal by a confusion with the second one (56). Rats reestablished three-fifths of their previous *ad libitum* intake within 10 days in a 2-hr daily meal versus four-fifths in a 4-hr daily meal (57). According to Balagura *et al.* (58), the level and speed of adjustment is better in a programmed nocturnal meal compared to a diurnal one. This effect, confirmed in mice (59), was not replicated in rats (60). In dogs, a meal intake in the period of low *ad libitum* intake led to a severe loss of intake and body weight compared to the daily free-feeding (61). Rats that easily adapted to one daily meal did not adapt to a meal every 48 or 72 hr, and, with this schedule, they progressively lost their weight (62). Rats that adapted to a 2-hr daily meal, either at night or during the day, were offered a supplementary, unexpected meal given either 6 or 14 hr after the fixed meal. Interestingly, intakes in those meals did not depend on the premeal time interval contrary to the response to food deprivation. This result will be discussed later in relation to the fact that the blood glucose level is maintained from day to day in rats adapted to one 2-hr daily meal (63).

Rats that were shifted from one daily meal to the *ad libitum* condition overate sometimes (64). Rats returning to a free-feeding condition also exhibited the "memory" of the previous meal time. In the *ad libitum* control, rats drank water only during the meal. *Ad libitum* they continued to drink at the previous meal time (65).

Convergent results show that the condition for the adjustment to a feeding schedule is a regular feeding. These predictable meals are also a condition of the metabolic adjustment to a feeding schedule detailed in Chapter 2.

By studying further underlying mechanisms, one will reach the conclusion that (1) the progressive adjustment to the scheduled meal (or meals) through an increase of MS is learned, precisely a relearning of conditioned stimuli, which determines the initial strength of both stimulation to eat and the satiating capacity of foods; and (2) through this learning and the metabolic adjustment, the rat in each huge daily meal anticipates for calories needed to maintain its body energy balance until the next daily meal.

V. *Changes in Feeding Patterns as a Function of Altered Meal Size, Food Utilization, and Food Demand*

Studying MS, MMI relationships in rats provided suggestions on the mechanism involved in meal initiations and MS. Such suggestions were strengthened in a series of experimental conditions in which either the MSs or the rate and nature of food utilization before and after meals or both were changed.

A. Increasing or Decreasing the Meal Size

The postprandial correlation suggested that an experimentally altered MS should induce a shorter postmeal satiety after a reduced meal and a prolonged one after a larger meal. Indeed, this is verified in *ad libitum* condition. When a meal is pharmacologically suppressed by cholecystokinin administration, the postprandial duration of satiety was proportionally reduced and, thus, the cumulative intake maintained (66). Vagotomized rats take small meals. Their meal frequency is proportionally augmented; thereby their daily intake is unchanged (67). Increasing the intake by infusing a liquid food in the stomach after the end of the meal proportionally prolongs postmeal satiety (68).

This suggests that the filling and storage of foods in the gastrointestinal tract and the rate of its further utilization through the intestinal absorption are keyfactors of meal patterning. Sham-feeding experiments strengthen this conclusion. When foods passing the mouth exit from the stomach through an open fistula, rats sham-fed big meals separated by periods of satiety. The length of these periods has no relationship at all with the size of the previous sham-feeding episode (69). The same conclusion may be drawn from the fact that the removal of foods from the stomach during or immediately after a meal is terminated leads to a rapidly resumed oral intake (70). In Chapter 2, we will further discuss the role of emptying the gastrointestinal tract in meal initiation. The apparently paradoxical effect of a blockade of intestinal absorption by mannitol will be discussed then.

Here we come to a decisive point represented by the response of rats and other animals to a caloric dilution of familiar diet. This phenomenon is called "caloric adjustment." Since the pioneering works of Gasnier *et al.* (71), the caloric adjustment has been extensively studied as the best evidence for a caloric regulation by food intake. In a detailed study, Janowitz and Grossman (72) showed that the rat offered *ad libitum* its normal diet adulterated by adding an inert material (e.g., cellulose) augmented the weight and volume of its meals. The adjustment occurred in 24 hr and up to a dilution of 33%. Then, reconcentrating the diet, the rat rapidly suppresses the volume intake and, after 2 or 3 days of caloric overeating, also reestablished the initial caloric intake. In another detailed study (73), the caloric density expressed in Kilocalories per volume unit was shown to be a better predictor of the adjustment than Kilocalories per weight of food. According to the author, the volume, not the weight, is the determining factor in the adjustment. Of course rats possess no means for sensing the weight of ingested food or calories *per se*. The volume can only be involved as oral and gastric stimuli. However, inasmuch as the rat is thus demonstrated to eat for calories, the terms of caloric intake and of caloric appetite are justified. The expression of the MS to postprandial satiety duration in terms of calories instead of gram of food eaten (onset ratio vs. satiety ratio) is also validated.

Here, the interesting fact is that the initial adjustment is not achieved by an augmentation of MS on the diluted diet nor by a reduction of this size on the enriched one. During the first hours of the *ad libitum* intake, the adjustment is operated through the meal frequency. The preceding volume intake is at first unchanged or slightly modified, perhaps due to the change of palatability by adding the inert material, but the postmeal interval is shortened on the diluted diet and prolonged on the reconcentrated one. Roughly, the caloric intake to the postmeal satiety ratio is maintained (74, 75). After some hours, and increasing during successive days, the adjustment is achieved through enhanced MSs. This change of the meal pattern will be interpreted as an effect of the delay of learning or relearning the palatability of the food.

Striking evidence for the role of MMIs in the possibility of adjustment is indicated by the fact that the regulatory response to a diluted food is absent or considerably impaired in feeding schedules (76, 77). In rats, no adjustment at all occurs in a 2-hr daily meal. In a pattern of four meals regularly spaced during the night, the adjustment (which can be achieved through MS only) is extremely low and "plateaues" maximally at a lowered level of cumulative caloric intake (Le Magnen, unpublished) (Fig. 1.12). The most classic experiment was performed in the dog (72). Dogs fed for 2 hr daily took weeks and months before adjusting their volume intake to the dilution of their familiar diet. The same long delay was observed in the reduction of this augmented intake on the reconcentrated diet. This same impairment in the human feeding schedule will also be interpreted as a cause of long delays in the human response to changes of caloric density.

B. Enhanced Food Utilization, Food Deprivation, and Fat Synthesis

The duration of satiety after a given meal and, thus, the occurrence of events presiding meal initiation depend on the rate of the meal-to-meal metabolic food utilization and of food demands. It is curtailed when a part of ingested food is utilized for fat synthesis in addition to oxidative metabolism and when this oxidative metabolism itself is augmented. It is prolonged when fat oxidation instead of, or in addition to, ingested foods is used in oxidative metabolism. This prolonging effect will be the same when continuous intragastric or intravenous administrations of nutrients will be substituted for or added to the oral intake. As detailed elsewhere (Chapter 8) in this volume, these phenomena are the basis for body energy balance and the so-called lipostatic mechanism.

As already seen, a shortening of the time preceding meal onsets relative to the previous meal is observed after a short-term food deprivation at night. The restoration of intestinal content and of body fats metabolized during the food removal accounts for this effect. As also already mentioned, meal initiations at night occur before the time that oxidative metabolism would only account for by the use of ingested foods. Onset ratios expressed in calories per minute are 1.5 or 2 times higher than the contemporary metabolic rate also expressed in calories per minute. This relative shortening of postmeal satiety is due to fat synthesis added to oxidative metabolism, as discussed later. This dual food utilization quantitatively accounted for meal frequencies, particularly at the beginning of the dark period. It is the basis for a nocturnal hyperphagia and positive energy balance (78). A similar trend is observed in rats with a hypothalamic ventromedial lesion (see Chapter 9). After lesioning, those rats, as is well known, are hyperphagic and become obese, i.e., accumulate fats. The study of their meal pattern showed that after big meals the postmeal interval was shorter than that after meals of the same size in intact rats.

C. Reduced Food Demand

1. *Fat Oxidation*   Utilizing fats mobilized from fat stores in oxidative metabolism suppresses and may entirely eliminate the oral intake. This call to an

internal fuel is subtracted from the call to an external one. This was confirmed by examining the normal meal pattern during the day in relation to the underlying metabolism at this time. During the day and contrary to the night, meals are initiated later than accounted for by ingested foods only. A study by various procedures (detailed in Chapter 2) demonstrated that this prolongation of postmeal satiety is an effect of fat oxidation. A saving of glucose utilization explains delays of meal initiations. These delays are immediately lifted when lipolysis was blocked either by a continuous insulin infusion or by a fat oxidation inhibitor (β-mercuro-acetate) (78–80). This is also the case when food deprivation at night, preventing the concurrent fat synthesis, also prevents the subsequent diurnal fat mobilization.

After a forced feeding and weight gain induced by various means (e.g., protamin-zinc-insulin treatment, gavage), rats are temporarily hypophagic and lose weight until they return to their initial body weight. These phenomena also will be examined in regard to the lipostatic mechanism. The important point here is that the residual feeding during this recovery of body weight occurs exclusively at night and reproduces exactly a normal diurnal pattern (Fig. 1.13).

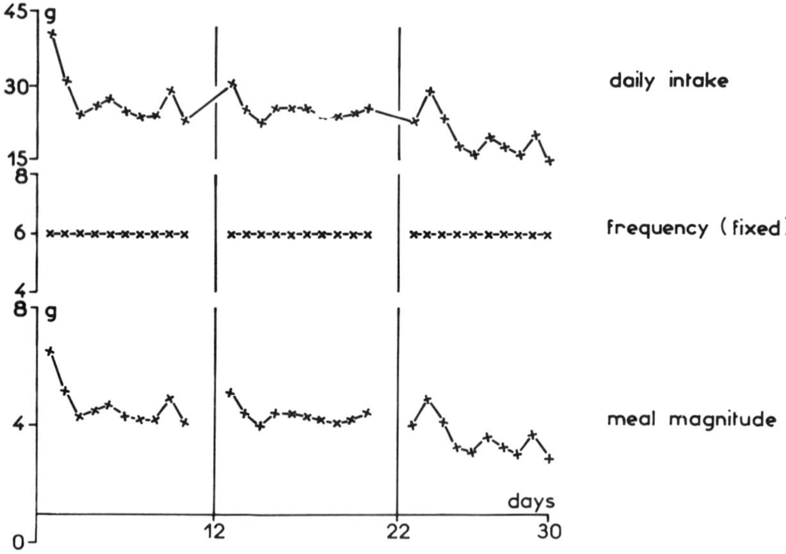

Figure 1.12 Deficient adjustment to a caloric dilution of the familiar food in a feeding schedule of six nocturnal meals.

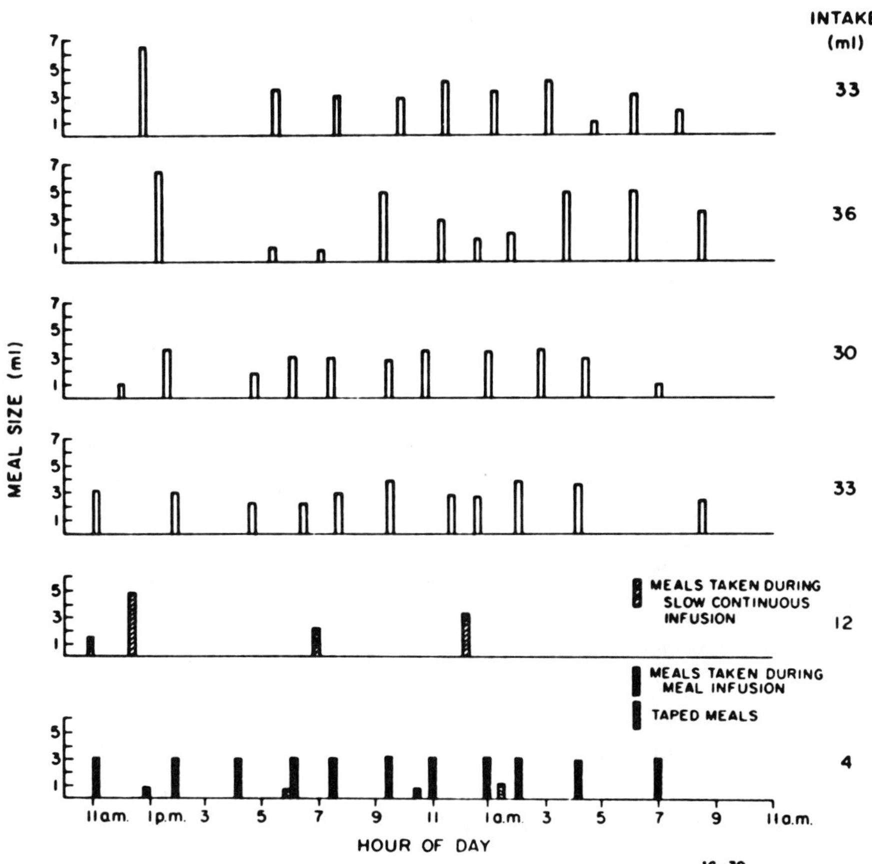

Figure 1.13 Suppression of food intake with intragastric loading: relation to natural feeding cycle. [From Quartemain *et al.* (83).]

2. *Intragastric and Intravenous Feeding*   Like the supply of an internal fuel to the pool of oxidized metabolites, the supply of an external fuel is compensated for by a change of the temporal patterning of meals.

Regarding intragastric feeding, misleading interpretations were introduced by the confusion of two very different experimental conditions. When a gastric load is given, sometimes before or during a meal, this load interferes with the satiating process and, consequently, with the MS (see Chapter 6). A different condition is realized when a gastric load is given some hours before or after a meal and also when a continuous gastric or intestinal infusion is associated to a concomitant free oral intake. In this case, satiety from meal to meal rather than satiation within meals is affected.

In two famous experiments, the enteral feeding effect on the temporal patterning of meals was demonstrated. In an experiment by Booth (81), the gastric load of a glucose solution was given some hours before the restitution of an *ad libitum* intake. For some hours of this free oral intake, a caloric compensation occurred by the prolongation of MMIs. These delays of meal initiations were observed as soon as the second meal. In experiments by Thomas and Mayer (68), rats took their liquid food (1 kcal/ml) by lever-pressing. A continuous and slow infusion of the same liquid food into the stomach at the level of 50%, and later 75%, suppressed proportionally the oral intake. This suppression was realized, although exclusively through delayed meals. MSs were slightly affected due to the effect of the continuous infusion during the meals (68).

Whereas the response to caloric dilution is deficient in a feeding schedule of one meal per day, the suppression of intake by intragastric chronic feeding is also impaired in this condition of feeding, excluding adjustments to the temporal patterning of meals. In dogs fed a daily single meal, gastric loads led to reduced meals after weeks or months only (82). A fascinating experiment illustrated the importance of the spontaneously established meal-to-meal patterning in a regulatory control of intake. Rats were intragastrically fed continuously and at a constant rate throughout the 24 hr by 100% of their *ad libitum* previous intake. Permitted to eat, their residual oral intake was 30–50% of their *ad libitum* previous intake, thus considerably augmenting their total intake (intragastric plus oral). When the intragastric feeding was distributed according to the preceding *ad libitum* pattern, the residual oral intake fell to 2–18%. When the intragastric feeding was given not at the time but 99 min prior to previous free meals, the compensation by the reduction of the oral intake was not better than in continuous perfusion (83).

Continuous intravenous infusions and their effects on the free oral intake were carried out by Nicolaïdis and Rowland (84), the first to use the chronically implanted intracardiac catheter. Infusions of carbohydrate solutions at 100% of the previous *ad libitum* intake induced only a partially compensatory reduction of the oral intake. Adding intermittent injections of insulin improved the compensation. The pattern of the residual intake was made of small meals separated by long MMIs. This pattern was identical to that observed during the day in untreated rats. These results were replicated and confirmed by Porte and Woods (85).

D. Conclusion

In free-feeding, the postprandial correlation was revealed only by spontaneous variations of MSs. Various experiments detailed in this section highlight and extend the significance of the phenomenon in the control of food intake and the regulation by food intake of body energy balance. Together, the results provide evidence indicating that the duration of satiety induced by meal intake, and thus the time of initiation of the subsequent meal, depend on the energy intake in the meal and on metabolic food utilization and demand following that meal. Thus, from meal to meal and between nocturnal and diurnal periods, the time patterning of meal initiations is adjusted to metabolic requirements. This adjustment and its mechanism will later be seen as the main factor in the regulation of the body energy balance by food intake.

VI. *Human Feeding Patterns*

How, how much, and when do humans eat? All have their own experience and a trivial knowledge of what is the most common food habit in their social environment. Partly for this reason, studies on human feeding patterns have developed slowly. They are still relatively limited. The study of human feeding has, for centuries, been obscured by the confrontation between subjective and objective approaches.

A. Hunger Feeling and Brain Hunger Arousal of Eating

For centuries, observations of food intake in humans have been biased by considering the hunger feeling only and by ignoring relations of this feeling with the choice and intake of foods. Generally, hunger is perceived in a state of short-term food deprivation. It is felt as a malaise or a desire for food. In acute food deprivation or energy deficits, the feeling may become a painful one. Of course everyone knows the components of this feeling, including the epigastric cramp. Questionnaires have been used on a large population of subjects to describe the various subjective aspects of hunger and their variations from subject to subject (86).

Stomach contractions have long been assumed to be causes of hunger. Such assertions and studies about relations between gastric contractions and hunger have also long hampered the progress of knowledge on human feeding. Indeed, it is well known that contractions occur in an empty stomach. Thus, subjectively, stomach contractions could be validly considered as a partial cause of the hunger feeling, but they could not be considered as a cause of food intake. Again, a confusion of thinking came from the ambiguity of the term "hunger," used to designate both the subjective feeling and the brain arousal of eating, whose feeling is an effect and not a cause.

B. Temporal Feeding Patterns in Humans

The dual prandial and diurnal periodicity of feeding is observed in humans. Even in exceptional conditions of *ad libitum* intake, feeding is never entirely scattered in a nibbling pattern. Subjects taking several snacks between their main meals maintain a pattern of discrete eating bouts. Despite free access to foods available at all times, they initiate and terminate separated and sized meals. In all cases, the diurnal, or circadian, periodicity is present in human feeding. Humans, whatever their feeding schedule, eat during the 12–15 hr of their wakefulness and do not eat (i.e., are fasting) for 8–10 hr during sleep.

The most common and universal feeding pattern of humans is a feeding schedule of three or four meals distributed at fixed hours between morning and evening times. This distribution and amounts of foods eaten in each meal vary widely according to sociocultural constraints. The Anglo-Saxon world is accustomed to a high-calorie breakfast, whereas in continental Europe breakfast is low in calories. The respective intake at lunch and dinner is also extremely different according to a series of socioprofessional and individual factors. We will return to this calorie difference in a comparison between normal weight and obese subjects.

The fixed meal pattern is now changing and differentiated in developed countries. Familial meals at fixed hours are strictly maintained in some countries. Elsewhere, the noon meal is collectively eaten in refectories in schools and at work. Generally, foods remain available at fixed hours only. When food dispensers or permanently opened cafeterias make food continuously available for the consumer, a feeding condition of free access may be established. It is of interest to note that an unexpected meal given to rats adapted to a feeding schedule (cited in Chapter 1, Section IV) is a trivial experience for humans. Aside from meals, cakes, candies, ice creams, and generally high-palatable foods can be occasionally eaten without any apparent relation to the time elapsed since the habitual meal and its size. This raises the question about stimuli acting to initiate such intakes.

Specialists, sociologists, and anthropologists are studying the social aspect of human feeding habits. This social context is of utmost importance. In all cultures, the meal is a basic social activity. Many studies have stressed the different eating patterns between meals eaten in isolation and collective meals and between self-served and served meals. In many populations, the meal is a ritual with esthetic and religious aspects.

At a biological point of view, two facts may be pointed out. The human feeding schedule, established in infancy, is both flexible and resistant to changes.

Inasmuch as it is not flexible, i.e., inasmuch as the feeding schedule is respected, the time of meal initiation is not freely determined. Consequently, no pre- or postprandial correlations should be observed. For this reason, any adjustment of intake to the requirement of body energy balance can be achieved only in a succession of days. We will see later that this is indeed observed. However, and surprisingly, the possibility of an adjustment from meal to meal within a day was hardly investigated. The presence of a negative correlation between amounts eaten at lunch and dinner was reported (87). The meal pattern of human subjects was studied and analyzed in a series of experiments by de Castro (88, 89) and de Castro and Elmore (90). From a complex statistical analysis of results obtained in various conditions of observation on subject groups, these studies concluded a human tendency to take more food in a meal as the time since the preceding meal is longer.

The fate of the human meal pattern in a free-running condition was studied. In one study (91), subjects lived for some days in a closed flat and entirely deprived of temporal cues. They asked for their meal by pressing a button. In this condition, the fixed meal pattern progressively disappeared and subjects eventually asked for their meal after a time proportional to the intake in the preceding meal. Thus, they exhibited a significant postprandial correlation. In another similar study (92), six isolated subjects increased their sleep–wakefulness cycle (34 hr). In half of the subjects, the meal frequency increased as a function of the prolongation of sleep. This suggests that in this condition sleep periodicity governs the meal periodicity. Subjects would be awakened by hunger and, thus, would take their meal in a state comparable to that of food deprivation.

C. Microstructure of Meals in Humans

An oscillographic recording of chewing and swallowing from the beginning to the end of test meals was used to study details of the intrameal eating pattern of human subjects (93). Normal weight subjects were instructed to take (at will and successively) pieces of bread of a constant volume and to eat them freely until satiation. The bread pieces were differently flavored by a layer of various items: cheese, jam, and so on. One of the flavored foods was offered in a "single-flavored meal." Various flavored foods were simultaneously offered in mixed varied meals, and these varied meals were compared to single flavored ones. Parameters recorded from the beginning to the end of the meal and statistically analyzed were MS, duration of mouthful, number and frequency of chewing each mouthful until swallowing, number of swallows, intervals between mouthful, and the mean eating rate. The evolution of these parameters from the beginning to the end of the meal and the effect of the palatability of different flavored foods were examined. The effect of palatability of the food on this pattern will be described elsewhere (see Chapter 4).

Figure 1.14 illustrates the chewing–swallowing pattern obtained in a single-flavor meal. The prominent fact considered here is that the duration and number of masticatory movements of each mouthful before swallowing and the duration of intervals are increasing from the beginning to the end of the meal. In other words, the eating rate is decreasing at the approach of satiety. MSs are greater in mixed-flavored meals than in single-flavor meals (Fig. 1.15).

The early work on the microstructure of human meals (94) was replicated and the result confirmed by Hill (95). An increase in chewing time and the number of masticatory movements of each mouthful at the end of the meal were again indicated as a feature of the approach of satiety. Using the same technique of chewing–swallowing recordings, Spiegel (96) also confirmed these results. Kissileff (97) found comparable data by using a less naturalistic test meal. Eight subjects of each sex took this test-meal 3 hr after their control breakfast. The food was a liquid food contained in a reservoir. The subject took this food at will through a mouthpiece. In this peculiar condition of testing, the cumulative intake curve also showed an initial rate of eating followed by a deceleration. Men exhibited a higher initial rate of eating than did women, but they also decelerated faster. A quadratic equation adequately fits this curve (Fig. 1.16).

The preceding studies on the microstructure of human meals in terms of masticatory movements and swallowing were carried out using food units of constant size, physical form, and caloric density. Even in this unique condition, chewing and the overall eating rate varied among individual subjects. An

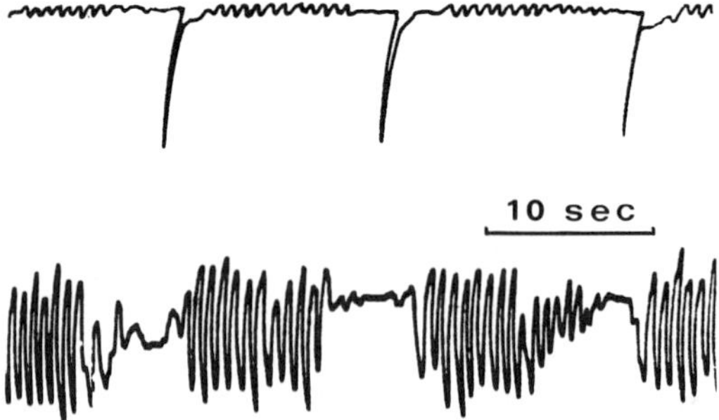

Figure 1.14 Chewing–swallowing patterns in humans.

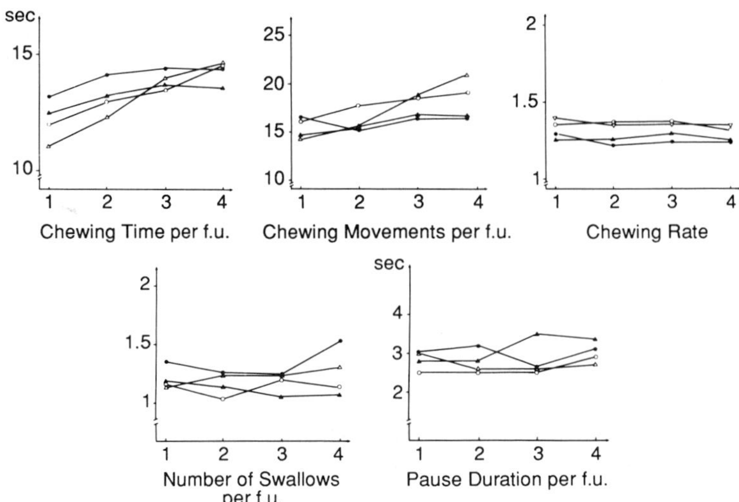

Figure 1.15 Evolution of five parameters from the first to the last quarter of meals. (SEMs are not represented for the sake of clarity.) △, high deprivation, high preference; ▲, high deprivation, low preference; ○ low deprivation, high preference; ●, low deprivation, low preference; f.u., food unit. [Bellisle *et al.* (98).]

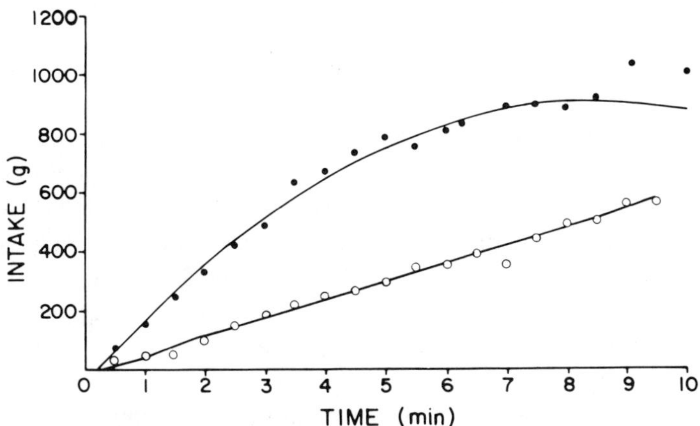

Figure 1.16 Cumulative curve of food intake in the human meal. ●, $I = -37.86 + 226.4t - 13.397 t^2$, $r^2 = .99$, 21h; ○, $I = -9.45 + 60.8t + .160t^2$, $r = .99$, 1h. [Kissileff *et al.* (97).]

individual eating style was apparent. These differences would presumably be greater in a comparison (which was never made) among foods of various physical forms, particularly of various textures. No comparative study exists on the unitary volumes of food taken in the mouth and their differences according to the food and personal habits. The chewing time of hard or soft solid foods, of solid and semiliquid foods, and again of an individual eating style have not been compared. It is only empirically known that a contrasting behavior exists between people swallowing rapidly after a minimal mastication and salivation of the food and others swallowing all foods (either hard or soft ones) after a slow and prolonged mastication and salivation. The role of such differences in the satiating process and also in the digestibility of foods is unknown. However, mastication *per se* (even of a calorically inert material) was reported to provide a relief of the hunger feeling. Prisoners coming back from deportation reported that they masticated leaves and pieces of woods, thus obtaining a transitory relief of their terrible hunger.

D. Short and Long Fast-Induced Feeding in Humans

That food deprivation increases the hunger feeling and food intake is common sense. It is also self-evident that the study of fast-induced feeding would be unuseful. This may explain the surprising lack of investigations on this basic phenomenon. However, science is made to quantify phenomena, even the most trivial ones, to find their determinants.

Within the human feeding schedule, omitting a meal acutely induces an increase in the size of the following meal. By recording the chewing–swallowing pattern in the conditions already described, subjects tested at noon after an overnight fast exhibited a typical change of their meal shape. Compared to the nondeprived condition, they showed, as expected, an increase in the initial rate of eating associated with a more accentuated contrast between the beginning and the end of the meal. The ratio of intake duration of the first to the last quarters of the amount eaten in the meal was 1.21 (98, 99). This ratio of durations of the first to the last quarters of the amount eaten in the meal was 1.21 following a 3-hr fast and 1.32 following a 15-hr fast on the high-palatability diet. This change of the intrameal structure of the meal is the acute effect of the elimination of the first morning small meal and therefore of the extension of the premeal time from about 5 to 17 hr. This effect would have to be compared to the effect of omitting lunch on the subsequent dinner and also of omitting the dinner on the overnight breakfast. Presumably, the same time of deprivation has different effects according to the time of day or of night. We eat our breakfast without a clear hunger feeling and we eat a small meal despite the 10-hr nocturnal fast. The same 10-hr fast created by omitting lunch would likely induce an acute hunger before dinner and would augment this meal. It is one of the surprising gaps in our knowledge of human feeding due to the lack of simple experiments, probably too simple to have been made.

The effect of omitting the three day meals, i.e., of a 12-hr deprivation, on hunger during the subsequent night and intakes in the subsequent day is of course suspected, but again it was not measured by anyone. Chronic effects and adjustments to changes of the normal schedule have also not been studied. Can humans, like rats, learn an anticipatory feeding in the first morning meal when repetitively deprived of the noon meal? Can humans eat every 48 hr only and still ensure their long-term body energy balance in this condition? Is it possible to eat only one meal per day? And, if so, would this be good or bad? What would be the best hour for such a daily meal? All these questions, important in common life, are without answers due to a lack of appropriate studies.

The effects of a long fast are better documented, unfortunately due to the increasing frequency of hunger strikes used as blackmail. The terrible experience of millions of human beings deported to concentration camps everywhere in the world also provides rich information (but is so horrible that their use is humanly difficult). Using volunteers, an extensive study about human starvation was carried out by Keys (99). The experience and studies of long fasting by individual volunteers have also been reported. Among others, a report by Langfeld (100) is an example. The subjects reported the disappearance of hunger after 3 days. Psychophysiological examination showed maintained sensory and mental functions, particularly of memory, during the first weeks. Activity is also maintained until the first impairment of neural function consisting of muscular weakness. This disappearance of hunger after some days of fasting was reported in many other observations on long fasting and on hunger strikes. Its mechanisms will have to be explained. It is of interest to note that in human subjects not entirely starved but profoundly restricted, hunger persists and increases over time. In undernourished people, a very painful hunger was reported to be a cause of cardiac failure.

Whereas human feeding measure in adulthood and in a metabolic steady-state condition have been reviewed here, the role of peripheral mechanisms in determining these feeding responses will be examined in Chapter 2 and the role of food palatability in Chapter 4. Its changes in various conditions of body energy expenditures will be detailed in Chapter 8 and obesity in Chapter 10. Finally, its ontogenesis from infancy to adulthood will be examined in Chapter 12.

VII. *Feeding–Drinking Relationships*

Feeding and drinking are very different and separately governed behaviors. Drinking water and saline solutions is involved in the regulation both of blood osmolality and body water content: They are not involved in body energy regulation. Internal and sensory stimuli as well as brain targets of these stimuli are different. In humans, feeding and drinking are, at least after deprivation, associated with two discriminated feelings: hunger and thirst. However, the two ingestive behaviors interact.

A. Feeding–Drinking Patterns in Animal Models

A comparative review of feeding–drinking relationships in various animal species (a very large and exciting topic) is beyond the scope of this volume. Again, only the rat will be compared to humans.

Rats, water-deprived for 22 hr/day, reduce their food consumption by 70%; food-deprived they reduce their water intake by 60%. The 21-hr water deprivation augments the 3-hr subsequent food intake (water being then available) by 9 g. The augmentation of intake is almost identical to that induced by 21 hr of food deprivation. Conversely, rats rapidly drink 8–10 ml of water after a food deprivation, almost the amount drunk following the same time of water deprivation (101). Water restriction reduces not only the simultaneous food consumption but also the metabolic efficiency for growth of the ingested foods (102).

Rats with diabetes insipidus (DI) offered an isotonic NaCl solution as the only source of fluid gain no water by drinking this solution inasmuch as their capacity to concentrate their urine is impaired. In this condition, they become totally aphagic (103). In the same DI rats, an old and admirable experiment (today forgotten) was carried out by Bruce and Kennedy (104). To match their polyuria, DI rats drink up to 200 ml of water per day to survive. If, instead of water, they are offered a 10% glucose solution, they are obliged to get 50% more calories than their *ad libitum* intake to maintain their 200-ml daily intake of fluid. They avoided the increase of caloric intake and obesity by drastically reducing their intake of the solution, and so they died. Other rats placed in the same condition survived after a ventromedial hypothalamic lesion and the elimination of the lock preventing hyperphagia and obesity.

In an instructive experiment (105), water-deprived rats received 4–8 ml of water or 50 ml of a 0.3 or 1.5 saline solution either orally or in the stomach before a choice between food and water. Results indicated that (1) the oral intake of fluid is more efficient in reducing water intake and in potentializing feeding than the gastric tubing is, (2) hypertonic saline solution is less efficient than a hypotonic one and water, and (3) the dual effect (reduction of drinking, augmentation of food intake) is greater after 8 than after 4 ml of water. Thus, both osmolality and volume are active in these effects. Convergent results suggested that hyperosmolality before meals inhibits food intake and meal initiation and that preprandial drinking would relieve this inhibitory factor. In water-deprived rats (as shown earlier), water ingestion immediately initiates an enhanced food intake. Adding NaCl to water increases the latency to overeat (106). This latency to eat and blood osmolality are negatively correlated to water drunk before food presentation (107). In a choice between drinking and feeding, the preference for eating increases with the degree of hydration.

A differential responsiveness to taste stimuli of water- and food-deprived rats substantiates (among other data) the distinction of the two brain behavioral arousals of eating and drinking. Rats made hungry either by food deprivation or by insulin administration and offered a choice among a saccharin solution, water, and an isotonic salt solution increased their drinking of the sweetened solution only. Rendered thirsty, they increased their intake of water and of the salty solution (108, 109). This increase of salty solutions in thirsty Na-deprived rats occurs acutely, even if the salty solution is not NaCl but made of another salt without sodium (110). The far-reaching significance of such anticipatory responses to taste will be commented upon elsewhere.

This role of oral stimuli causes some problems for the rat when fed liquid versus solid food. Rats overeat a liquid diet. In one experiment (111), rats given their solid food diluted with 50–90% water overate and gained weight. The bitterness of the liquid food adulterated by sucrose octo-acetate eliminate the preference for the liquid food in a choice between the liquid and solid food. After some weeks of habituation to this adulterated diet, rats nevertheless overate this diet *ad libitum*.

B. Meal-Associated Drinking

The relationships between feeding and drinking within meals was extensively studied in a series of experiments conducted with simultaneous recordings of feeding and drinking patterns (112). Results of these investigations led to the conclusion that the normal drinking of fed rats is a food-associated drinking in which, in the absence of current osmolar or hypovolemic stimuli to drink, rats

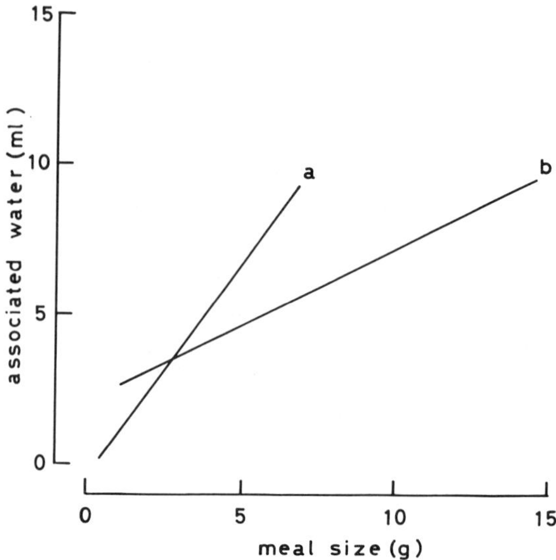

Figure 1.17 Positive correlation in rat between periprandial drinking and amount of solid food eaten during the meal. [Fitzsimons and Le Magnen (112).]

anticipate their further water requirements on the basis of eating food. Oral food stimuli then act as "conditioned stimuli" to drink (112) (Fig. 1.17). Various reports confirmed and extended these results (113, 114). A striking confirmation of meal-associated drinking and of its independency of internal stimuli was provided in another species—the pig. Blood samples were taken and blood osmolality, hematocrite, and blood cell volumes determined in four different conditions: no feeding and no drinking, drinking just before eating, drinking during eating, and drinking without eating. The only difference in blood parameters found in the four conditions was an increase of cell volumes during the meal. The authors concluded that neither hypertonicity nor hypovolemia stimulated the normal drinking of pigs associated to meals (115).

Through conditioning and based on food stimuli, water requirements, and therefore thirst, vary with the composition of foods. Proteins produce a water loss due to the osmotic load and diuresis and starches due to the intestinal need of water by hydrolysis. On the contrary, the water requirement after fat intake is minimal due to the gain of internal water by fat oxidation. Consequently, the water : food ratios (milliliters per gram) exhibited by rats eating high-protein, high-carbohydrate, and high-fat diets are 1.9, 1.2, and 0.8, respectively.

C. Prandial Drinking

Meal-associated drinking differs from so-called "prandial drinking." In the preceding experiments, the microstructure of alternate eating and drinking within the meal was not examined. It is assumed that rats drink during pauses recorded during their normal meal-eating pattern. This microstructure of feeding–drinking relationships within meals of humans will be described in the following section.

In rats, a specific type of these relationships was described in lateral hypothalamic-recovered rats (116). After recovering from aphagia, lesioned rats exhibited a nibbling pattern. In very prolonged meals, they alternated short bouts of eating and drinking. This was reproduced in surgically desalivated rats, and the hypothesis was presented that the lateral hypothalamic lesion impaired salivation and that the dry mouth caused both the nibbling pattern and the rapid alternation between eating and drinking. However, this nibbling pattern is also produced by olfactory bulbectomy and by surgical dysconnections between amygdala and the hypothalamus (117). It was assumed that the nibbling pattern in these various conditions, including the lateral hypothalamic lesion, was due to the loss of olfactory projections to the lateral hypothalamic feeding system and to a resulting impairment of the salivation process. The alternating drinking would be the result of the crumbled meal, perhaps analogous to the schedule-induced polydipsia. In this feeding schedule forcing the rat to nibble its food intake, an alternation between feeding and drinking is also observed.

D. Feeding–Drinking Relationships in Humans

Humans drink when thirsty or not thirsty, like they eat either hungry or not. Humans (it is our common experience) drink avidly pure or flavored water in large amounts when they are thirsty under the effects of two thirst stimuli: hyperosmolality (e.g., after a salt intake) or hypovolemia, mainly due to basic or occasional water losses (e.g., sweating). In water deprivation, these two internal stimuli of thirst act synergistically to induce the response to deprivation (118). This acute water intake after exercise (e.g., in high ambient temperature) generally occurs aside from meals. In humans, like in other mammals, the urge to drink is greater than the urge to eat as an effect of the absence of body water reserves stamped however by the saving of water through oliguria. Conversely, humans (like other species) can cope with excessive drinking due to water diuresis. In the limits of this regulatory action of the kidney, humans can over-drink considerably (e.g., flavored or alcoholic beverages).

In addition to this extraprandial and occasional fluid intake, the daily water intake of humans is, like that in the rat, a periprandial drinking. Using the technique of recording the chewing–swallowing pattern (described in Section VI), the intrameal drinking of human subjects was investigated. Subjects allowed to drink freely during experimental meals spent 7.7% of meal durations to drink. The number of swallows and swallowing frequencies was found correlated to durations of drinks. Contrary to what happens in the rat, water intake was not correlated to the same amount of food eaten during the meal; this is true in lean subjects only. Obese subjects drank more during their meal than their lean counterparts, and their volume of water drunk was correlated to MS. Figure 1.18 illustrates the distribution of drinks and water volumes in the four quarters of meal duration. Frequencies of drinking episodes interrupting the chewing of solid food progressed as the meal progressed over time. At least 50% of drinks occur in the last quarter of meal. In the mixed meal, during which various flavored foods were offered, 76% of water drafts occurred when subjects, after chewing a particular food, shifted to another food. Thus, it was suggested that a part of the prandial drinking in humans is used to rinse the mouth and, in so doing, to exalt the differential palatability of food (93) (Fig. 1.18).

Temporal and quantitative relations between water and food intake and the effect of water restriction on solid intake were also reported by Engell (119) and de Castro (120). Ninety-eight percent of the daily water intake occurs at meal

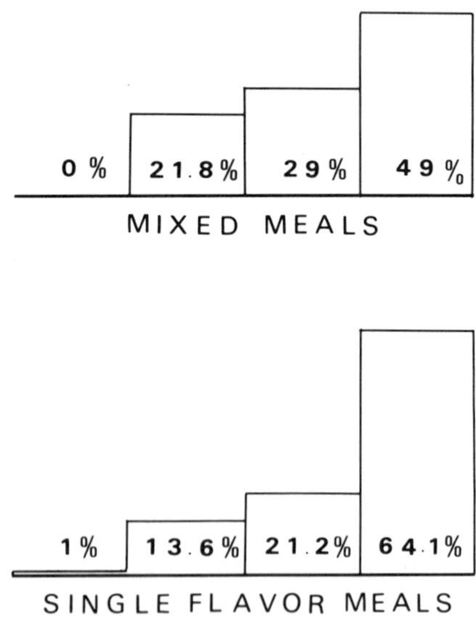

Figure 1.18 Pattern of prandial drinking in humans.

time. A 40% limited water availability reduces food intake, but food acceptability, as estimated by a preference rating, is not affected by water restriction. A subjective rating of thirst intensity is negatively correlated to food intake. As in the rat, hyperosmolality and a resulting thirst before the meal seem to antagonize feeding. Coming to the table both hungry and thirsty, humans begin to drink. The urge to drink overcomes the urge to eat.

However, water is not the most common fluid drunk by humans. Various flavored beverages are consumed apart from as well as within the meal. Humans made thirsty after water loss in warm ambient temperature or after exercise prefer cool and flavored beverages (e.g., peppermint, anis, bitter). The same flavors are disliked when associated with foods during the meal. One well-known American beverage, cola, appetitive when humans are thirsty, is considered awful (and even scandalous, at least in France) when associated with various palatable dishes. Such associations between food and fluid flavors are based on the rules of food esthetics. These "rules" also mandate that a dry white wine must accompany seafoods, and not meats; red wine meats, and not fish; sweet white wine the dessert only. In some countries, such as France, these rules are respected perhaps even more than the law.

VIII. *Sleep and Food Intake*

A. Sleep and Meal Pattern

The best and most extensive study of sleep–feeding relationships was carried out by Danguir *et al.* (7). For the first time, he recorded simultaneously the 24-hr electro-encephalogram (EEG) and meal patterns in normal and ventromedial and lateral hypothalamus-lesioned rats. The EEG was analyzed in terms of slow waves (SWs) and rapid eye movement (REM) sleep times, the feeding pattern in terms of MSs and MMIs. In normal rats, MS was found to be correlated to the duration of both SW and REM sleep in the subsequent postmeal intervals, as it is correlated to the duration of this interval. But, surprisingly, a higher correlation was also found between the size of the same meal and the duration of sleep in the interval following the subsequent meal.

B. Effects of Food Deprivation on the Sleep–Wakefulness Pattern

The effects of the total food deprivation on the sleep–wakefulness pattern were studied in rats. The respective effects on the SW and REM sleep were recorded and analyzed by a simultaneous 3-hr EEG, electroolfactogram (EOG), and electromyogram (EMG) recordings. The amount of wakefulness increased as a function of a positive accelerated curve with time of food deprivation, while sleep times diminished. Sleep totally disappeared after 9–11 days of fasting. The REM sleep disappeared before the SW sleep. However, the time course of this evolution was different for the two types of sleep. The SW sleep disappeared gradually. On the contrary, the REM sleep increased on the first day of deprivation, was maintained at control level the 2 following days, and later totally disappeared. At restoration of food access, REM sleep was initially above and SW below the control level (121). In cats, the duration of the REM sleep during 12 hr of free food intake predicts the food consumption during the 12 subsequent hr. It is a better predictor of intake in this period than the total sleep and wakefulness and the previous consumption. However, in cats, food-deprived at night, the REM sleep during the night is no longer a predictor of the subsequent *ad libitum* intake (122). In rats, REM sleep deprivation produces a reduction of MSs at night, compensated for by larger meals during the day becoming identical to the reduced nocturnal meals (123).

References

1. Kleiber, M. (1961). *The fire of life: An introduction to animal energetics*. New York: Wiley, 428 pp.
2. Le Magnen, J., and Tallon, S. (1966). La périodicité spontanée de la prise ad libitum d'aliments du rat. *J. Physiol. (Paris)* **38,** 327–349.
3. Le Magnen, J., and Devos, M. (1980). Parameters of the meal pattern in rats. *Neurosci. Biobehav. Rev.* **4** (Suppl. 1), 1–11.
4. Strubbe, J. H., Keyser, J., Tjalling Dijkstra, and Alingh Prins, A. J. 1986. Interaction between circadian and caloric control of feeding behavior in the rat. *Physiol. Behav.* **36,** 489–493.
5. Armstrong, S., Clarke, J., and Coleman, G. (1978). Light–dark variation in laboratory rat stomach and small intestine content. *Physiol. Behav.* **21,** 785–788.
6. Kissileff, H. R. (1970). Free feeding in normal and "recovered lateral" rats monitored by a pellet-detecting eatometer. *Physiol. Behav.* **5,** 163–174.
7. Danguir, J., Nicolaïdis, S., and Gérard, H. (1979). Relations between sleep and feeding patterns in the rat. *J. Comp. Physiol. Psychol.* **93,** 820–830.
8. Davis, J. D. (1973). The effectiveness of some sugars in stimulating licking behavior in the rat. *Physiol. Behav.* **11,** 39–45.
9. Davis, J. D., and Smith, G. P. (1988). Analysis of lick rate measures the positive and negative feedback effects of carbohydrates on eating. *Appetite* **11,** 229–238.
10. Levitsky, D. A. (1974). Feeding conditions and intermeal relationships. *Physiol. Behav.* **12,** 779–788.
11. Temple, B., Shor-Posner, B., and Leibowitz, S. (1989). Temporal pattern of mac-

ronutrient intake in freely feeding and deprived rats. *Am. J. Physiol.* **256,** R541–548.

12. Snowdon, C. T. (1969). Motivation, regulation and the control of meal parameters with oral and intragastric feeding. *J. Comp. Physiol. Psychol.* **69,** 91–100.

13. Thomas, D. W., and Mayer, J. (1968). Meal taking and regulation of food intake by normal and hypothalamic hyperphagic rats. *J. Comp. Physiol. Psychol.* **66,** 642–653.

14. Balagura, S., and Coscina, D. V. (1968). Periodicity of food intake in the rat as measured by an operant response. *Physiol. Behav.* **3,** 641–643.

15. Davies, R. F. (1977). Long–short term regulation of feeding patterns in the rat. *J. Comp. Physiol. Psychol.* **91,** 574–585.

16. de Castro, J. M. (1975). Meal pattern correlations: Facts and artefacts. *Physiol. Behav.* **15,** 13–17.

17. de Castro, J. M. (1978). An analysis of the variance in meal patterning. *Neurosci. Biobehav. Rev.* **2,** 301–309.

18. Kraly, F. S. (1981). Pregastric stimulation and CCK are not sufficient for the meal size–intermeal interval correlation in the rat. *Physiol. Behav.* **27,** 457–462.

19. Rosenwasser, A. M., Boulos, Z., and Terman, M. (1981). Circadian organization of food intake and meal patterns in the rat. *Physiol. Behav.* **27,** 33–39.

20. Panksepp, J. (1973). Reanalysis of feeding patterns in the rat. *J. Comp. Physiol. Psychol.* **82,** 78–94.

21. Wolf, L. L., and Hainsworth, F. R. (1977). Temporal patterning of feeding by hummingbirds. *Anim. Behav.* **25,** 976–989.

22. Duncan, I. J., Horne, A. R., Hughes, B. O., and Wood-Gush, D. G. (1970). The pattern of food intake in female brown leghorn fowls as recorded in a Skinner box. *Anim. Behav.* **18,** 245–255.

23. Marcilloux, J. C., Simon, J., and Auffray, P. (1985). Effects of VMH lesions on plasma insulin in the goose. *Physiol. Behav.* **35,** 725–728.

24. Ardisson, J. L., Dolisi, C., Ozon, C., and Crenesse, D. (1981). Caractéristiques des prises d'eau et d'aliments spontanées chez des chiens en situation ad lib. *Physiol. Behav.* **26,** 361–371.

25. Sanderson, J. D., and Vanderweele, D. A. (1975). Analysis of feeding patterns in normal and vagotomized rabbits. *Physiol. Behav.* **15,** 357–364.

26. Prudhon, M. M. (1972). Evolution au cours de la croissance des caractéristiques des consommations liquides et solides du lapin nourri ad libitum. *Ann. Zootechnie* **24,** 451–460.

27. Hansen, B. C., Jen, K. L. C., and Kalnasy, L. W. (1981). Control of food intake and meal patterns in monkeys. *Physiol. Behav.* **27,** 803–810.

28. Hirsch, E., Dubose, C., and Jacobs, H. L. (1978). Dietary control of food intake in cats. *Physiol. Behav.* **20,** 287–295.

29. Auffray, P., and Marcilloux, J. C. (1983). Etude de la séquence alimentaire du porc adulte. *Reprod. Nutr. Dévelop.* **23,** 517–524.

30. Hirsch, E. (1973). Some determinants of intake and patterns of feeding in the guinea pig. *Physiol. Behav.* **11,** 687–704.
31. Weijs, W. A. (1975). Mandibular movements of the albino rat during feeding. *J. Morph.* **145,** 107–124.
32. Hiiemae, K., Thexton, A. J., and Crompton, A. W. (1978). Intra-oral food transport: The fundamental mechanism of feeding. In D. S. Carlson and A. MacNamara, Jr. (eds.) *Muscle adaptation and the maxiofacial region* (pp. 181–208). Center for Human Growth and Development, University of Michigan, Ann Arbor, Michigan.
33. Epstein, A. N., and Teitelbaum, P. (1962). Regulation of food intake in the absence of taste, smell and other oro-pharingeal sensations. *J. Comp. Physiol. Psychol.* **55,** 753–759.
34. Holman, G. L. (1969). Intragastric reinforcement effect. *J. Comp. Physiol. Psychol.* **69,** 432–441.
35. Nicolaïdis, S., and Rowland, N. (1977). Self-intraveinous feeding: Long-term regulation of body energy balance in rats. *Science* **195,** 589–590.
36. Varner, L. H. (1928). The study of hunger behavior in the rat by means of the obstruction method. *J. Comp. Physiol. Psychol.* **8,** 273–300.
37. Miller, N. E. (1955). Shortcomings of food consumption as a measure of hunger: Results from other behavioral techniques. *Ann. N.Y. Acad. Sci.* **63,** 141–143.
38. Heron, W. T., and Skinner, B. F. (1937). Changes in hunger during starvation. *Psychol. Rec.* **1,** 51–60.
39. Bare, J. K. (1959). Hunger, deprivation and the day/night cycle. *J. Comp. Physiol. Psychol.* **52,** 129–131.
40. Bare, J. K., and Cicala, G. (1960). Deprivation and time of testing as determinants of food intake. *J. Comp. Physiol. Psychol.* **53,** 151–154.
41. Bellinger, L. L., and Mendel, V. E. (1975). Effect of deprivation and time of refeeding on food intake. *Physiol. Behav.* **14,** 43–46.
42. Tagliaferro, A. R., and Levitsky, D. A. (1982). Overcompensation of food intake following brief periods of food restriction. *Physiol. Behav.* **29,** 747–750.
43. Le Magnen, J., Devos, M., and Larue-Achagiotis, C. (1980). Food deprivation induced parallel changes in blood glucose, plasma free fatty acids and feeding during the two parts of the diurnal cycle in rats. *Neurosci. Biobehav. Rev.* **4** (Suppl. 1), 17–23.
44. Larue-Achagiotis, C., and Le Magnen, J. (1982). Effects of short-term nocturnal and diurnal food deprivation on subsequent feeding in intact and VMH lesioned rats: Relations to blood glucose level. *Physiol. Behav.* **16,** 245–248.
45. Le Magnen, J. (1968). Eating rate as related to deprivation and palatability in normal and hyperphagic rats. International Congress on the Physiology of Food and Fluid Intake (ICPFFI), Haverford, Pennsylvania.
46. Le Magnen, J., Devos, M., Gaudillière, J. P., Louis-Sylvestre, J., and Tallon, S. (1973). Role of a lipostatic mechanism in regulation by feeding of energy balance in rats. *J. Comp. Physiol. Psychol.* **84,** 1–23.

47. Armstrong, S., Coleman, G., and Singer, G. (1980). Food and water deprivation: Changes in rat feeding, drinking, activity and body weight. *Neurosci. Biobehav. Rev.* **4,** 377–402.

48. Levitsky, D. A., Faust, I., and Glassman, M. (1976). The ingestion of food and recovery of body weight following fasting in the naive rat. *Physiol. Behav.* **17,** 575–580.

49. Le Magnen, J., and Tallon, S. (1968). L'effet du jeûne préalable sur les caractéristiques temporelles de la prise d'aliments chez le rat. *J. Physiol. (Paris)* **60,** 143–154.

50. Finger, F. W. (1951). The effect of food deprivation and subsequent satiation upon general activity in the rat. *J. Comp. Physiol. Psychol.* **54,** 557–564.

51. Ghent, L. (1951). The relation of experience to the development of hunger. *Can. J. Psychol.* **5,** 77–81.

52. Le Magnen, J. (1959). Etude d'un phénomène d'appétit provisionnel. *C.R. Acad. Sci. (Paris)* **249,** 2400–2402.

53. Bolles, R. C., and Thichler, B. (1977). Deprivation, weight loss and intake in the rat as a function of age: Evidence for an obligatory growth factor. *Behav. Biol.* **1,** 207–212.

54. Pénicaud, L., and Le Magnen, J. (1980). Recovery of body weight following starvation or food restriction in rats. *Neurosci. Biobehav. Rev.* **4** (Suppl. 1), 47–52.

55. Andik, I., Donhoffer, S. Z., Moring, I., and Szentes, J. (1951). The effect of starvation on food intake and selection. *Acta Physiol. Acad. Sci. Hung.* **2,** 363–368.

56. Baillie, P. (1977). Patterns of ingestion of rats adjusting to a controlled-feeding schedule of four hours per day. *J. Physiol. (London)* **273,** 35P.

57. Booth, D. A. (1977). Caloric compensation in rat with continuous and intermittent access to foods. *Physiol. Behav.* **8,** 891–901.

58. Balagura, S., Harrell, L. E., and Roy, E. (1975). Effect of the dark–light cycle on neuroendocrine and behavioral responses to schedule feeding. *Physiol. Behav.* **15,** 245–248.

59. Nelson, G., Scheving, L., and Halberg, F. (1975). Circadian rhythm in mice fed a single daily meal in different stages of lighting regimen. *J. Nutr.* **105,** 171–184.

60. Valle, C., and Peleo, Y. (1976). Body weight in rats: Effects of night versus light meals. *Physiol. Psychol.* **4,** 494–496.

61. Ozon, C., Dolisi, C., Crenesse, D., and Ardisson, J. L. (1986). Effects of feeding time upon daily intake and body weight. *Physiol. Behav.* **36,** 583–586.

62. Duffort, R. H. (1965). The rat's adjustment to 23, 47 and 71 hour food deprivation schedules. *Psychol. Res.* **15,** 663–679.

63. Lima, F. B., Hell, N. S., Timo-Iaria, C., Dolnikoff, M. S., and Pupo, A. A. (1982). Carbohydrate metabolism and food intake in food restricted rats: Effects of an unexpected meal. *Physiol. Behav.* **29,** 931–937.

64. Lawrence, D. H., and Mason, W. A. (1955). Intake and weight adjustment in rats to changes in feeding schedule. *J. Comp. Physiol. Psychol.* **48,** 43–46.

65. Mori, T., Nagai, K., and Nakagawa, H. (1983). Dependence of memory of meal time upon circadian biological clock in rats. *Physiol. Behav.* **30,** 259–265.

66. Kraly, F., Carlson, J., Resneck, T., and Smith, G. P. (1988). Effects of cholecystokinin on meal size and intermeal intervals in sham-feeding rats. *J. Comp. Physiol. Psychol.* **94,** 597–607.

67. Snowdon, C. T., and Epstein, A. N. (1970). Oral and intragastric feeding in vagotomized rats. *J. Comp. Physiol. Psychol.* **71,** 59-67.

68. Thomas, D. W., and Mayer, J. (1978). Meal size as a determinant of food intake in normal and hypothalamic obese rats. *Physiol. Behav.* **21,** 113–117.

69. Kraly, F. S., Carty, W. J., and Smith, G. P. (1978). Effect of pregastric food stimuli on meal size and intermeal interval in the rat. *Physiol. Behav.* **20,** 779–784.

70. Snowdon, C. T. (1969). Regulation and the control of meal parameters with oral and intragastric feeding. *J. Comp. Physiol. Psychol.* **69,** 91–100.

71. Gasnier, A., Gompel, M., Hamon, F., and Mayer, A. (1932). Regulation automatique de l'ingestion et de l'absorption des aliments en fonction de la teneur en eau du régime. *Ann. Physiol. (Paris)* **8,** 870–890.

72. Janowitz, H. D., and Grossman, M. I. (1949). Effect of variations in nutritive density on intake of food of dogs and rats. *Am. J. Physiol.* **158,** 184–193.

73. Peterson, C. D., and Baumgardt, B. R. (1971). Food and energy intake of rats fed diets varying in energy concentrations and density. *J. Nutr.* **101,** 1057–1068.

74. Le Magnen, J. (1969). Peripheral and systemic actions of food in the caloric regulation of intake. *Ann. NY Acad. Sci.* **268,** 3107–3110.

75. Kanarek, N. L. (1976). Energetics of meal pattern in rats. *Physiol. Behav.* **17,** 395–399.

76. Jacobs, H. L., and Sharma, K. N. (1969). Taste versus calories: Sensory and metabolic signals in the control of food intake. Ann. NY *Acad. Sci.* **157** (Art. 2), 1084–1125.

77. Smith, M., Pool, R., and Weinberg, H. (1962). The role of bulk in the control of eating. *J. Comp. Physiol. Psychol.* **55,** 115–120.

78. Le Magnen, J., and Devos, M. (1970). Metabolic correlates of the meal onset in the free food intake of rats. *Physiol. Behav.* **5,** 805–814.

79. Larue-Achagiotis, C., and Le Magnen, J. (1983). Effects of a daytime infusion of exogenous insulin on subsequent nocturnal food intake and body weight in rats. *Physiol. Behav.* **30,** 573–576.

80. Langhans, W., and Scharrer, E. (1987). Role of fatty acid oxidation in the control of meal pattern. *Behav. Neural Biol.* **37,** 7–16.

81. Booth, D. A. (1972). Satiety and behavioral caloric compensation following intragastric glucose load in the rat. *J. Comp. Physiol. Psychol.* **78,** 412–432.

82. Share, I., Martyniuk, E., and Grossman, M. I. (1952). Effect of prolonged intragastric feeding on oral food intake in dogs. *Am. J. Physiol.* **169,** 229–235.

83. Quatermain, D., Kissileff, R., Shapiro, R., and Miller, N. E. (1971). Suppression of food intake with intragastric loading: Relation to natural feeding cycle. *Science* **173,** 941–943.

84. Nicolaïdis, S., and Rowland, N. (1976). Metering of intravenous versus oral nutrients and regulation of energy balance. *Am. J. Physiol.* **231**, 661–668.

85. Porte, D. Jr., and Woods, S. C. (1981). Regulation of body weight and food intake by insulin. *Diabetologia* **20**, 274–280.

86. Monello, L. G., and Mayer, J. (1967). Hunger and satiety sensations in men, women, boys and girls. *Am. J. Clin. Nutr.* **20**, 253–261.

87. Jiang, C. L., and Hunt, J. N. (1983). The relation between freely chosen meals and body habitus. *Am. J. Clin. Nutr.* **38**, 32–40.

88. de Castro, J. M., McCormick, J., Pedersen, M., and Kreitzman, S. N. (1986). Spontaneous human meal patterns are related to preprandial factors regardless of natural environment constraints. *Physiol. Behav.* **38**, 25–29.

89. de Castro, J. M. (1988). Physiological, environmental, and subjective determinants of food intake in humans: A meal pattern analysis. *Physiol. Behav.* **44**, 651–659.

90. de Castro, J. M., and Elmore, D. K. (1988). Subjective hunger relationships with meal patterns in the spontaneous feeding behavior in humans. *Physiol. Behav.* **43**, 159–165.

91. Bernstein, I. L. (1981). Meal patterns in "free-running humans." *Physiol. Behav.* **27**, 621–624.

92. Green, J., Pollak, C. P., and Smith, G. P. (1987). The effect of desynchronization on meal patterns of humans living in time isolation. *Physiol. Behav.* **39**, 203–209.

93. Bellisle, F., and Le Magnen, J. (1981). The structure of meals in humans: Eating and drinking patterns in lean and obese subjects. *Physiol. Behav.* **27**, 649–658.

94. Pierson, A., and Le Magnen, J. (1969). Etude quantitative du processus de regulation des réponses alimentaires chez l'homme. *Physiol. Behav.* **4**, 61–67.

95. Hill, S. W. (1974). Eating responses of humans during dinner meals. *J. Comp. Physiol. Psychol.* **86**, 652–657.

96. Spiegel, T. A. (1990). Regulation of energy balance and control of food intake in humans. *Appetite* **14**, 67–69.

97. Kissileff, H. R., Thornton, J., and Becker, E. (1982). A quadratic equation adequately describes the cumulative food intake curve in man. *Appetite* **3**, 255–272.

98. Bellisle, F., Lucas, F., Amrani, R., and Le Magnen, J. (1984). Deprivation, palatability and the micro-structure of meals in human subjects. *Appetite* **5**, 85–94.

99. Keys, A. (1950). *Biology of human starvation.* Minneapolis: University of Minnesota Press.

100. Langfeld, H. S. (1914). The psychophysiology of a long prolonged fast. *Psychol. Monogr.* **16**, 61–62.

101. Verplank, W. L., and Hayes, J. (1953). Absorption de nourriture et d'eau en fonction du régime alimentaire. *J. Comp. Physiol. Psychol.* **46**, 327.

102. Crampton, E. W., and Lloyd, L. E. (1957). The effect of water restriction on food intake and food efficiency of growing rat. *J. Nutr.* **57**, 213.

103. Kakolewski, J. W., and Deaux, E. (1972). Aphagia in the presence of drinking an isosmotic NaCl solution. *Physiol. Behav.* **8,** 623–630.

104. Bruce, H. M., and Kennedy, G. C. (1951). The central nervous control of food and water intake. *Proc. R. Soc. (London)* **138,** 528–543.

105. Hsiao, T., and Smutz, E. (1976). Thirst reducing and hyperinducing effects of water and saline by stomach tubing versus drinking in rats. *Physiol. Psychol.* **4,** 111–113.

106. Kakolewski, J. W., and Deaux, E. (1970). Initiation of eating as a function of ingestion of hypoosmotic solutions. *Am. J. Physiol.* **218,** 590–595.

107. Deaux, E., and Kakolewski, J. W. (1971). Character of osmotic changes resulting in the initiation of eating. *J. Comp. Physiol. Psychol.* **74,** 248–253.

108. Le Magnen, J. (1953). Activité de l'insuline sur la consommation spontanée des solutions rapides. *C.R. Soc. Biol. (Paris)* **147,** 1753–1757.

109. Le Magnen, J. (1953). Régulation immédiate de la prise spontanée d'eau et de sel chez le rat blanc dans des états imposés de déséquilibre hydrominéral. *C.R. Soc. Biol. (Paris)* **147,** 619–623.

110. Schulkin, J. (1982). Behavior of sodium deficient rats: the search for the salty taste. *J. Comp. Physiol. Psychol.* **96,** 628–634.

111. Ramirez, I. (1986). Feeding a liquid diet increases caloric intake and weight gain without producing obesity. International Symposium on the Physiology of Food and Fluid Intake (ICPFFI), Seattle, Washington, p. 62. [Abstract.]

112. Fitzsimons, J. T., and Le Magnen, J. (1969). Eating as a regulatory control of drinking in the rat. *J. Comp. Physiol. Psychol.* **67,** 273–283.

113. Lucas, G. A., Timberlake, W., and Gawley, D. J. (1989). Learning and meal-associated drinking: Meal-related deficits produce adjustments in postprandial drinking. *Physiol. Behav.* **46,** 361–367.

114. de Castro, J. M. (1989). The interaction of fluid and food intake in the spontaneous feeding and drinking patterns of rats. *Physiol. Behav.* **45,** 861–870.

115. Houpt, T., and Anderson, C. (1990). Spontaneous drinking: Is it stimulated by hypertonicity or hypovolemia? *Am. J. Physiol.* **258,** R141–144.

116. Epstein, A. N. (1972). The lateral hypothalamic syndrome: Its implications for the physiosiological psychology of hunger and thirst. In J. M. Sprague and A. N. Epstein (eds.), *Progress in psychobiology and physiological psychology,* Vol. 4. pp. 263–317. New York: Academic Press.

117. Larue, C. (1975). Prandial drinking and the disruption of meal patterns in olfactory bulbectomized rats. *Physiol. Behav.* **15,** 491–493.

118. Kutscher, C. L. (1972). Interaction of food and water deprivation on drinking: Effect on body water losses and characteristics of solution offered. *Physiol. Behav.* **9,** 753–758.

119. Engell, G. (1988). Interdependency of food and water intake in humans. *Appetite* **10,** 133–141.

120. de Castro, J. M. (1988). A microregulatory analysis of spontaneous fluid intake by

humans: Evidence that the amount of liquid ingested and its timing is mainly governed by feeding. *Physiol. Behav.* **43**, 705–714.

121. Jacobs, H. L., and MacGinty, H. G. (1971). Effects of food deprivation on sleep and wakefulness in the rats. *Exp. Neurol.* **40**, 212–222.

122. Siegel, J. P. (1975). REM sleep predicts food intake. *Physiol. Behav.* **15**, 399–403.

123. Elomaa, E. (1981). The light/dark difference in meal size in the laboratory rat on a standard diet is abolished during REM sleep deprivation. *Physiol. Behav.* **26**, 487–493.

Chapter Two

The Stimulus to Eat

As normal feeding processes are observed and measured in the rat model and in humans, the question of causal determinants accounting for these processes arises.

Altogether, considerable evidence indicates that the overall stimulation to eat results from the combination of systemic and sensory stimuli. Two types of information are transmitted and integrated into the brain. A blood-borne or humoral signal reflects a state of energy or, more generally, a nutritional deficit to be identified. This signal acts intermittently on brain targets and gives rise to the "hunger arousal of eating." This is the "systemic stimulus to eat." The other information concerns food. Sensory afferent pathways coming from the oral cavity and activated by foods in the mouth provide the sensory stimulation to eat or not to eat. This sensory stimulus of eating is called "the palatability" of a food. The combination of the systemic and sensory stimulations to eat a food is called "the appetite" for a food, in relation to some aspects of its nutritive properties.

I. *Meal Initiation in Undeprived Animals*

Analysis of the normal free-feeding pattern led to the conclusion that two distinct systems are responsible for cumulative intake over time. The first one periodically initiates the eating of the offered food. Because this meal initiation is also the end of satiety developed by the preceding meal, this system might be called "the hunger–satiety system." The other system, apparently independent of the former, governs the amount eaten within the meal until an abolition of its initial stimulation, i.e., until satiation.

A tremendous confusion was introduced in the literature by the synonymous English meaning of two words: satiety and satiation. Eating a food induces the onset of satiety, i.e., eliminates the stimulation to eat. This process is not "the mechanism of satiety" proper but, rather, the process of its induction. For a long time, this author proposed to designate this process satiation. When a meal is terminated, the absence of stimulation until the initiation of a new meal is satiety. The mechanism by which the subject is not stimulated to eat for a time is the mechanism of satiety (sometimes called "persistent satiety").

What is the systemic stimulus to eat or not to eat (i.e., of the hunger–satiety system in the *ad libitum* condition of feeding)?

A. Theories

Until the recent discovery of the exact nature of the hunger-state signal and of its brain targets, investigations on food intake and body energy balance produced a field of theories.

Everyone agrees that a continuous outflow of energy is the "fire of life." Everyone expresses this outflow in terms of power. Computed from oxygen consumption or directly measured by calorimetry, this energy outflow or expenditure is commonly expressed in Kilocalories, or Kilojoules, per day. These joules divided by the number of seconds in a day become watts. The human body with expenditures of the order of 10,000 kJ/day develops a power of about 110 watts.

The three parameters of energy outflow are heat production, oxygen consumption proportional to the overall oxidative metabolism, and caloric intake postulated to be equal to the energy output. It was not surprising that, on the basis of these three parameters and their relationships, various theories proposed that food intake is controlled either by heat production, oxygen consumption, or the nutrient as substrates of the energy production.

The thermostatic theory was first presented by Brobeck (1), who, with Hetherington and Ranson (2) and later with Anand (3), discovered the effects of hypothalamic lesions on food intake. The theory was based on the changes of food intake observed in cold or heat exposure (1). Cold exposure enhances food intake, and this enhancement matches the increase in energy losses. Heat exposure somewhat reduces intake, parallel to a reduced expenditure-limiting thermolysis. It was thus hypothesized that humans are hungry when they are cold, and satiated when warm. Specifically, it was proposed that some sensors of heat outflow, the skin or the brain temperature, provide the stimulus to eat or not to eat. Unfortunately, this attractive theory was ruled out. No clear-cut relations were found between core, skin, or brain temperatures and eating initiation (4).

In energostatic theories (5, 6), the flow of energy in the brain, whatever the macronutrients providing this energy in the whole body, is hypothesized to be responsible for the control of food intake. Some specialized brain neurons would be witnesses of the all-body energy expenditures inasmuch as they would reflect the whole-body oxygen consumption by their own oxygen uptake. A decrease in oxygen consumption, not due to activity, observed in rats prior to meal was presented in support of this notion (6).

A final set of theories, supported by much preliminary evidence as well as common sense, proposed that the control of food intake was operated by substrates of energy production, i.e., nutrients. The only substrate whose concentration in the blood is closely regulated is glucose. Its continuous supply to the brain is a vital requisite. Based on these facts, Mayer (7) proposed that glucose tissue availability and tissue glucose uptake were the stimuli to eat or not to eat. At the time, his theory was based on the intake-promoting action of insulin and on the hyperphagia observed in diabetic animals and humans. No technique was then available for animal models to test a correlation or a causal relationship between changes in blood glucose or insulin levels and feeding patterns; however, in humans, the arteriovenous differences of blood glucose exhibited some correlations with hunger and satiety feelings (see Chapter 2, Section VII). The main support of this theory was the first finding of the presence of glucosensitive sites in the ventromedial hypothalamic (VMH) nuclei selectively lesioned by gold-thioglucose. Mayer (7) and his co-workers hypothesized that these cells were the sensors of brain glucose availability. While the hypothalamic nuclei was thought to be involved in satiety, they suggested that glucose availability was a positive stimulus of satiety rather than glucopenia, a stimulus of hunger. This interpretation was later ruled out by the demonstration that the VMH is neither involved in satiety nor in other feeding mechanisms proper (see Chapter 9). However, this glucostatic theory was a decisive step toward the now-achieved demonstration that this is no longer a theory but the fact.

B. A Transient Fall of Blood Glucose Level as the Systemic Stimulus to Eat, or Hunger

Three new techniques—chronically implanted intracardiac catheter, permitting easy blood samplings and intravenous infusions; the automated glucose-oxidase measure of blood glucose; the radioimmunoassay of plasma insulin—allowed Steffens (8) and his co-workers to carry out the first pioneering works on the correlation between blood parameters and feeding patterns in rats. By intermittent blood samplings, Steffens found a constancy of blood glucose level from meal to meal and its postabsorptive prandial augmentation. Insulin concentration, at a mean level from meal to meal, fell just before meal onset (9). However, injecting glucose between meals, thus inducing hyperinsulinemia, did not change the time and size of meals. Plasma insulin concentration thus would not be the stimulus to eat (10). Using the same blood sampling technique every 10 min from the end of a meal to the onset of the subsequent one, the authors tested the blood glucose level and the time of meal initiation. The mean intermeal level was averaged in a group. The last point averaged was the level between 10 and 0 min before the onset. In this condition, Strubbe *et al.* (10) failed to find any drop of blood glucose at the approach of the meal initiation; therefore, they claimed that blood glucose level was not linked to this initiation.

The opposite was demonstrated by Louis-Sylvestre and Le Magnen (11) and their pupils Campfield, Bradon, and Smith (12) by continuously monitoring blood glucose level in free-moving and -eating rats.

In the initial experiment (11), rats freely moving in their home cages were offered their familiar food *ad libitum*. The free intake was continuously recorded by a food cup weighing device as rats were slightly heparinized, a microflow of blood (25 µl/min) was drawn continuously via an implanted intracardiac catheter. Their blood glucose level was also continuously measured by a glucose analyzer (Yellow Spring). Recordings began 1 hr after light on or light off and lasted 3–5 hr. A fall of blood glucose preceded every meal in the night and during the day. This fall began an average of 5 min before the start of the meal. The maximal amplitude of this preprandial hypoglycemia was −6.5%. This fall occurred after a constant level following the prandial rise associated with the preceding meal. From this constant level of 32 mg%, the lowest value obtained during the preprandial drop was 92 mg%. The amplitude of the drop had no relation to the size of the initiated meal (Fig. 2.1).

This experiment and its findings were fascinating. Rats were closed in their chambers. Two screens allowed the experimenter to follow the blood glucose from one of them, the weight of the food cup and, thus, the rat's intake on the other. Each time the first screen indicated an initial fall of blood glucose, it was 100% predictable that the rat would begin to eat some minutes later.

With the support of this author (J. L. M.), this initial experiment was renewed and admirably developed by Campfield and Smith (13, 14). The technique was improved by computerized data processing, which allowed the minimal resolution time. They confirmed the typical shape of the preprandial fall of the blood glucose level. This fall began an average of 12.3 min prior to the meal initiation, reached a nadir 5.4 min before this initiation at a level that averaged 11.6% lower than the constant intermeal level. Then, blood glucose increased toward baseline level at the time of meal onset (Fig. 2.2). No correlation existed

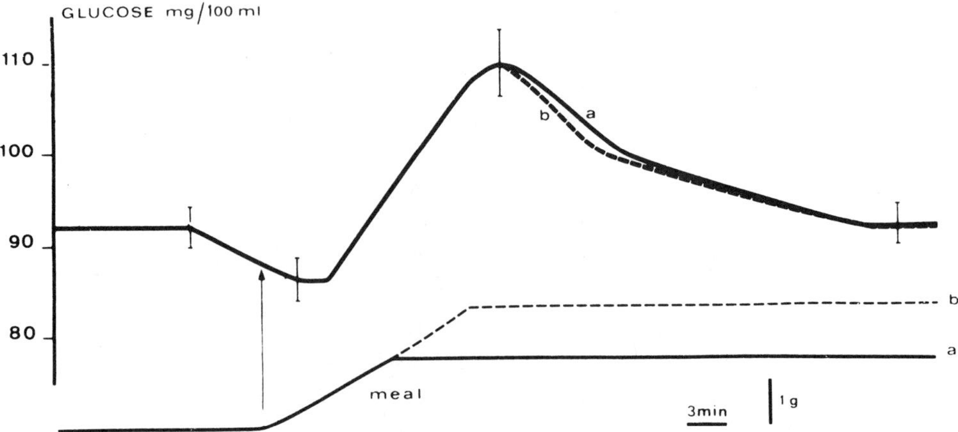

Figure 2.1 In rats, a slight fall of blood glucose precedes the initiation of every meal. K_a, 1.7; K_b, 2.2.

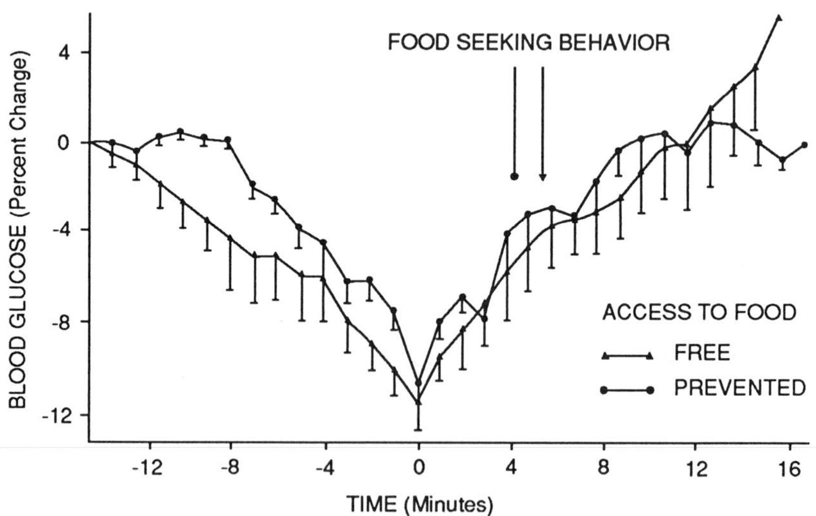

Figure 2.2 Percentage of change of blood glucose associated to meal initiation. [Campfield and Smith (13).]

between the nadir level and the size of the subsequent meal. No fall without initiation of a meal and no meal without a preprandial hypoglycemia were observed. The transient fall of blood glucose level prior to meal initiation was observed in hyperinsulinic VMH-lesioned and obese Zucker rats. Hypoinsulinic diabetic rats also showed a 10% decline of the basal hyperglycemia preceding every meal.

This transient premeal fall of blood glucose level is preceded by a brief peak of plasma insulin concentration (+50% from base line). This 50% peak of insulin produced by the injection of an acetylcholine analogue during a long diurnal meal-to-meal interval elicits both the transient fall of blood glucose level and a subsequent meal initiation. Infusing 20 mg of glucose at the beginning of the drop of blood glucose level partly stamped this drop. This suppression of the signal retards the meal by 300 min. A new meal occurs at this time after renewed slight hypoglycemia. When the food cup is removed or closed, the normal fall of blood glucose level and the return to base line occur. If the food access is reestablished 2–8 min before this return to base line, the rat eats. After this delay, the rat does not eat and eats again 70 min later, after a new normal fall of blood glucose level. In vagotomized rats, meal initiation occurs after a slower and longer fall but at the same amplitude as that in intact rats. In these vagotomized rats, the verified absence of the initial peak of insulin may cause the particular shape of the transient fall of the blood glucose level that precedes meals. A comparison between fructose and glucose infusions during the premeal decline of blood glucose was instructive. Infusing fructose instead of glucose did not retard the meal initiation. At the beginning of the night, a low dose of infused fructose induces hypoglycemia and a high dose hyperglycemia. Whatever the dose, the initiation of the first meal was neither precipitated nor retarded. At the beginning of the day, a dose-dependent fall of blood glucose was produced by fructose. This transient hypoglycemia produced three different shapes. A meal was initiated only when this shape was identical to that preceding spontaneous meals.

The causal relationship between the premeal hypoglycemia and meal initiation was also confirmed by the fact that the suppression of this hypoglycemia or hyperglycemia induced at the time of meal onset retards the meal like the small infusion does. Glucagon injection and a small epinephrine injection, both inducing hyperglycemia, retard the meal (15; Chabert, unpublished).

Thus, it seems fully demonstrated that a transient fall of blood glucose level is the necessary signal for meal initiation.

Another type of experiment further validated this conclusion. Forty-eight rats were assigned to eight different groups (16). Their free-meal patterns were at first recorded. Then, they were deprived of food at various times of the day according to their group: from the first to the third hour of the day (Group 1), from the third to the sixth hour (Group 2), etc. Groups 5–8 were night groups. Later, the same rats were submitted again to the same time-dependent 3-hr deprivation, and blood samples were taken before and at the end of the 3-hr fast. Blood glucose level was determined. These determinations confirmed other data about the time-dependency of the deprivation-induced hypoglycemia throughout the dark–light cycle. At the beginning of the day (Groups 1 and 2), no fall of blood glucose was induced by the 3-hr deprivation. This fall was present at other times of the day and night. However, the fall was not significantly different in the four night groups. The *ad libitum* meal pattern of initially fed rats was analyzed in terms of cumulative 3-hr intake (feeding rate). This *ad libitum* feeding rate and its variations according to the time of day were plotted against the deprivation-induced hypoglycemia in the corresponding 3-hr period in the eight groups. The correlation between the two parameters was shown to be highly significant (Fig. 2.3) (r = 0.65). When day and night groups were separately plotted, the same high correlation appeared during the day but was absent in the night groups. In the group deprived for the last 3 hr of the day, a very high interindividual

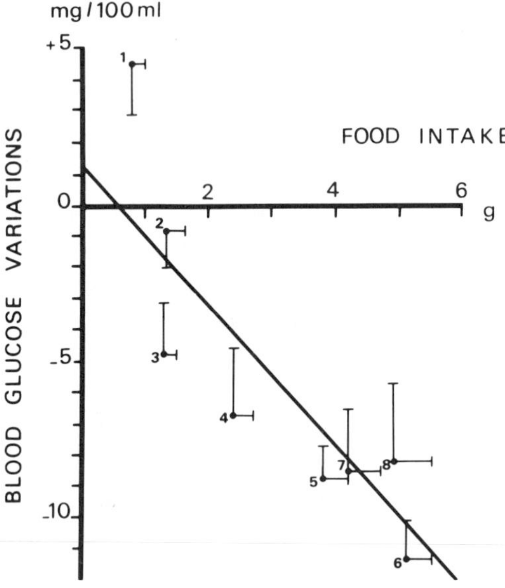

Figure 2.3 Correlation throughout the day between 3-hr free food intake and the fall of blood glucose level when the food is removed during the same 3 hr. See text for descriptions of groups 1–8.

correlation was present between the deprivation fall and the preceding *ad libitum* intake during the same hours ($r = 0.94$).

The cumulative intake during 3 hr is the product of meal size and meal frequency during the period, and it varies between night and day, mainly according to the frequency of meal initiations. The fall of blood glucose in deprived rats is an indication of the frequency at which the meal is initiated during free access and of its variations throughout the circadian cycle. Thus, it is assumed that the 3-hr free intake is the intake needed to prevent the fall of blood glucose for the 3 hr. The rat throughout the day eats at a rate that just maintains its blood glucose level because the slightest fall of this level initiates a meal. Therefore, one can also assume that stimulation of feeding is a component of the counterregulation of hypoglycemia. This counterregulation by stimulation of feeding precedes the intervention of glucagon and later of catecholamine releases.

The effects of glucoregulatory system manipulations (insulin, tolbutamide or glucose antimetabolite administrations and intravenous vs. intraportal administrations of glucose) on feeding patterns will be discussed in Section V of this chapter. Already, the above-mentioned experiments by Nicolaïdis and Rowland (17) and by Porte and Woods (18) show that continuous intravenous infusions of glucose solutions strongly reduce oral intake. The experiment by Porte and Wood showed that this suppression was not due to hyperglycemia but, rather, to the suppression of the hypoglycemic stimulus to eat. Finally, the medical literature reports pathological cases of chronic hypoglycemia in humans, always associated with a chronic hunger feeling and obesity (19).

C. Meal Initiation and the Emptying of the Gastrointestinal Tract

The time of meal initiation (end of satiety) is correlated to both the caloric intake in the previous meal and the rate of food utilization since this previous meal (see Chapter 1, Section V). Thus, the premeal fall of blood glucose level must be considered as also dependent on these premeal events and as a result of some degree of exhaustion of the gastrointestinal tract filled by the previous meal. This is comparable to gasoline in the car and its timing: The car depends on gasoline, which the motor consumes until the red signal on the dashboard indicates a degree of emptiness of the tank.

The physiological signal represented by slight hypoglycemia is necessarily the result of a decreased hepatic glucose production, which suddenly fails to match tissue uptakes. This decrease may be postulated to be in turn the result of a declining intestinal absorption that downstream feeds the liver. The relations between liver input to and output from the liver are not well known.

Simultaneous recordings of the glucose content in the portal and the jugular or hepatic veins (20, 21) before, during, and after a meal do not provide a clear indication of a parallel decrease of glucose concentration upstream and downstream before the meal. A chronic measurement of the same parameters from meal to meal would be needed, and unfortunately this was never carried out. Another postulate must be added: Because the gastric load of a meal apparently feeds the liver during a time varying with the peripheral tissue utilization of metabolites, it is necessary to postulate that both the gastric emptying in the intestine and the intestinal absorption are in some manner regulated as a function of the rate of hepatic glucose production. To continue the image of the car, the feeding of the carburator (equivalent to the liver) and its accelerated or decelerated oil flow into the motor is automatically regulated downward by the pump drawing the gasoline from the gasoline tank. Some indications support this postulate. Newman and Booth (22) showed that the rates of gastric emptying and of intestinal absorption were equal; therefore, the dry matter and carbohydrate intestinal content were constant over time. Then, the emptiness of the stomach would lead to a drop of intestinal absorption. Interestingly, the authors showed that this rate of intestinal absorption, and thus of the gastric emptying, was at night twice that of the day, contemporary with the high nocturnal rate of feeding and food utilization. Vachon and Savoie (23) confirmed both the elevated gastric emptying and intestinal absorption at night. During the day, intestinal absorption progressively declines. The rapid repetition of a meal following the withdrawal of foods from the stomach, i.e., the gastric emptying, also argues for this mechanism. In addition, Snowdon (24) showed that the duration of the gastric emptying of a meal is exactly the same (110 min) as the mean duration of postprandial satiety following an identical meal. The initiation of the subsequent meal at the time of complete gastric emptying was so suggested.

The question then arises as to the mechanism by which the rate of hepatic glucose production may rule the rate of intestinal absorption. Two possibilities exist for such a mechanism and should be explored experimentally. The presence of vagal fibers responding to carbohydrates and proteins in the duodenum was demonstrated by Mei (25) and his coworkers. In addition, they showed that the glucose stimulation of these chemosensors could produce a pancreatic insulin release via a vago-vagal loop. The control of hepatic glucose production by pancreatic hormones is well known. Through these efferent–afferent pathways, a signal of some degree of emptiness of the small intestine would cause the premeal fall of hepatic glucose production. The brief peak of insulin shown to precede the premeal fall of blood glucose could be this signal, brought about by the so-called entero–insular axis. Another possibility is based upon the finding by Niijima (26) that the firing rate of the hepatic branch of the vagus nerve varies inversely with the concentration of hexoses in the portal vein. Contrary to various assertions (see Chapter 3, Section I), the involvement of this system and of putative hepatic glucoreceptors is unlikely. Rather, it could be involved in a hepatic vagal afferent and abdominal efferent loop by which the rate of hepatic glucose production could regulate the flow of metabolites from the gastrointestinal tract. Such a system was not investigated and is only plausible.

Finally, the paradoxical effect of the blockade of intestinal absorption must be explained. Mannitol added to food or to saccharin or glucose solutions blocks intestinal absorption. The solid food or solution are emptied from the stomach but accumulate in the intestine. The apparent paradox is that this mannitol effect does not change the meal size but does introduce a dose-dependent delay of subsequent meals. During this delay, a fall of blood glucose identical to that in a short-term deprivation should, on the contrary, stimulate meal onset. This meal initiation does not occur, perhaps because the specific and correct signal is not generated due to the lack of the initial insulin release. Another possibility should be the absence of a permissive action exerted by the normal lowering of the intestinal content, which, through vagal afferents, would be necessary for the understanding of the humoral signal by its brain targets (27).

II. *The Systemic Stimulus of Food Deprivation-Induced Feeding*

As shown in preceding chapters, in the *ad libitum* condition of feeding, the timing of meal initiation depends on the previous caloric intake and on the metabolic utilization of this intake and is caused by a transient fall of blood glucose. It was also shown that, in this condition of free access, the size of the initiated meal is not affected by the previous caloric intake and its utilization nor by the amplitude of the initiating signal. In Chapter 1, Section III, we have seen that, by contrast, a short-term food deprivation increases the subsequent meal size and precipitates the initiation of following meals. In addition, this effect of food deprivation was shown to be lower during the day than at night. What are the metabolic effects of food deprivation accounting for these effects; i.e., what are the stimuli, or the stimulus, of food deprivation-induced feeding?

A. Metabolic Effects of a 12-hr Fast Compared during the Two Parts of the Diurnal Cycle

Figure 2.4 illustrates the evolution of blood glucose and PFFA levels during a 12-hr fast during the nocturnal and diurnal periods, both compared to

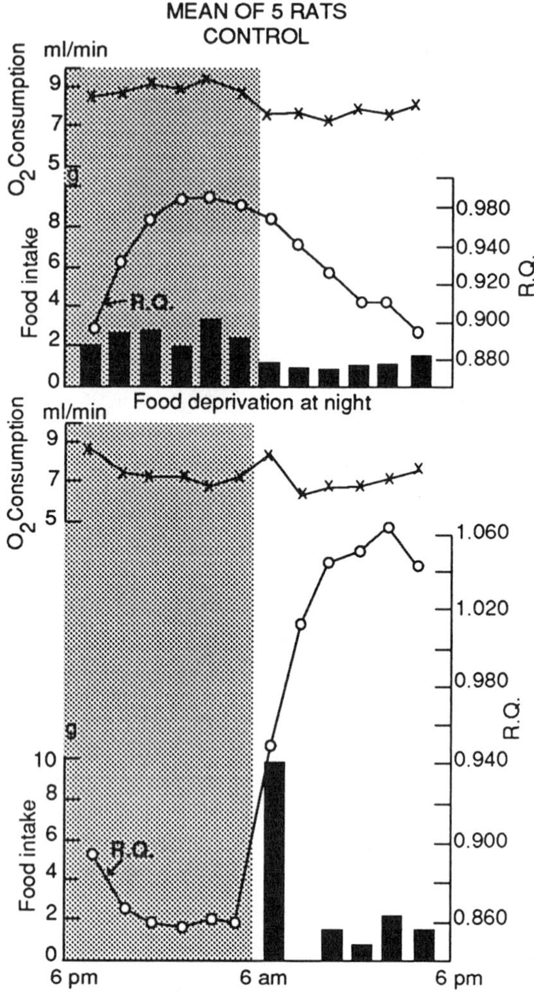

Figure 2.4 Metabolic rates (O_2 consumption) and respiratory quotients (R.Q.) during a nocturnal fast and the subsequent diurnal refeeding—food consumption during this 12-hr diurnal refeeding.

levels in fed rats. At night, a progressive linear decrease of plasma glucose and increase in PFFA are induced by 0–12 hr of food deprivation. A regression analysis shows that at night a fall of 2.8 mg of blood glucose level per hour is significantly correlated to an increase of the PFFA level by 14 micro-equivalents/hr. During the day, the analysis of variance displays that a quadratic equation fits the evolution of the fasting blood glucose over time. From 0 to 8 hr of fast, a mean linear blood glucose level decrease of 1.2 mg/hr (vs. 2.8 at night) is observed. After 8 hr of fast, a steeper fall of blood glucose level is induced (3 mg/hr), thus comparable to values at night. This diurnal fasting induces a progressive elevation in PFFA level; it is slower than at night (8.5 microequivalents versus 17 at night).

The evolution of O_2 consumption and CO_2 expiration during the same 12-hr dark and light fasting was studied. Rats were placed in the calorimetric chamber of an open-circuit device in which their free-food intake could be recorded. After habituation, O_2 consumption and the respiratory quotient (RQ; volume of expired CO_2 to volume of O_2 consumption) were recorded in free-fed and then in 12-hr deprived rats, during either dark or light. Figure 2.5 illustrates the evolution of O_2 consumption and the RQ averaged on 2 successive hr during dark and light times of fasting. In comparison with fed rats, deprived rats exhibit at night a slight decline of O_2 consumption and a marked drop of the RQ, indicating the occurrence of a time-dependent lipolysis. In fed rats, both O_2 consumption and the RQ are lower during the day than at night. Under fasting, both are decreased further; however, the RQ remains higher during the diurnal than the nocturnal fasting. Rats refed during the day after the nocturnal fast show a very elevated RQ, indicating an active fat synthesis; refed at night following a diurnal fast, they show an unchanged RQ level.

B. Feeding Responses to Fasting

Short-term feeding responses to 2–10 hr of food deprivation recorded either at night or during the day were reported in Chapter 1, Section III. Following the same time of deprivation, the increase in the first meal size was 2.4 times lower during the day compared with that at night. As indicated elsewhere, the amount eaten within a meal in deprived rats may be considered as a measure of the strength of the stimulation to eat present at meal initiation. Thus, apparently the same time of fasting induces an enhancement in the stimulation to eat, which is 2.4 times less during the day than at night. It is assumed that the difference is based on the parallel difference observed in the fall of blood glucose level by fasting during the two periods. At night, the fast-induced decrement of blood glucose level plotted against the fast elevation in the first meal shows that a decrement of 2.8 mg/hr of blood glucose level is associated to an increment of 0.5 g/hr of the first postfast meal. During the first hours of the day, a fall of blood glucose level of 1.2 mg/hr of fast is correlated to a rise of 0.23 g of the first meal. Thus, it is striking to observe that during the two periods an identical increase of the strength of stimulation to eat at the beginning of the first postfast meal is

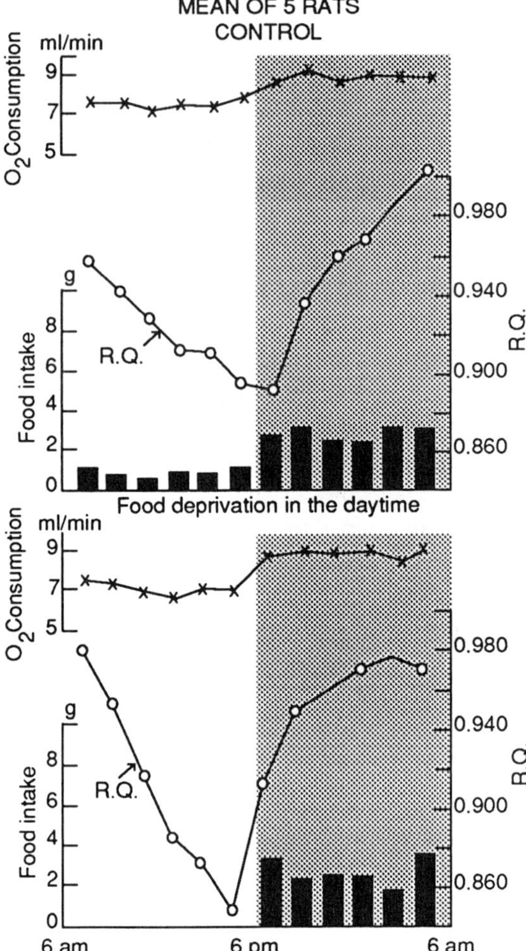

Figure 2.5 Metabolic rates (O_2 consumption) and respiratory quotients (R.Q.) during a diurnal fast and the subsequent nocturnal refeeding—food consumption during this 12-hr nocturnal refeeding.

induced by the same fall of blood glucose level: 2.7 versus 3.0 g added to the meal per 10 mg of fall of blood glucose.

From these results, it was concluded that (1) contrary to the free-feeding condition, in which a transient fall of blood glucose as an all-or-none stimulus initiates the meal without affecting its size, a fall of blood glucose level in a short-term food deprivation is a graded stimulus to eat, which increases the size of postfast meals; (2) the same decrement of blood glucose provides the same strength of stimulation of meal intake during the two parts of the diurnal cycle; and (3) the lower response to food deprivation during the day is parallel to the lower rate of decline of blood glucose level by fasting during this period (Figs. 2.6, 2.7, 2.8).

In a complementary experiment, the relationship between the fast-induced fall of blood glucose and the shortening of the meal-to-meal interval relative to meal size in refed rats was investigated. Six hours of food intake at night and during the day following 0–6 hr of food removal prior to the refeeding were recorded. Changes of the first meal and of the first postmeal intervals in refed rats

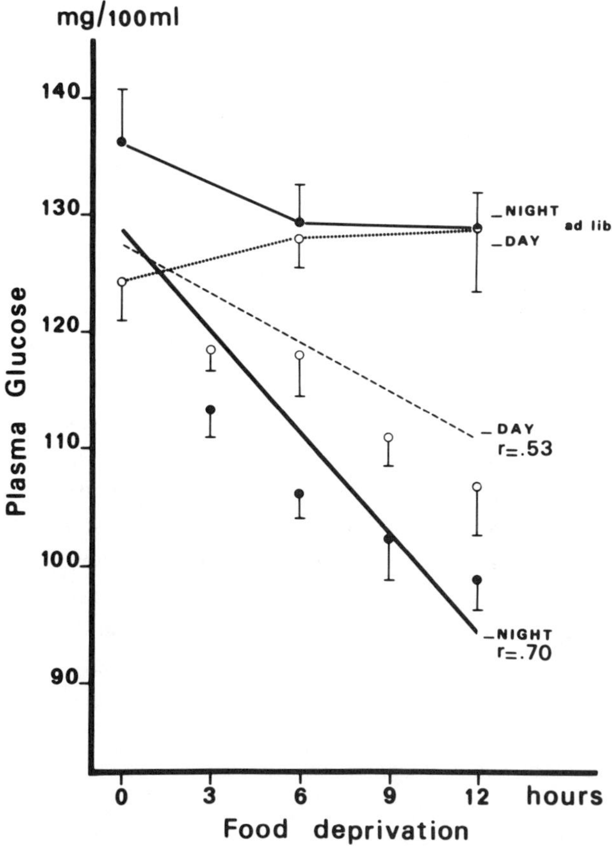

Figure 2.6 Evolution of blood glucose level during a 12-hr fast compared between night and day to control fed rats.

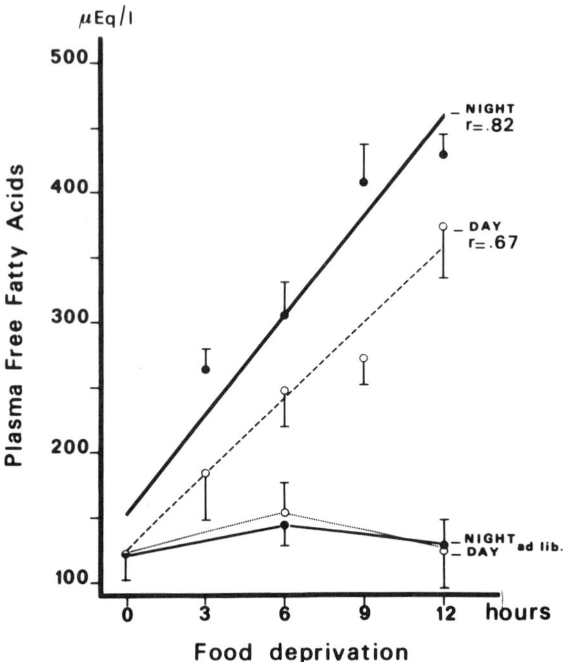

Figure 2.7 Evolution of plasma free fatty acid level during a 12-hr fast compared between night and day to control fed rats.

were correlated to blood glucose and PFFA levels prior to and 30 min after the end of the first meal. At night, the relative shortening of the postmeal satiety was shown associated with the fast-induced fall of blood glucose level and rise of PFFA recorded prior to the meal. By contrast, blood glucose levels 30 min after the end of the first meal were not different from the mean level in the *ad libitum* condition; rather, the PFFA level at this time increases as a function of fast duration. During the day, the 2–6 hr of food deprivation at the beginning of the period did not produce a significant fall of blood glucose level, and no relations appeared between this level at the sixth hour and both the first meal and the postmeal satiety; they were unchanged (28–31).

Thus, it seems clearly demonstrated that the decrease in blood glucose level induced by a short-term nocturnal or diurnal fasting is the physiological systemic stimulus of the feeding response to such fast durations. No explanation is available for the deficiency of the feeding response to 24 and 48 hr of food deprivation. The fasting metabolism established at this time, including lipolysis

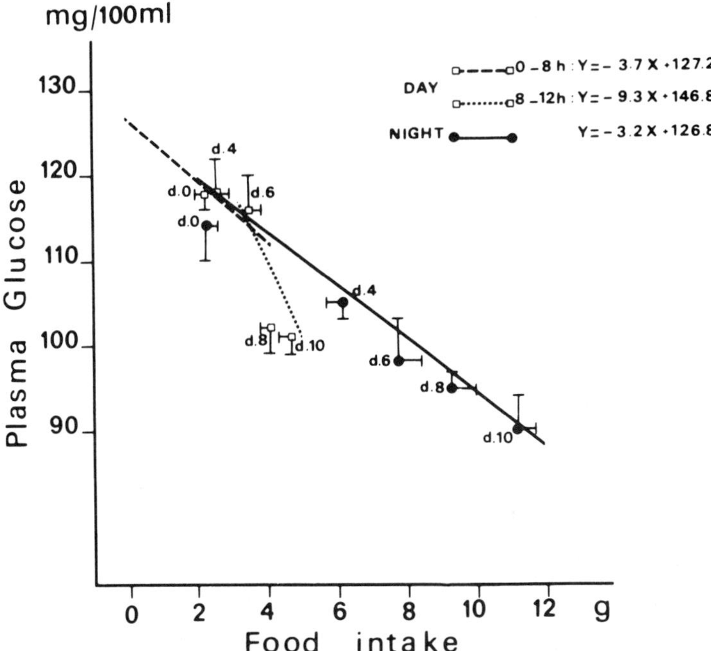

Figure 2.8 First meal following short-term food deprivation as a function of the fall of blood glucose induced by the deprivation at night and day.

and fat oxidation, might be suspected as the cause of this deficient response in refed rats.

III. *Metabolic and Neuroendocrine Bases for the Diurnal Periodicity of Feeding*

As previously described, a diurnal periodicity of feeding in free-fed rats results from large and frequent meals at night and small and infrequent meals during the day. The same caloric intake in a meal was shown followed at night by a postprandial satiety two to three times shorter than that during the day. The postprandial satiety terminated by the onset of the subsequent meal was found correlated to previous caloric intake at night and poorly or not at all during the day. Moreover, under food deprivation, the same response to a short-term food removal was obtained after a deprivation two to four times longer during the light period than during the dark period. These and other facts suggested that meal onset may depend on the metabolic food utilization since the previous meal.

Hence, the control of meal frequency was thought to be sensitive to metabolic events and to the steady-state or shifts of body energy balance.

A. The Lipogenesis–Lipolysis Diurnal Cycle

The investigations of the lipogenesis–lipolysis diurnal cycle and of its role in the diurnal periodicity of feeding necessitated a simultaneous measurement of the meal pattern and of metabolic events in unanesthesized rats. This was achieved in the 1960s by the simultaneous recordings of the free-feeding and respiratory exchanges of rats during one or a succession of days. Using this technique, a series of works (32–34) demonstrated (1) the existence of a nocturnal lipogenic and a diurnal lipolytic cycle, (2) that fat synthesis at night was a cause of fat mobilization and oxidation during the day, (3) that the former was a cause of the high frequency of meals at night, the latter of prolonged satiety during the day, and (4) that neuroendocrine mechanisms governing insulin releases and glucose utilization underlied this metabolic feeding cycle to achieve the daily energy balance.

Ten adult rats were used in the initial works (32). They were closed in a metabolic chamber in which a dark–light cycle (light on at 5 A.M.) was monitored. Fourteen cycles were obtained and analyzed. It is well known that the RQ (volume of expired CO_2 to volume of O_2 consumption) averaged for sufficient long periods of time reflects the ratio of CO_2 release to O_2 uptake involved in fat or protein synthesis and glucose and fat oxidations. A RQ lower than the RQ of food oxidation (0.94) indicated lipolysis and fat oxidation. A higher value indicated a fat synthesis added to the oxidative metabolism. In addition, equations derived from the biochemical bases of fat synthesis and oxidation permitted calculation of the amount of calories either used in fat synthesis or yielded by fat oxidation. The 10 rats exhibited a clear lipogenetic–lipolytic diurnal cycle. During the dark period, after 2 hr of low RQ, the 10 subsequent hr indicated a high fat synthesis. On average, this fat synthesis was 0.67 kcal/hr and 8 kcal/hr during the 12 hr. During the day, a fall of RQ from the fourth to the twelfth hours indicated a lipolytic fat oxidation phase. On average, 6.6 kcal was derived from fat oxidation and contributed to oxidative metabolism. Comparison between feeding and metabolic rates (O_2 consumption) showed that at night the feeding rate was 41% higher than the metabolic rate; thus, a strongly positive energy balance was present. During the day, the feeding rate was 52% lower than the metabolic rate; thus, a strongly negative energy balance was observed. The mean rate of lipogenesis during the night represented 24% of the metabolic rate, whereas the mean rate of calories supplied by fat oxidation during the day represented 24% of the current metabolic rate. Thus, the excess of nocturnal intake over the metabolic rate was higher than accounted for by fat synthesis. This confirmed the suggestion that, in addition to fat storage, a storage of food in the gastrointestinal tract (particularly at the end of the night) participated in this nocturnal energy storage. During the day, the intake deficit also exceeded the rate of fat oxidation. These initial data were later reexamined and confirmed in other groups of rats. However, it was shown that the amplitude and even the presence of the diurnal lipogenetic–lipolytic cycle were variable among rats. In 30 rats, the mean cycle is illustrated in Figs. 2.9 and 2.10. Note in this figure that the PFFA concentration cycle follows this metabolic cycle with low values during fat synthesis and high values during the diurnal lipolysis.

The discovery of the fat synthesis–fat oxidation diurnal cycle was fully confirmed by other investigators. Measurements of carcass fats in rats sacrificed every 2 hr at night showed increasing fat body mass along the period (35). ^{14}C-acetate injected at night and not during the day labeled fats (36). A weight gain–weight loss cycle parallels the fat storage–fat mobilization cycle. At the end of the night, rats gain some gram, partly lost at the end of the subsequent day (37). Other investigators also confirmed by various means the high fat and glycogen synthesis and low PFFA level at night, and lipolysis and fat oxidation and increasing PFFA during the day (38–40). The study by Cornich and Cattene (38) showed that fat synthesis at night was higher in the liver and the rest of the carcass than in the adipose tissue. They suggested that the rest of the carcass is the intramuscular depots of fats.

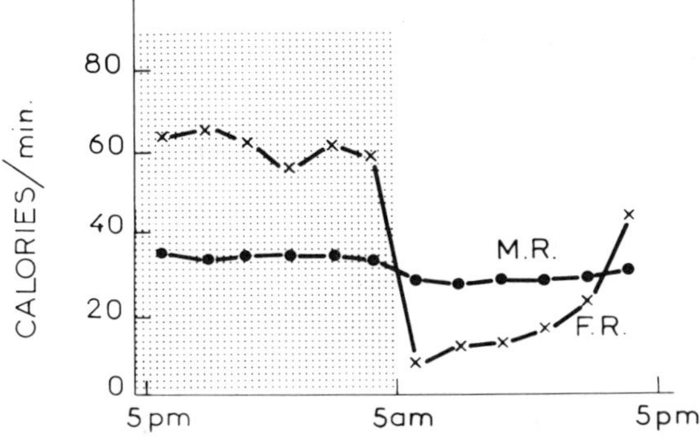

Figure 2.9 Evolution of feeding rates (F.R.) and metabolic rates (M.R.) during the diurnal cycle (average of 30 rats).

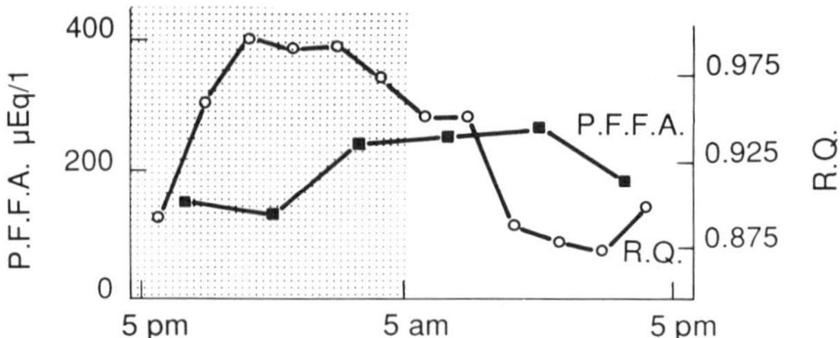

Figure 2.10 The lipogenesis–lipolysis diurnal cycle. Evolution of the respiratory quotient (R.Q.) and of plasma free fatty acid (P.F.F.A.) levels throughout the diurnal cycle (average of 30 rats).

B. Nocturnal Energy Intake and Storage as a Cause of Low Intake and Fat Oxidation during the Day

Dependency of low intake and fat oxidation during the day on the previous high energy intake and storage at night was investigated. As already illustrated in Fig. 2.4, food deprivation at night leads to lipolysis and fat oxidation during the period. This prevention of the nocturnal energy storage immediately leads to a drastic change of feeding and metabolic pattern during the subsequent day, with a mean feeding rate and high fat synthesis reproducing a normal nocturnal meal pattern. In the initial 10 rats, feeding in the diurnal period was predictable based on feeding and metabolic events measured in the previous nocturnal period. In other studies, a negative significant correlation was found between the total nocturnal and subsequent diurnal intakes. The more the rat eats at night, the lower during the day. The total intake at night (sum of meal sizes) is positively correlated to the sum of times of satiety (meal-to-meal interval) during the subsequent day. Thus, in addition to the correlation between meal sizes and postmeal satiety at night, apparently a comparable correlation exists between night intake and day satiety. It was a new, clear indication that prolonged satiety between short meals during the day was an effect of the metabolic utilization, of the energy stored in the previous nocturnal period.

Among others the following experiment demonstrated that substrates oxidized during the day were derived from fats and foods stored during the previous night. Rats were presented a ^{14}C-labeled food (1.8 mC/100 g) at night and the same unlabeled food the subsequent day, or, conversely, the labeled food offered during the day and the unlabeled one the subsequent night. The total CO_2 and $^{14}CO_2$ expirations were continuously recorded along with the meal pattern. The disintegration per minute was calculated. When the labeled food was ingested during the night, only 62.6% of the ingested radioactivity was recovered in $^{14}CO_2$ expired during the period. The remaining 24.4% as expired during the subsequent light period and beginning of the following night (Figure 2.11), while rats ate the unlabeled food at this time. By contrast, when the labeled food was eaten during the day, 93.2% of ingested radioactivity was recovered during the period. Strong evidence thus indicated that energy stored at night was actually oxidized during the subsequent day, but the most interesting fact was that the specific radioactivity of expired CO_2 during the day appeared to be synchronized with the meal pattern. When the unlabeled food was eaten during the day after the labeled one the previous night, the $^{14}CO_2$ output progressively increased through the long meal-to-meal intervals, reaching the higher level at the end of this interval with the meal onset. When the labeled food was eaten during the day, the highest $^{14}CO_2$ expiration occurred not immediately before but immediately after the intake of meals. It was a striking indication that fat oxidation during the day was a cause of prolonged satiety from meal to meal and of the hypophagic pattern at this time (Figs. 2.12 and 2.13).

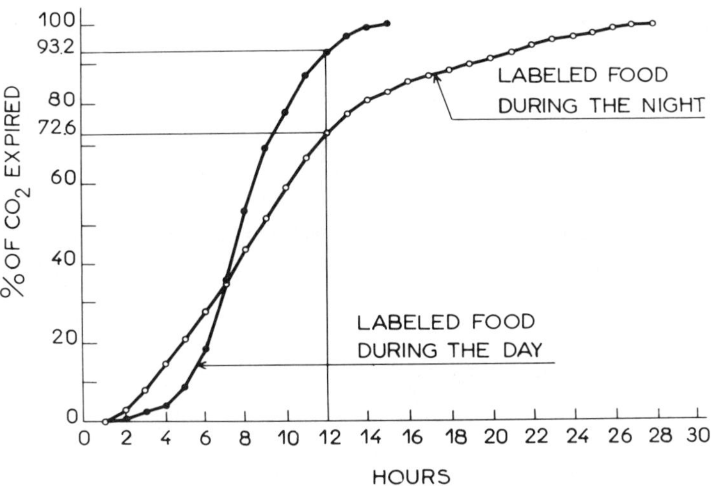

Figure 2.11 Expiration of $^{14}CO_2$ during the night and the subsequent day from ^{14}C-labeled food eaten at night.

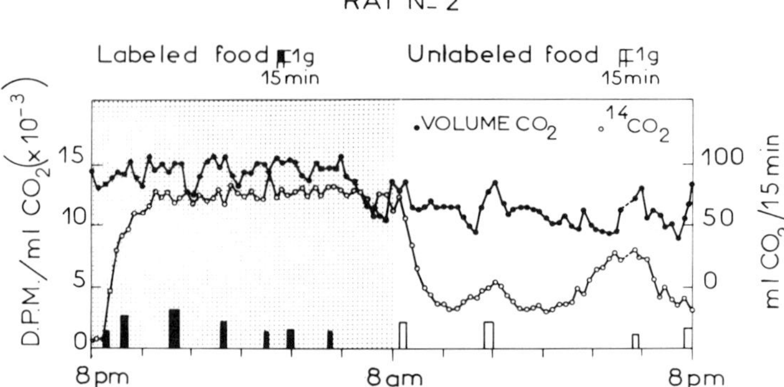

Figure 2.12 Expiration of $^{14}CO_2$ during the day in relation to the meal pattern from ^{14}C-labeled food eaten the preceding night.

C. Fat Oxidation as a Cause of Prolonged Postmeal Satiety

The prevention of fat oxidation by various means during the day and its absence in hyperphagic hypothalamic rats readily abolishes the typical diurnal meal pattern and establishes a nocturnallike pattern. This is a clear-cut demonstration that fat oxidation and the internal supply of fuel are responsible for the delayed meal initiations.

The effects of the food deprivation at night already provided evidence for this fact. In hypothalamic hyperphagic rats following VMH lesion, it was demonstrated that lipolysis and fat oxidation during the day were absent (see Chapter

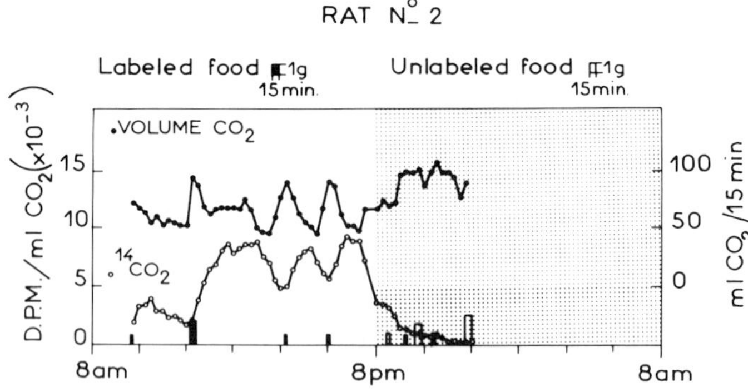

Figure 2.13 Expiration of $^{14}CO_2$ during the day in relation to the meal pattern from ^{14}C-labeled food eaten during the same day.

9, Section III). In contrast to intact rats the RQ was found during the day at the same level as that observed at night. This 24-hr fat synthesis was claimed to be the cause of their weight gain leading to obesity. As an effect of this nocturnallike fat synthesis during the day, their meal pattern by this time exactly reproduces the normal nocturnal meal pattern. The postprandial correlation, normally nul or low at this time, becomes highly significant, as at night (41).

Injecting long-acting insulin (protamin-zinc-insulin) prevents diurnal lipolysis and accentuates nocturnal fat synthesis. Consequently, onsets of meals are precipitated at night and the nocturnal meal pattern was introduced during the day. Injecting continuously through the implanted cardiac catheter $0.1-0.6\mu/h$ regular insulin during the day also established the nocturnallike meal pattern. Moreover, this fat synthesis, experimentally induced during the day, produced a compensatory weight loss and a daylike meal pattern during the subsequent night (42). Finally, a blockade of fat oxidation by β-mercuro-acetate was shown to increase the daily food intake, mainly by suppressing the prolonged postmeal intervals during the day (43). This effect, erroneously called "lipoprivic feeding," was in fact an effect of the alleviation of satiety induced by fat oxidation.

D. Saving of Glucose Utilization as a Cause of Prolonged Satiety by Fat Oxidation

We have seen that a transient fall of blood glucose triggers the start of a meal in the night as well as in the day. How does fat oxidation act in the latter period to delay this event? It is suggested to be due to the decreasing use of foods stored in the gastrointestinal tract by the previous meal intake. As indicated earlier, the rate of intestinal absorption is twice the diurnal level at night. On the other hand, elevation of PFFA as well as a pharmacologically induced increase in fat oxidation produces a parallel lowering of glucose utilization, i.e., a sparing of the utilization of carbohydrates supplied by the food. In dogs and humans, injecting ^{14}C-glucose after an elevation of PFFA, induced either by the ingestion of high fat diet or by heparine administration, demonstrated a parallel reduction of glucose oxidation and of hepatic glucose production (44, 45). The percentages of glucose converted in CO_2 and of $^{14}CO_2$ derived from glucose fall by 11 and 21%, respectively. The blood glucose level was maintained despite the elevation of plasma insulin concentration. Thus, a relative tissue insulin insensitivity was associated to fat oxidation and the saving of glucose utilization. The same metabolic pattern is present in rats during the day. It is also found in humans during the night (see Chapter 2, Section II). Presumably, fat oxidation is that of peripheral tissue only, not of the brain, which does not oxidize fats and requires a continuous glucose supply. The sparing of glucose utilization by peripheral tissues permits the maintenance of the blood glucose level and the vital supply of glucose to the brain. The low intake during the day, manifested by the delayed drop of blood glucose initiating meals, is a contribution to the vital maintenance of the brain glucose feeding.

E. Neuroendocrine Cycle Associated to the Metabolic Feeding Cycle

Surprisingly, a neuroendocrine cycle associated to the metabolic feeding cycle (glucose utilization, insulin release and responsiveness) was more extensively studied and earlier in humans than in rats. The rate of glucose utilization is higher at night than during the day. In one experiment, 0.5 g/kg body weight was intravenously injected after 5 hr of food removal either at night or during the day. Blood samples taken before and six times after the glucose load tested the rate of glucose utilization. The straight line of the glucose disappearance rate on semi-log coordinates demonstrated the different rates in the two periods with a slope of 3.65 and 2.7, respectively (Fig. 2.14). This was revealed in 5-hr deprived rats. Conflicting results were reported regarding the presence of a diurnal cycle of blood glucose level (46–48). Presumably, the claimed cyclic variations of blood glucose were biased because blood samplings and determinations were not referred to the meal pattern. The probability of samplings to fall during a postprandial hyperglycemia was higher at night than during the day. However, the modern techniques ascertained, as already mentioned, an identical intermeal level during the two periods. However, convergent reports demonstrated the insulin diurnal cycle. Plasma insulin concentration is lower during the day and increases in the last hours of the day in rats (49, 50). In the latter work, rats were fed for 6 hr at night or 6 hr during the day. Identical cycles of insulin responsiveness and glucose disappearance rates were also clearly demonstrated in mice (51–53). In the mouse, the interaction between changes of feeding and the neuroendocrine background was studied by Petersen (54). This elegant study yielded convincing evidence that, at the transition between light and dark periods, a positive feedback between stimuli associated with feeding and insulin release bring about both the high feeding rate and fat synthesis and the hyperinsulinemia present at night.

A seasonal cycle of weight gain–weight loss was extensively studied in hibernators: hamsters, ground-squirrels, dearmice, etc. (53). The neuroendocrine background of these cycles was shown generally to be analogous to that observed in diurnal cycles. A beautiful review of genetically programmed aphagia or hypophagia in many animal species from fishes to mammals was published by Mrosovsky and Sherry (55). These long-lasting hypohagias are comparable to the

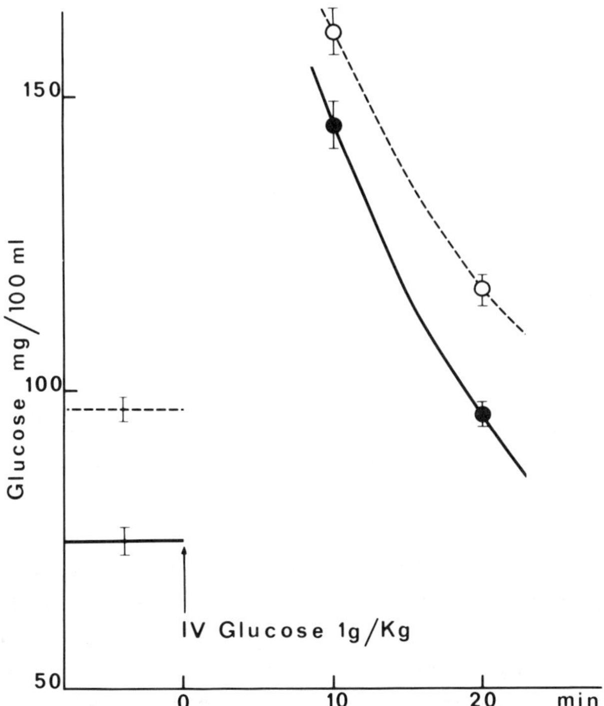

Figure 2.14 Difference of glucose tolerance between night and day in normal rat. ●, night; ○, day.

rat diurnal hypophagia with generally the maintenance of scarce meals. In migrating birds, hyperphagia and an enormous increase in body fats precede the migration. The distance of the migration is proportional to this load of fuel before starting. It is comparable to the night–daytime correlations in rats (56).

IV. *Initiation of Scheduled Meals*

Rats shifted from a free access to foods to a 2-hr daily meal progressively augment their intake in the single meal, at which they eat up to 10–12 gram. In so doing, they maintain their weight (see Chapter 1, Section IV). The progressive and enormous increase of the meal size and therefore of the satiation process suggested a relearning of responses to the food by which the size of a meal is determined (see Chapter 6). How are such scheduled meals initiated?

A. Metabolic Adjustments to Feeding Schedules

One or two "served" daily meals are followed by a strong lipogenesis. It is impossible to cite and detail a great number of works in which this postmeal fat synthesis was recognized. Among them, however, the work of Fabry (57) must be acknowledged. This postmeal fat synthesis is a first indication of the mechanism initiating the meal. This initiation and the meal size are not a response to food deprivation nor to the stimulus seen to be involved in this condition. The meal is not a response to an energy deficit present at the time of food presentation. Rather, the meal is a storing of food, anticipating the energy expenditures until the next meal. A process identical to that seen in "provisional appetite" (Chapter 1, Section III) occurs in the development of such an anticipatory response.

The body energy balance is achieved in a feeding schedule despite the number of fixed daily meals. Two groups of rats, one on a 2-hr meal, the other on 10 meals/day, had, after 4 mo, identical intake, body weight, energy expenditures, and food efficiency. But all these parameters were lower than they were in *ad libitum*-fed rats (58). Consistent with this result, a group of rats pair-fed three times per day in the stomach the same amount of food taken in a single daily meal by another group maintained the same body weight and particularly the same body fats as controls (59). Hepatic glycogen during the meal as well as the insulin concentration 60 min following the start of a single 2-hr daily meal increases. Hyperinsulinemia is higher when the meal is presented during the light period (60). In addition to the high glycogen and fat synthesis associated to the meal, a prominent fact is the maintained euglycemia for the 22 hr separating the daily meals (61). This is new evidence indicating that a fall of blood glucose level, like that induced by food deprivation, is not the stimulus of meal initiation. This maintained blood glucose also explained the response to unexpected meals described elsewhere (62). As an effect of the elevated lipid metabolism, the single daily meal leads to relative glucose intolerance (63). Finally, a big daily meal is associated, as expected, to an augmented prandial thermogenesis as measured by direct and indirect calorimetry (64).

B. Conditioned Stimuli to Eat in Feeding Schedules

External stimuli temporarily associated to the food presentation are demonstrated to be conditioned stimuli (CS) of scheduled meals. The convincing evidence for this mechanism came from experiments by Woods (65). Rats were fed twice daily. After some time of adjustment, an augmentation of plasma insulin concentration was observed at the habitual time of the omitted food presentation

(Fig. 2.15). An identical insulin release occurred under the stimulation by an olfactory cue regularly associated with previous deliveries of foods. The time became a CS to eat only when the meal was offered at fixed regular hours, and the odor only if temporarily associated with food presentation (Fig. 2.16). In another series of experiments, Woods *et al.* (66) demonstrated that external stimuli, repetitively paired with an insulin administration or with a glucose load producing an endogenous insulin release, became CS of an insulin release and of a resulting hypoglycemia. The habitual place of the daily feeding also provided a CS. The rat suddenly fed in a new environment reduced its intake, which was further augmented as a function of the exposure time to the new environment (67).

C. Conditioned Alterations of Meal Patterns

Balagura and Harrell (68) and Balagura (69) showed that hypoglycemia induced by insulin or tolbutamide administrations, or hyperglycemia induced by glucagon, might be an external CS. After repetitions of insulin or glucagon injections, the injection procedure alone produced the response. Feeding was stimulated by insulin and retarded by glucagon. These feeding responses also were stimulated by the external stimulus repetitively associated to insulin or glucagon administrations.

Thus, the initiation of regularly programmed meals apparently depends on

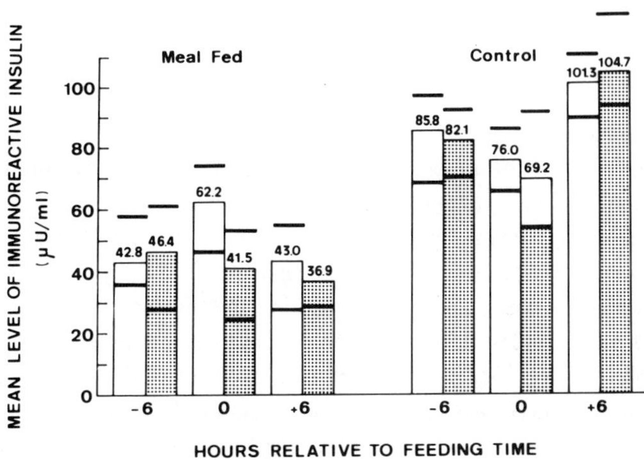

Figure 2.15 Mean levels of immunoreactive insulin for various groups of rats. □, saline; ▨, atropine.

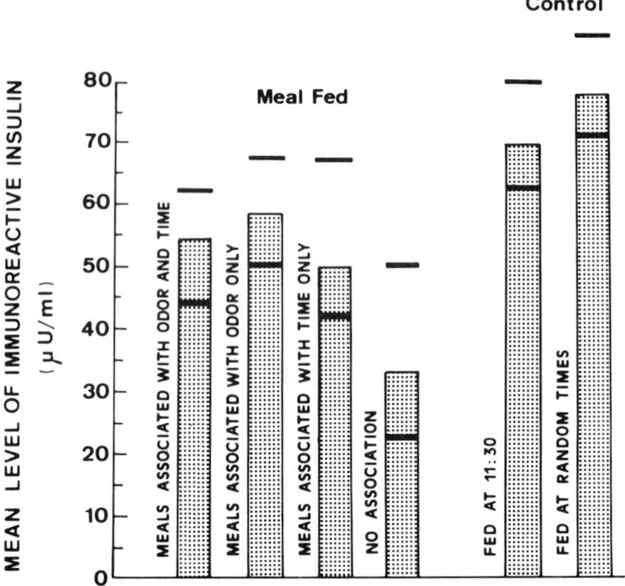

Figure 2.16 Mean levels of immunoreactive insulin for various groups of rats.

external stimuli, mainly the time, acting as a CS to eat, and an associated conditioned insulin secretion and resulting hypoglycemia are presumably a cause of the meal initiation.

D. Conditioned Feeding

Thus, rats adapted to a feeding schedule become spontaneously conditioned to the time of the meal and environmental cues. Is the conditioning of meal initiation in the *ad libitum* condition possible? The positive answer was provided by an experiment by Weingarten (70) (Figure 2.17). Rats were on a schedule of six regularly separated daily meals. During 11 days, a tone was given at the start of every meal, another tone during the intervals. Tested *ad libitum*, rats initiated meals under the stimulation by the tone previously paired with the scheduled meal. Such meals may be provoked by the CS during the intervals between spontaneous meals. These extra meals could reach 21% of the free intake. Investigating whether or not a C.S. of meal initiation was also a C.S. of a fall of glycemia, was of importance. In contrast to previous results (71), Weingarten, Campfield, and Smith (personal communication), using the tech-

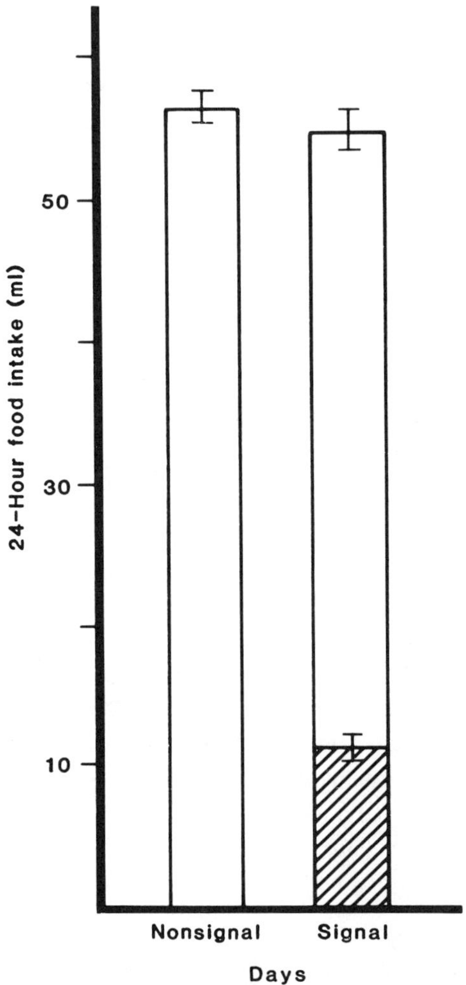

Figure 2.17 Daily food intake of rats on nonsignal and signal days. Hatched bar, intake during a signaled meal; open bar, intake from free-feeding food bottle. Vertical lines indicate SEM.N,7. Weingarten (70).

nique of continuous determination of blood glucose, have shown that an arbitrary chosen stimulus becomes a C.S. of the initiation of a meal and also becomes a C.S. of a fall in blood glucose level. When the C.S. is delivered in the absence of the food, the blood glucose is reduced to lower level. When food is offered, the shape of the fluctuation of blood glucose is similar to the pattern (discussed earlier) to ad lib meals and occurs between the C.S. delivery and the start of the conditioned meal.

Thus, external stimuli repeatedly associated to intake become conditioned stimuli to eat, both in scheduled meals of the Woods experiments cited above and in the Weingarten scheduled meals. In the two conditions, this stimulation of eating by the external stimulus seems to depend on a rise of insulin (Woods), or a fall of blood glucose (Weingarten). We shall see later that the stimulation to eat by these sensory properties of foods, in other words, the palatability of that food is also mainly learned and is also associated to a reflexely induced insulin release known as the "cephalic phase of insulin release."

V. *A Behavioral Self-Regulation of Blood Glucose Level by Rats*

The preceding chapters have shown that the rat regulates its blood glucose level by feeding because minimal hypoglycemia stimulates eating. Rats eat to correct hypoglycemia. Base on these results, the possibility to learn a behavioral response by the reinforcing effects of changes of blood glucose levels and of their association to meals was hypothesized. To test this hypothesis, a series of experiments was carried out to see whether or not a rat could learn a behavioral response other than feeding to regulate its blood glucose level. This learned behavior was the self-intravenous injection of insulin, glucagon, and glucose, alone or combined. Jouhaneau and Le Magnen (72) demonstrated that indeed rats trained to self-inject the glucoregulatory hormones and glucose perform the self-injections in a significant relation to the meal-induced changes of their blood glucose level. The far-reaching significance of this finding (later replicated by other investigators) justifies a brief review of the original work in this chapter.

A. Insulin Self-Injection

Rats with a chronically implanted intracardiac catheter were allowed, by lever-pressing, to self-inject a solution of 5 IU of regular insulin per milliliter at a rate of 1.25 IU/min (Fig. 2.18). Rats learned rapidly to press the lever to obtain insulin. Their daily self-injection, six times that of a saline control, increased progressively during 5 days and then reached a plateau. The daily self-injection, almost exclusively performed at night, averaged 7.7 IU/kg (2–18 IU/kg according to the rat). While rats were allowed to eat freely and their meal patterns were recorded, the periprandial distribution of self-administrations was analyzed. Four phases were considered: the preprandial (10 min before the start of meals), prandial, postprandial (20 min after the end of meal), and extraprandial. Three controls of the periprandial distribution of self-injections were used: (1) the computed probability of random distribution, (2) the distribution in a free-operant condition (no injection by lever-pressing), and (3) saline self-injection. A random distribution would have given 20, 5, 20, and 55% in the preprandial, prandial, postprandial, and extraprandial times, respectively; the distribution in free-operant and saline conditions was not significantly different from this random one. The periprandial distribution of self-injected insulin was considerably different: 55% of the daily injections occurred during the postprandial period, and 79% in the first half of this period. The extraprandial distribution fell to 13%, whereas the preprandial and prandial distributions were not different from those of the control. Thus, rats had learned to self-inject insulin massively during the postprandial hyperglycemia.

The glucagon self-administration was tested in two groups with a rate of 25 μg/min and 10 μg/min, respectively. The training of rats to self-inject glucagon alone failed entirely. In the two groups, an initial self-injection lower than that of saline fell progressively to zero.

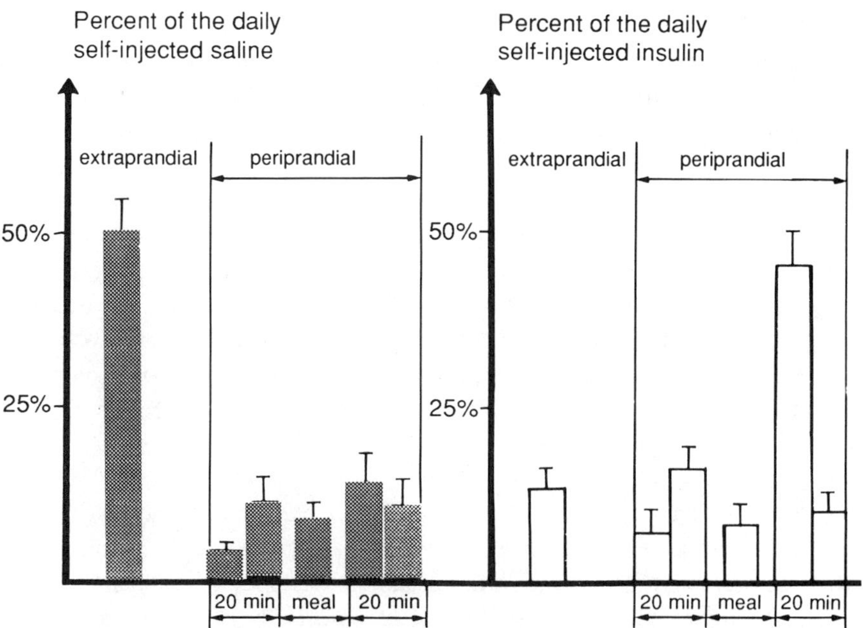

Figure 2.18 Periprandial self-injections of insulin.

B. Combined Insulin and Glucagon Self-Administration

In a combined insulin and glucose self-administration experiment, two levers were available: one glucagon injection, the other for insulin. In the combined administrations, rats self-injected daily 12 times the amount of glucagon self-injected alone, and insulin 4 times less than insulin alone. A relatively constant ratio of the two self-injections was recorded: 9 μg glucagon to 1 IU insulin.

The respective periprandial distribution of the self-injection of the two hormones was very suggestive: insulin versus glucagon—preprandial, 28 versus 30%; prandial, 18 versus 34%; postprandial, 44 versus 27%; extraprandial, 10 versus 19%.

C. Combined Insulin–Glucose Intravenous, or Intragastric Self-Injections

In combined insulin–glucose self-injection experiments, two levers were again available in the cage. Pressing one of them injected a 5% glucose solution intravenously at a rate of 12.5 mg of glucose per minute; the other one injected insulin at a rate of 5 IU/min. Paired with glucose self-injection, rats self-injected three times the daily amount of insulin they injected when paired with glucagon (Fig. 2.19). This daily amount increased progressively from 0.28 IU the first day to up to 3.6 IU after 8 days. The paired glucose self-administration, averaging 35 mg/day, also increased progressively from 4.6 to 49 mg/day, parallel to the insulin increment. During the rise, the ratio between the two self-administrations was very constant (15.8 mg of glucose per 1 IU insulin). Insulin and glucose self-administrations were significantly correlated across days and across rats. The periprandial distribution of insulin was then 30.3, 37.0, and 30.0%, respectively, and of glucose 41.0, 37.7, and 21.3%, respectively. The two distributions were inverse of each other with a maximal self-administration of glucose before the meal and a maximum self-administration of insulin after the meal. A dose of 17.3 mg of glucose per IU was self-injected before the meal, 9.1 mg/IU after the meal.

In another group, rats with chronically implanted intravenous or intragastric catheters were allowed to self-inject glucose intragastrically and insulin as before intravenously. The intragastric self-administration was considerably higher than the intravenous administration of glucose in the preceding group and reached 800 mg/day. A relatively constant ratio of glucose to insulin self-administration was again observed (160 mg glucose per 1 IU of insulin). The peripran-

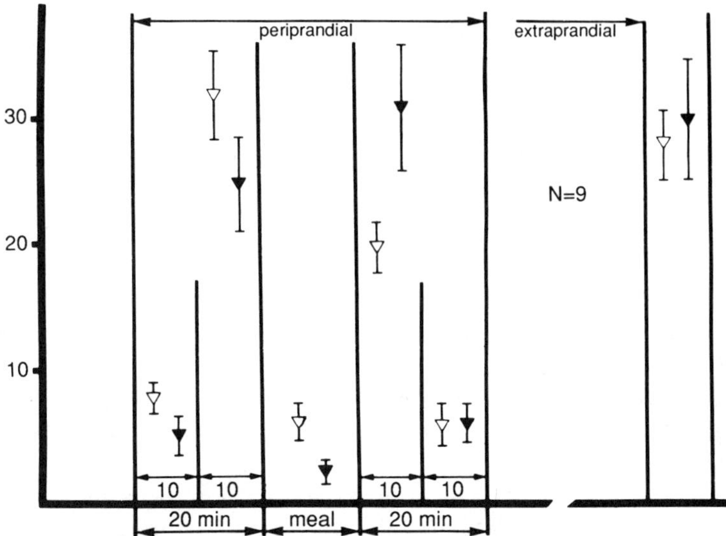

Figure 2.19 Percentage of the daily amount of self-injected insulin and glucose (I.V.).

dial distribution was comparable to that observed when insulin self-administration was paired with intravenous glucose. Again, glucose was maximally self-injected before meals, insulin in the 20 min following the meal (Fig. 2.20).

Consequently, in the two conditions of either intravenous or intragastric self-administration of glucose paired with insulin self-administration, the ratio of

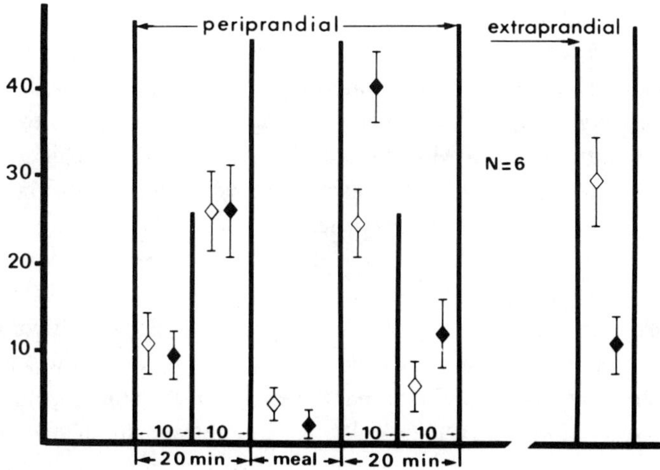

Figure 2.20 Percentage of the daily amount of self-injected insulin and glucose (i.g.).

the glucose to insulin self-administration was exactly inverted from the preprandial to the postprandial period. Rats before every meal are hypoglycemic and press for glucose. Overcompensating for hypoglycemia, rats also press for a substantial amount of insulin. Rats with hyperglycemia press for high amounts of insulin after every meal. As shown in a complementary experiment, rats, overcompensating for hyperglycemia, also press for minute amounts of glucose. In the two conditions of these paired glucose–insulin self-administrations, lever-pressing in the extraprandial periods was extremely low.

Various other complementary experiments in this series confirmed these main findings. Note that in these various insulin, glucagon, or glucose self-administration paradigms neither the size, frequency of meals, nor daily food intake were significantly affected. Even in the condition of high self-administration of insulin alone, the daily intake was not modified and the body weight was maintained.

Thus, the ability of rats to learn a self-regulatory behavior of its blood glucose level was clearly demonstrated. This supported the notion that meal initiation after a slight fall of blood glucose and satiety during intermeal euglycemia may involve a learning process.

VI. *Metabolic and Feeding Effects of Alteration in Glucoregulation and Glucose Utilization*

A. Insulin Administration

The stimulation of food intake and of hunger in humans by insulin administration was observed as soon as insulin was available in the 1920s. Longlasting studies on this effect of insulin administration were limited to its effect on stomach contractions because, at this time, hunger caused stomach contractions. Based on the initial work of Morgan and Morgan (73), using regular insulin in the rat, and of Mackay *et al.* (74), using long-acting insulin in the rabbit, a considerable number of works were carried out. Generally speaking, they are of little interest. The purpose of studying the effects of injecting exogenous insulin—to investigate a possible role of endogenous insulin and of its various physiological activities in the feeding mechanism—was apparently forgotten. In this extrapolation from exogenous to endogenous actions (when it was investigated), researchers overlooked the fact that the effects of injecting insulin are quite unphysiological effects. Physiologically, insulin release stimulated by hyperglycemia counteracts this hyperglycemia, suppresses hepatic glucose production, increases tissue glucose uptake, and mainly induces a lipogenesis. Also, insulin is never released in euglycemia, and it induces hypoglycemia like insulin administration does. Thus, hypoglycemia induced by insulin administration is an artificial situation that provides almost no information about the role of endogenous insulin in feeding. In fact, physiological or pathological states of chronic hyperinsulinemia exist. They are particularly realized in obesity but are not associated with chronic hypoglycemias. Only the hyperinsulinemic–euglycemic clamp, and not acute or chronic insulin administration, can reproduce this state, thus allowing the study of its effects on food intake. Another neglected point is the fact that the effects of endogenous insulin are mainly (if not exclusively) peripheral ones. Convergent and definitive data ascertain that the brain glucose uptake is not insulin-dependent, with the only apparent exception being specific glucosensitive sites in the VMH (see Chapter 9). This possible and site-limited activity of insulin in the brain potentially involved in feeding is mediated either by blood-borne insulin or through cerebrospinal fluid. However, Porte and Woods (75) showed that fluctuations of insulin concentration in the cerebrospinal fluid did not reflect the time course of fluctuations in the plasma.

Mistakes are still caused by experimental procedures that lead to comparisons of noncomparable things, or by the absence of necessary comparisons. The acute effects of injecting acutely or chronically regular or long-acting insulin (protamin-zinc-insulin) are not comparable without caution. The acute effects of insulin tested during the day in rats cannot be extrapolated to its noncomparable effect during the night. The injection time related to the temporal patterning of food intake is a crucial consideration in the interpretation of results. Taking into account the plasma insulin level from meal to meal and its rise during the meal, injecting insulin will have very different effects on this endogenous pattern according to the injection times. Finally, the effects of chronic exogenous administration of insulin, like those of other physiological agents, may not be directly used to understand the physiological effect of the same agent. Injecting insulin chronically adds insulin to its physiological level; therefore, it counteracts its pancreatic insulin release in a compensatory way. Moreover, this chronic administration creates an unresponsive insulin tissue or artificially increases this condition of insulin resistance.

Together, these difficulties have hampered knowledge on the exact role of endogenous insulin *per se* in feeding.

Short-acting normal insulin injected out of meal time in *ad libitum* or slightly deprived animals produces strong hypoglycemia and stimulates eating. In all animals studied including humans, the feeding response occurs after a delay of 20–40 min; however, this delay, generally tested during the day, was not compared between night and day and in undeprived rats according to the timing of injection within the meal-to-meal interval. During this delay, the blood glucose level drops sharply, reaching the unphysiological level of 51 mg/100 ml with the efficient feeding doses. The feeding response occurs after this lower blood glucose level is reached and on a plateau of this level. The response is dose-dependent with an optimal considerable dose of 20 IU/kg (76) (Fig. 2.21). During the day, the response is an increase of meal frequency during some hours, later compensated by a reduced meal frequency. Both plasma insulin and induced blood glucose level are unphysiological effects. A continuous insulin infusion through the intravenous catheter of 1.5 IU/hr for 9 hr produces the increase of cumulative intake by augmenting the meal frequency only. Blood glucose level

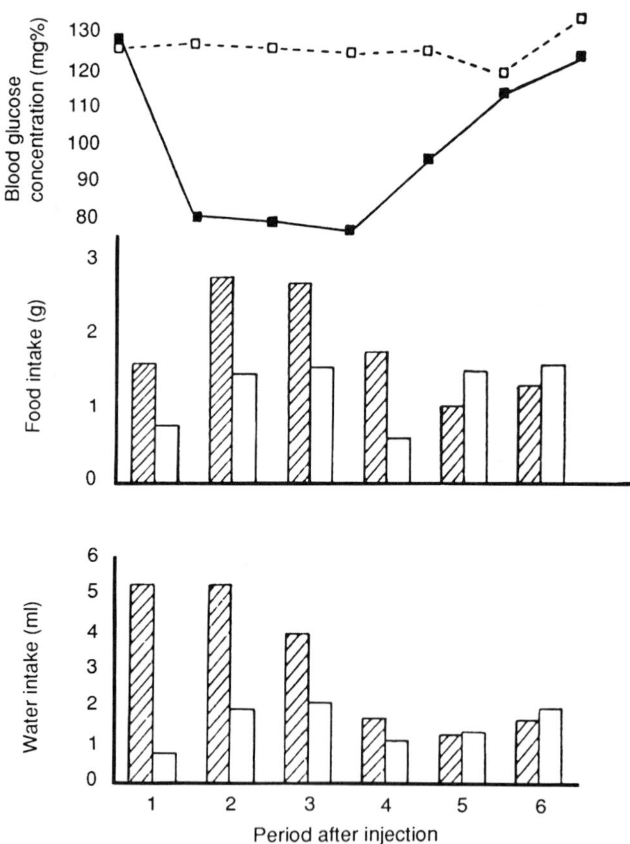

Figure 2.21 Role of blood glucose level in insulin-induced feeding. [Booth and Brookover (76).]

falls sharply (50 mg%) between meals, without reaching a normoglycemic level during meals (77). Intragastric or intravenous infusions of glucose, maintaining the blood glucose level following the insulin administration, prevents its effect. As shown later, the hyperinsulinic–euglycemic clamp by which a high plasma insulin and normoglycemia are simultaneously maintained does not stimulate eating. Thus, feeding is clearly not a response to insulin level *per se* but to induced hypoglycemia.

The use of the chronically implanted intracardiac catheter revealed the different actions of a slow insulin infusion during the two parts of the diurnal cycle. At night, the promoting feeding activity is modest (20%) and is not dose-dependent. During the day, the promoting feeding action of regular insulin is dose-dependent. As already shown, insulin infusion at this time produces a dose-dependent increase of the cumulative intake achieved by an increase of meal frequency. Under the highest dose, a three-to fourfold increase of the diurnal intake is achieved (Fig. 2.22 and 2.23). This diurnal-induced overeating is compensated for by strong hypophagia in the subsequent night. The 24-hr feeding pattern of the light period under insulin and subsequent night, free of insulin

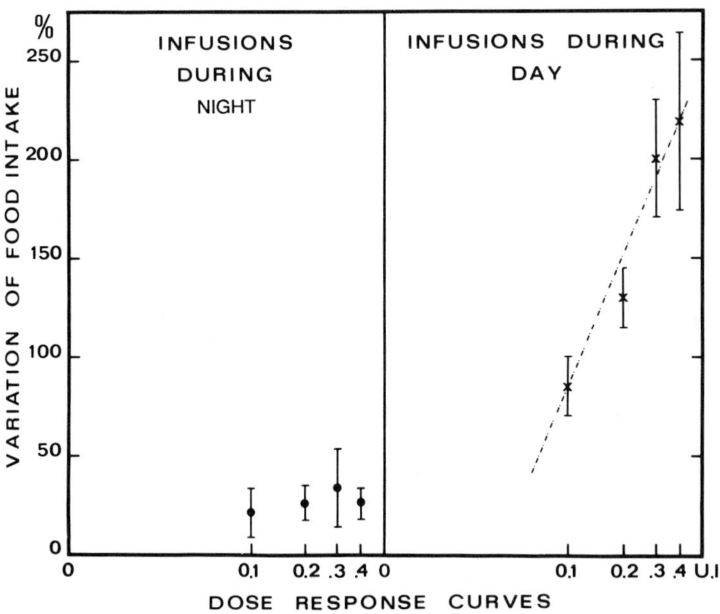

Figure 2.22 Dose-dependent changes of food intake under insulin infusion during the day or night.

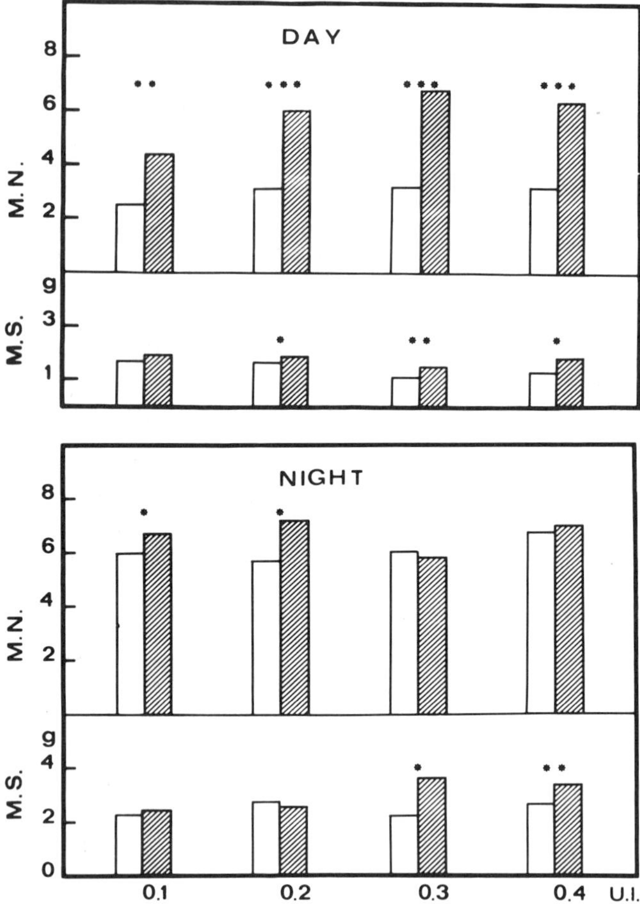

Figure 2.23 Changes in meal size and meal number during a continuous perfusion of regular insulin during night and day: dose–response relationship. ▨, insulin; □, saline.

produces an exactly inverted diurnal feeding pattern (78, 79). The use of long-acting insulin, like that of continuous infusions of regular insulin, abolishes the circadian feeding pattern. In a calorimetric chamber, it was seen that a diurnal-induced high RQ and diurnal fat synthesis are responsible for the observed 24-hr hyperphagia (see Fig. 8.3 in Chapter 8, Section II) (80). During the day and with the optimal dose, the antilipolytic–lipogenic action of insulin not only eliminates the decreased intake by fat oxidation but also affects the condition of fat storage and consequent hyperphagia normally prevailing at night. Thus, insulin *per se* does not promote eating but, like endogenous insulin, acts through the insulin-dependent glucose tissue uptake. This diurnal-induced hyperphagia will be seen also in VMH-lesioned rats and, in these rats, is caused by endogenous hyperinsulinism. These facts proved that insulin administration, in addition to its acute action by the induced hypoglycemia, acts chronically throughout the diurnal cycle by interacting with the lipogenetic–lipolytic diurnal cycle. Under a hyper-insulinic–euglycemic clamp, by which the glucose uptake of insulin-sensitive tissues is compensated for by the glucose infusion, hyperinsulinemia did not induce overeating (Marfaing and Penicaud, personal communication).

This was strikingly confirmed by Towby *et al.* (81). Rats were given a protamin-zinc-insulin treatment and were pair-fed with saline-treated controls to prevent overeating. After some days, insulin-treated rats and controls had the same body weight, but the former showed increased carcass fats achieved at the expense of lean tissue and without hyperphagia.

Some subsidiary facts must be mentioned. Food being removed, a delayed effect of an acute regular insulin administration was observed (82). It is easily explained by the insulin-induced increase of glucose uptake added to the effect of food deprivation. Similarly, cold exposure and insulin administration act synergistically to increase intake (83).

Another significance of insulin given during the meal will be discussed in Chapter 6 in relation to systemic determinants of meal size. The effect of intra-ventricular or local infusions in the brain will also be examined elsewhere in relation to brain functions in the so-called lipostatic mechanism (see Chapters 8 and 9).

B. Glucagon and Epinephrine Administrations

The effects of glucagon administration on spontaneous food intake were also studied. The purpose (often forgotten) was not the study of exogenously administered glucagon but of pancreatic release of glucagon and of its metabolic effects on feeding mechanisms. Here again, a distinction has to be made between a possible action of endogenous glucagon and the effect of injecting exogenous glucagon. The role of endogenous glucagon in the determination of the meal sizes will be discussed in Chapter 6.

Metabolically, glucagon is the opponent of insulin. The pancreatic glucagon released in hypoglycemia counteracts this lowering of blood glucose by augmenting hepatic glucose production and glycogenolysis and, most of all, by mobilizing body fats. Injected in a state of euglycemia, it produces hyperglycemia, a condition as extraphysiologic as the hypoglycemia induced by insulin administration.

Glucagon injected intraperitoneally during the day to 8-hr deprived rats immediately reduced intake by delaying meals. At this time, hyperglycemia and enhanced food utilization were observed. Three hours later, this reduction of intake was compensated for by an augmentation of the cumulative intake due to elevated meal sizes (84). This retardation of meal initiation by glucagon administration was already witnessed by Balagura *et al.* (85). As already mentioned, inasmuch as the injected glucagon counteracts the premeal fall of blood glucose, this retarding effect is new evidence for the role of transient hypoglycemia in meal initiation. In rats fed from 11 A.M. to 4 P.M., glucagon infused at 2 P.M. augmented hepatic glucose production whatever the diet: high-fat, high-carbohydrate, or high-protein diets. But the intake was reduced only in rats fed the high-carbohydrate diet, indicating that glycogenolysis and hyperglycemia were not the only factors involved (86, 87). An acute or continuous intraportal infusion of glucagon also reduced intake during some hours (88).

More convincing was an experiment by MacLaughlin (89). Rats were autoimmunized against glucagon by a glucagon plus an adjuvent administration. A 30% reduction of free glucagon in the plasma, but a 120% increase in total glucagon, was observed, plausibly due to a compensatory mechanism. As an effect of this increase in circulating glucagon, rats drastically reduced their food intake and lost weight.

Thus, the conclusion may be drawn that the glucoregulatory hormone acts on the hunger–satiety system in the opposite way of insulin. Insulin augments intake by lowering blood glucose and inducing fat synthesis, thus precipitating meal initiations. Glucagon reduces intake by counteracting hypoglycemia and presumably by inducting fat mobilization, both leading to prolonged meal-to-meal satiety.

Injecting epinephrine, like glucagon, produces glycogenolysis, hyperglycemia, and lipolysis. In addition, epinephrine administration induces an inhibition of pancreatic insulin release. Also like glucagon, the injection inhibits food intake. Continuously infused at night, it eliminates the nocturnal hyperphagia. The residual meal pattern of small meals and long meal-to-meal intervals resembles the normal diurnal pattern. This nocturnally induced hypophagia is compensated for by a considerable increase in intake on the subsequent light period. This is the opposite of the diurnal insulin-induced hyperphagia followed by the subsequent nocturnal hypophagia (90).

C. Glucose Antimetabolites

In the late 1960s, the latent hypothesis that tissue and particularly brain tissue glucose utilization had something to do with feeding led to experimenting with the effects of a series of molecules interfering with the cellular glycolysis: 2-deoxyglucose (2-DG), 5-thioglucose, and 3-methylglucose.

Intraperitoneally or intravenously administered 2-DG induces hyperglycemia and strongly elevates PFFAs, indicating lipolysis and fat oxidation. The same injections, at first performed during the day, stimulate food intake during some hours by increasing mainly meal sizes in various species: rats, mice, rabbits, and monkeys. In rats, this effect was obtained in a narrow range of doses (250–500 mg/kg intraperitoneally). Later, a contrasting effect between the nocturnal and diurnal periods was pointed out. 2-DG injected at the beginning of the night inhibits feeding for some hours, this early inhibition being partly compensated for during the following hours (91, 92). As with insulin, the stimulating effect during the day is inhibited by a concomitant glucose administration. Convergent results provided by intracerebral administrations (see Chapter 2, Section VIII) provided evidence indicating that both metabolic and feeding effects of 2-DG are due to a central and not to a peripheral activity of the compound. Another assessment of this central action was done by Stricker and Rowland (93). Glucose and mannose but not fructose injections inhibited the feeding effect of 2-DG. While fructose is not accessible to the brain but is peripherally metabolized, new evidence indicated a central activity of the glucose antimetabolite.

Together, the narrow ranges of low doses of 2-DG active to promote feeding during the day are seemingly higher than the threshold of the central action of 2-DG necessary to induce both feeding and lipolysis. But the same doses are apparently less than the threshold of a general blockade of peripheral glucose utilization. Beyond the intraperitoneal dose of 751 mg/kg in the rat, 2-DG exhibits its high toxicity. This high degree of toxicity results from the peripheral and generalized brain glucopenia. Therefore, the data provide evidence for the presence in the brain of specific glucosensitive sites on which 2-DG peripherally injected gives rise to both metabolic changes and an altered feeding pattern. Effects of centrally administered 2-DG will confirm this notion.

Nevertheless, Smith and Epstein (94) proposed the surprising notion that 2-DG, like insulin administration, demonstrated the existence of a specific glucoprivic feeding for them without any relation to the stimulation of a normal feeding. Later works on the brain activity of glucose antimetabolites ruled out this notion, which again was the result of the difficult extrapolation between the effects of exogenous administrations to endogenous physiological mechanisms.

D. Diabetes Mellitus

Injecting alloxan or streptozotocin in rats induces a diabetic state comparable to the human disease. In this state, the animal lacking insulin can no longer metabolize glucose, which is massively lost in urine. It oxidizes fats, proteins, and ketone bodies.

The diabetic rat is at first hypophagic (-30%). After 7–10 days, a hyperphagia is manifested ($+50\%$). Researchers disagree regarding the change of meal pattern through which hypophagia and hyperphagia are realized. According to Booth (95), hypophagia results from prolongations, hyperphagia from a curtailing of meal-to-meal intervals. The postprandial correlation initially present disappears afterward. According to de Castro and Balagura (96), elevated meal sizes and relatively shortened postmeal intervals account for the hyperphagia while the postprandial correlation and the diurnal cycle are maintained. Based on the same observations, Panksepp and Ritter (97) concluded that this altered pattern indicates the reduced postingestive satiety effect of the food. Insulin injection reduces hyperphagia. Intragastric glucose loads still reduce intake but by far less than in controls (98). During the hypophagic initial phase, PFFA, glycerol, and ketone bodies are at a high level, indicating lipolysis and fat oxidation. A return of these parameters to normal coincides with the appearance of hyperphagia. Thus, the initial hypophagia seems to be new evidence indicating that lowering food intake is due to fat oxidation (99).

In a series of works, Friedman and his co-workers (100–103) elucidated the origin of diabetic hyperphagia. Diabetic rats are no longer hyperphagic on a high-fat diet. They respond to a carbohydrate diet exactly like normal rats respond to a caloric dilution of the diet by inert materials. In other words, diabetics are hyperphagic in volume eaten because the carbohydrate content of the food becomes an inert nonmetabolizable material and is lost in urine. They continue to respond to a caloric dilution but in the high-fat diet only. Long-acting insulin injection suppresses the hyperphagia on the low-fat and produces it on the high-fat diet. This insulin-induced hyperphagia on the high-fat diet is greater than it is in the normal rat.

Richter and Schmidt (104) were the first to show that diabetic rats offered a choice between the three macronutrients shifted their choice from carbohydrates to fats. This classical finding was reexamined by various investigators. According to Kanarek and Ho (105), this choice of fats occurs after 2 wk and according to Bartness and Rowland (106), only with slight diabetes. By contrast, the reduction of intake during the initial phase is realized by a low fat intake. Finally, diabetic rats exhibit an aversion to a saccharin solution during initial hypophagia.

However, this response seems to be a trivial taste aversion conditioned by the pairing of taste with general malaise (107).

As already mentioned, in diabetics, like in intact rats, a slight fall of blood glucose, here from the permanently elevated level, stimulates the meal. During chronic hyperglycemia, an inhibition of transport system of glucose through the blood–brain barrier is well documented. Thus, the diabetic rat is not hyperglycemic in the brain, and the peripherally recorded hypoglycemia preceding meal initiation is presumably equivalent to that of intact rats at brain level.

VII. *Metabolic Bases for Hunger and Eating in Humans*

Finally, we come to humans—a difficult animal. Unfortunately, humans speak. They tell when they are hungry or satiated, what they like or dislike and, finally and objectively, they eat. Again the mixture of subjective and objective parameters poses hard problems in examining the present knowledge about relationships among hunger feeling, eating, and various metabolic events.

As already discussed, eating in a feeding schedule of temporarily fixed and permanently spaced meals creates a peculiar condition of feeding in humans like in other animals. The timing of meal initiations is that imposed by the schedule. We take our breakfast, lunch, or dinner not when and because we are hungry but when it is time and because the meal is served. Coming freely to the table, however, we are free to engage in the meal (or not) and to eat (or not) generally various courses until satiety. To clarify the matter, determinants of the amount eaten within the meal and the particular role of likes and dislikes as one of these determinants will be treated in other relevant chapters. Here the question is whether or not some metabolic parameters susceptible to generate a blood-borne or neural signal to the brain are correlated to both hunger and the onset of eating. Inasmuch as the three parameters are measured, a triangular covariance must be examined. Is hunger a prerequisite of eating or not? Is a particular metabolic condition, for instance, blood glucose level, significantly correlated or not to hunger and to its intensity? Are the same events or others correlated or not to the fact that we eat or not?

A. Tissue Glucose Disposal as a Correlate and a Possible Cause of Hunger and Eating

In the frame of the glucostatic theory, pioneering works attempted to relate a measure of tissue glucose uptake to both a scored hunger sensation and eating. In these works, an implicit postulate was that hunger and eating had a common cause; however, this apparently self-evident postulate may be opposed to the common observation that humans can eat their meal with a good appetite in due time without any previous hunger feeling.

In the 1950s, the first experiments used the arteriovenous difference of blood glucose as an index of the general body glucose disposal. Blood samples were taken from the cubital vein and finger arterioles. The difference in blood glucose levels measured in the two samples, called Δ-glucose, was considered as an indicator of the whole body (including the brain) glucose disposal. In the first extensive work, Van Itallie *et al.* (108) assumed the presence of a correlation between the Δ-glucose and the declaration of hunger or satiety by their subjects. A Δ-glucose of the order of 25 mg, i.e., a blood glucose level 30% lower in venous compared to arterial blood, was associated to satiety. A Δ-glucose approaching zero was associated with hunger before a meal or in deprived subjects. The latter fact was surprising: It is difficult to believe that even in a fasting condition the glucose uptake of the forearm, although immobilized, is reduced to zero. To extend their finding, the authors examined the fate of these relations between hunger and Δ-glucose after epinephrine, insulin, and cortisone administrations, in diabetic and hyperthyroidic conditions.

In the most advanced study at this time, the PFFAs high level was taken as an indication of the exhaustion of carbohydrate stores and of a parallel reduction of glucose utilization. Therefore, in three subjects, the correlation between the two parameters with hunger and satiety and, finally, the effect of meals of various compositions were tested in three subjects (108, 109). A rise of glycemia after the free meals was observed. A further rise in the arterial blood resulted in a postmeal rise of the Δ-glucose during 90 min. By the same time, PFFA declined. By 120 min after the meal, the Δ-glucose returned to the premeal level, while PFFA increased to a level sometimes higher than the premeal level. Satiety was declared during 3 and 4 hr until hunger returned. This was concomitant with a low absolute level of blood glucose, of arteriovenous differences, and of the high level of PFFA. In subjects fed a diet lacking carbohydrates, the premeal–postmeal differences of PFFA were maintained, but the parallel variations of Δ-glucose were minimal so that a correlation between Δ-glucose and hunger and satiety was no longer observed. These results remain of high interest after 30 years (Fig. 2.24).

However, a clear-cut relation among Δ-glucose, stomach contractions, and hunger was not confirmed by Stunkard and Wolff (110). The relation was manifested only by some of the subjects examined. Injecting glucose to hungry subjects elevated the Δ-glucose without suppressing hunger, the desire to eat, or the actual eating. Bernstein and Grossman (111) extended the negative result. Two hours after their breakfast, subjects received 200 ml of a 25 or 10% glucose

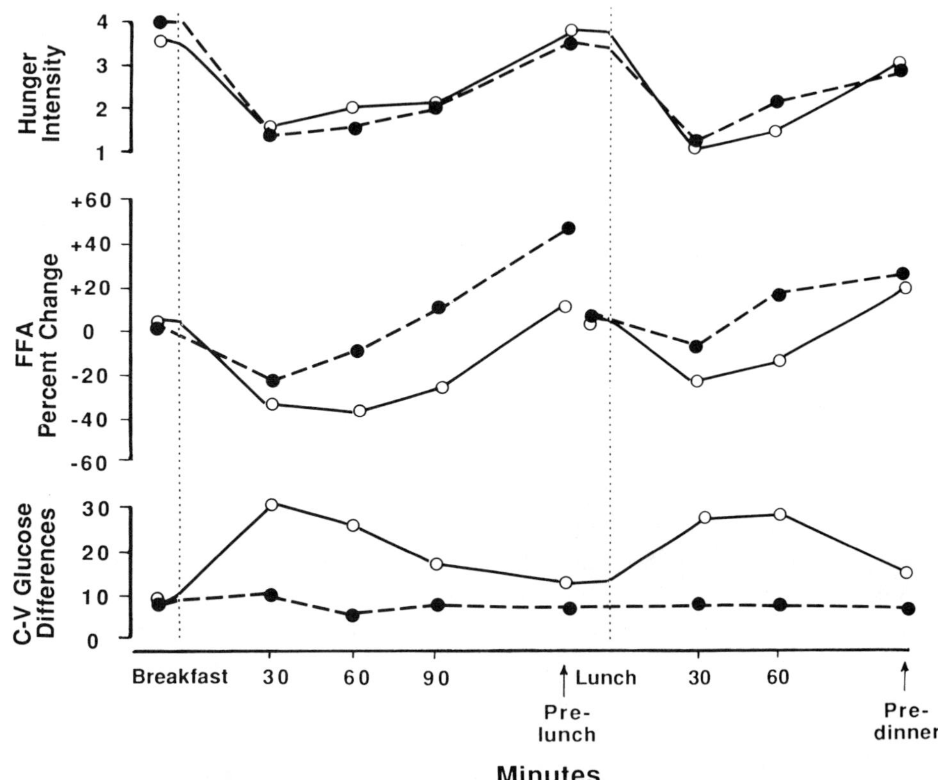

Figure 2.24 Evolution of blood parameters throughout the scheduled meal pattern in humans. ○—○, normal diet (9); ●– –●, CHO-poor diet (7). [Van Itallie and Hashim (109).]

solution intravenously or intragastrically, respectively. Arterial and venous blood glucose was followed until noon. No effect of alterations of blood glucose level was observed due to glucose administrations between the two meals. The most interesting result of the work was to provide, for the first time, a curve of the evolution of the blood glucose level from meal to meal in untreated controls.

Comparable data were provided later by Thornston and Horvath (112). Blood glucose was followed after a breakfast of 750 kcal and compared in adolescent and adult subjects. In adolescents, blood glucose level returned to the preprandial level after 2.3 hr and later fell below this level. In adult subjects, and contrary to adolescents, the blood glucose level was found to be higher than the fasting level during the 6 hr following breakfast.

Almost 25 years passed before a new approach was made. In relation to the finding of the premeal fall of blood glucose level in rats, a similar event at the time of the human scheduled meal was sought. From seven subjects isolated from temporal cues, blood samples were taken at random, averaging every 20 min. Samples taken from 120 to 90 min after the free demand of their meal were also analyzed. Among nine meals studied, five were preceded by a rise and four by a fall of blood glucose level (113). As acknowledged by the authors, this inconclusive result could be due to the 20-min periodicity of blood sampling, presumably too long to see transient hypoglycemia just preceding the meal. Excluding this possible bias, Lucas *et al.* (114) took a continuous microflow of blood through a permanent intravenous catheter from five subjects before a programmed meal in a laboratory setting condition. A decline of blood glucose level during the 20 min preceding the meal was observed in 5 among 18 sessions only. In an unpublished and similar study (cited by Polak *et al.*), subjects asked for their meal within 20 min from the start of a blood glucose level decline or some minutes after the end of this decline (Chabert, unpublished). Finally, a clear cut positive result has been provided by Smith, Rosenbaum, Campfield, and Hirsch (personal communication). They used the technique of prolonged drawing of a microflow of blood and continuous blood glucose determination applied from rats to human subjects. Subjects were placed in controlled conditions in the absence of temporal cues and blood glucose was recorded for 3 to 6 hours. Meal requests followed a pattern of decline in blood glucose in 9 out of 12 subjects.

Meanwhile, a new and different insight on the role of glycemia and insulinemia and on their relationships in responses to foods was provided by Rodin *et al.* (115). The fine technique of insulin and/or glycemic clamp was used in four groups of subjects tested after an overnight fast. A control group and three others were examined with the following clamps: hyperinsulinemia–hyperglycemia, hyperinsulinemia–hypoglycemia, and eu-insulinemia–hyperglycemia. The hunger rating, food pleasantness rating, and intakes of subjects were tested in these four different conditions. The two groups maintained with either hyperglycemia or hypoglycemia, showed an elevation of the amount of food and of feeding responses. In these two hyperinsulinemic conditions, the increase of glucose utilization derived from the glucose infusion needed to maintain the hyperglycemic or hypoglycemic level was augmented six and three times, respectively. This result suggests that under this experimental condition of chronic hyperinsulinemia the increased glucose uptake by endogenous energy stores, either over- or undercompensated for by an external glucose supply, may be surprisingly correlated to a stimulation of eating. The possibility that a chronic hyperinsulinism might lead to an understimulation of brain glucosensitive sites in a large range of glycemia will have to be examined and discussed.

Finally, the emptiness of the stomach also seems to be a factor in human meal onset. In the experiments by de Castro and Kreitzman (116), the stomach content added to the time elapsed since the last meal improved the predictability of the start of a new meal in subjects almost completely free to eat at will.

The parenteral feeding of patients who were or were not allowed to eat orally provides information on humans equivalent to that obtained by the intravenous feeding of rats (117). During a total parenteral feeding, a moderate hunger is maintained. The same was observed in experiments of total self-intragastric feeding (118). Patients fed parenterally but who can eat orally nevertheless eat small meals and declare a slight hunger prior to these consumptions.

B. Conditioned Feeding in Humans

Like in rats, strong evidence indicates that humans eat at the habitual time of their meals conditioned to the time and to environmental stimuli; thus, again like the rat on a feeding schedule, humans engage eating and eat their meal not to repair a previous deficit but rather anticipate further expenses by this intake to remain satiated until the time of the next habitual meal. In Chapter 4, the-so called cephalic phase, or preabsorptive insulin release, will be described in rats and in humans. The sight, odor, and flavor of a food at the start of a meal reflexly induce a peak of elevated plasma insulin. This insulin release is palatability-dependent. It is a conditioned response like palatability is. Presumably, it is involved together in the start of eating and in determining meal size among other factors.

In two experimental studies, the reality of a true conditioned feeding was demonstrated in humans like it was previously in rats. Satiated subjects were exposed to two foods (pizza or ice cream) and then were free to eat or not eat these foods. Following this pairing, the sight of the food actually eaten augments both the score of the desire to eat and the consumption of that food (119). In another study (120), children received a food in the presence of an auditive or a visual stimulus or at a given place (CS+). They received no food in the presence of other discriminative stimuli (CS−). Tested when satiated, children had a shorter latency to eat and ate more food associated with the CS+ than with the CS−. The bell in Pavlov's experiments brings about the salivation of the dog. The bell of the refectory gives us "l'eau à la bouche" and . . . (it is another French logo) "l'appétit vient en mangeant" (Appetite grows as one eats).

C. Metabolic Correlates of Responses to Food Deprivation in Humans

The fact that food deprivation produces hypoglycemia in humans like in animals is not contested. In the already mentioned experiment of recorded chewing–swallowing patterns, the blood glucose level was compared in the same subjects either eating their lunch after a morning breakfast or this first morning meal being omitted. The blood glucose level was found to be 82 mg/100 ml after the overnight fast versus 90 mg in the undeprived condition (121). According to Thornton and Horvath (122), a breakfast of 750 kcal containing 20 g proteins would be needed to maintain the blood glucose level above the fasting level for 3–4 hr. A decline below this level is aggravated by the omission of the midday meal. A lunch of 750 kcal at noon maintains blood glucose above the prebreakfast level for 5 hr. However, this comparison of one and two meals was made in acute conditions. Results in rats suggest that a metabolic adjustment occurs in chronic conditions leading to long-lasting normoglycemia adapted to the habitual feeding schedule. The common sense suggests that the fast-induced hypoglycemia and induced lipolysis have something to do with the acute hunger and the high consumption of food experienced by everyone after food deprivation. But no experimental data are available demonstrating the quantitative and causal relationships between these parameters in short- and long-term fastings. The effect of cold exposure and exercise, which create a condition of energy debt comparable to that of a fast through an enhanced energy expenditure will be examined elsewhere (see Chapter 8).

D. Neuroendocrine Correlates of the Diurnal Periodicity of Feeding in Humans

Humans eat their three or four daily meals during the day and spend 10–12 hr without eating and without hunger from dinner to breakfast on the other day. Thus, in terms of energy balance and inasmuch as this balance is achieved daily, humans consume twice the value of their 24-hr expenditures in their daily meals. The facts that, except in some eating disorders, humans are not hungry at night, they are not awakened by hunger, and breakfast is eaten without a severe hunger and is the smallest daily meal, suggest the presence in humans, like in rats, of a lipogenetic–lipolytic diurnal cycle and of its neuroendocrine background. Also like in rats, these diurnal metabolic cycles are suggested to be causes of a diurnal hunger–satiety cycle. A considerable number of experimental works gave definite evidence for these notions. They can be summarized as follows.

Human subjects tested by indirect calorimetry exhibit a diurnal cycle of O_2

consumption and of the RQ, indicating the presence of a lipogenetic–lipolytic cycle. Both VO_2 and RQ are high during the day and low during the night. Both rise at the beginning of the day before the first morning meal. These cycles are attenuated but maintained in subjects restricted to a protein diet. A parallel insulinemic and glucagonemic cycle is observed and maintained, although reduced in its amplitude during restriction (123). This diurnal cycle of insulinemia, the nocturnal lipolysis, and a body weight cycle were fully confirmed by many studies (e.g., 124, 125). In addition, convergent and abundant results showed a higher insulinemic response to oral or intravenous glucose and higher hypoglycemic response to insulin administration during the day compared to the evening or to the night. Finally, and consequently, a clear-cut diurnal cycle of glucose tolerance was unanimously demonstrated. At night, humans with a high PFFA level, lipolysis, and glucose intolerance are in a state of acute diabetes (e.g., 126, 127).

E. Insulin, Glucagon, and 2-Deoxy-D-Glucose Administration

Early on the hunger-promoting action of insulin administration was experimented with in humans (128). Hunger is declared by subjects 40 min post injection and 27 min after the lowest point of the blood glucose level. The sensation persists beyond the return to normoglycemia. It is not affected by a glucose injection. The most interesting aspect of this old experiment was the demonstration that this insulin-induced hunger was maintained in vagotomized patients. Stomach contractions being eliminated by the vagotomy, this evidence, among others, proved that stomach contractions were not causes of hunger.

The intravenous injection of 2 mg of glucagon abolishes hunger in fasting subjects (129). Glucagon administration reproduces the hyperglycemic prandial pattern without a meal. Hunger reappeared following a return of blood glucose to the normal level.

Injecting 2-DG (250–750 mg) increases the 2-hr subsequent intake in humans. It is later compensated for by a reduction of intake (130). Interestingly, in subjects suffering a spinal cord section who had lost catecholamine-induced hyperglycemia under 2-DG, the feeding response to the glucose antimetabolite was augmented.

References

1. Brobeck, J. R. (1955). Neural regulation of food intake. *Ann. N.Y. Acad. Sci.* **63** (Art. 1), 44–55.

2. Hetherington, A. W., and Ranson, S. W. (1940). Hypothalamic lesions and adiposity in rats. *Anat. Rec.* **78,** 149–172.

3. Anand, B. K., and Brobeck, J. R. (1951). Localisation of a "feeding center" in the hypothalamus of the rat. *Proc. Soc. Exp. Biol. Med.* **77,** 323–324.

4. Grossman, S. P., and Rechtschaffen, A. (1967). Variations in brain temperature in relation to food intake. *Physiol. Behav.* **2,** 379–383.

5. Booth, D. A. (1972). Postabsorptive induced suppression of appetite and the energostatic control of feeding. *Physiol. Behav.* **9,** 199–202.

6. Nicolaïdis, S. (1974). Short and long-term regulations of energy balance. XXVIth International Congress of the Union of Physiological Sciences (IUPS), New Delhi, pp. 122–123.

7. Mayer, J. (1953). Glucostatic mechanism of regulation of food intake. *N. Engl. J. Med.* **249,** 13–16.

8. Steffens, A. B. (1969). Blood glucose and FFA levels in relation to the meal pattern in the normal and VMH-lesioned rats. *Physiol. Behav.* **4,** 215–225.

9. Steffens, A. B. (1970). Plasma insulin content in relation to blood glucose level and meal pattern in the normal and hypothalamic hyperphagic rats. *Physiol. Behav.* **5,** 147–152.

10. Strubbe, J., Steffens, A. B., and De Ruiter, L. (1977). Plasma insulin and the time pattern of feeding in the rat. *Physiol. Behav.* **18,** 81–86.

11. Louis-Sylvestre, J., and Le Magnen, J. (1980). A fall in blood glucose level precedes meal onset in free feeding rats. *Neurosci. Biobehav. Rev.* **4** (Suppl. 1), 43–46.

12. Campfield, L. A., Brandon, P., and Smith, F. J. (1985). On-line continuous measurement of blood glucose and meal pattern in freely feeding rats: Role of glucose in meal initiation. *Brain Res. Bull.* **14,** 605–616.

13. Campfield, L. A., and Smith, F. J. (1986). Functional coupling between transient declines in blood glucose and feeding behavior; temporal relationships. *Br. Res. Bull.* **17,** 427–433.

14. Campfield, L. A., and Smith, F. J. (1990). Systemic factors in the control of food intake. In E. Stricker (ed.), *Neurobiology of food and fluid intake. Handbook of behavioral neurobiology,* Vol. 10. New York: Plenum Press.

15. Balagura, S., Kanner, M., and Harrell, L. E. (1975). Modification of feeding pattern by glucodynamic hormones. *Behav. Biol.* **13,** 457–466.

16. Larue-Achagiotis, C., and Le Magnen, J. (1985). Feeding rate and responses to food deprivation as a function to fasting-induced hypoglycemia. *Behav. Neurosci.* **99,** 176–180.

17. Nicolaïdis, S., and Rowland, N. (1976). Metering of intravenous versus oral nutrients and regulation of energy balance. *Am. J. Physiol.* **231,** 661–668.

18. Porte, D., and Woods, S. C. (1981). Regulation of body weight and food intake by insulin. *Diabetologia* **20,** 273–280.

19. Goldzierer, M. A. (1936). Chronic hypoglycemia. *Endocrinology* **20,** 86–89.

20. Strubbe, J. H., and Steffens, A. B. (1977). Blood glucose levels in portal and peripheral circulation and their relation to food intake in the rat. *Physiol. Behav.* **19,** 303–308.

21. Langhans, W., Geary, N., and Scharrer, E. (1982). Liver glycogen content decreases during meals in rats. *Am. J. Physiol.* **243,** R450–453.

22. Newman, J. C., and Booth, D. A. (1981). Gastrointestinal and metabolic consequences of a rat's meal on maintenance diet ad libitum. *Physiol. Behav.* **27,** 929–940.

23. Vachon, C., and Savoie, L. (1987). Circadian variation of food intake and digestive tract contents in the rat. *Physiol. Behav.* **39,** 629–632.

24. Snowdon, C. T. (1969). Motivation, regulation and the control of meal parameters with oral and intragastric feeding. *J. Comp. Physiol. Psychol.* **69,** 91–100.

25. Mei, N. (1982). Sensory structures in the viscera. In *Progress in sensory physiology,* Vol. 4, H. Autrum, D. Ottoson, E. R. Perl, R. F. Schmidt, I. Shimazu, and W. D. Willis (ed.), (pp. 1–42). Berlin: Springer-Verlag.

26. Niijima, A. (1969). Afferent discharges in chemoreceptors in the liver of the guinea-pig. *Science* **104,** 151.

27. Bernstein, I. L., and Vitiello, M. V. (1978). The small intestine and the control of meal patterns of the rat. *Physiol. Behav.* **20,** 417–422.

28. Le Magnen, J., Devos, M., and Larue-Achagiotis, C. (1980). Food deprivation induced parallel changes in blood glucose, plasma free fatty acids and feeding during two parts of the diurnal cycle in rats. *Neurosci. Biobehav. Rev.* **4** (Suppl. 1), 17–23.

29. Larue-Achagiotis, C., and Le Magnen, J. (1980). Changes of meal patterns induced by food deprivation: Metabolic correlates. *Neurosci. Biobehav. Rev.* **4** (Suppl. 1), 25–27.

30. Larue-Achagiotis, C., and Le Magnen, J. (1982). Effects of short-term nocturnal and diurnal food deprivation on subsequent feeding in intact and VMH-lesioned rats: Relations to blood glucose level. *Physiol. Behav.* **16,** 245–248.

31. Larue-Achagiotis, C., and Le Magnen, J. (1983). Fast-induced changes in plasma glucose, insulin and free fatty acid concentration compared in rats during the night and day. *Physiol. Behav.* **30,** 93–96.

32. Le Magnen, J., and Devos, M. (1970). Metabolic correlates of the meal onset in the free food intake of rats. *Physiol. Behav.* **5,** 805–814.

33. Le Magnen, J., Devos, M., Gaudillière, J. P., Louis-Sylvestre, J., and Tallon, S. (1973). Role of a lipostatic mechanism in regulation by feeding of energy balance in rats. *J. Comp. Physiol. Psychol.* **84,** 1–23.

34. Le Magnen, J., and Devos, M. (1984). Meal to meal energy balance in rats. *Physiol. Behav.* **32,** 39–44.

35. Panksepp, J. (1973). Reanalysis of feeding patterns in the rat. *J. Comp. Physiol. Psychol.* **82,** 78–84.

36. Kimura, T. T., Maji, T., and Ashida, K. (1970). Periodicity of food intake and

lipogenesis in rats subjected to two different feeding plans. *J. Nutr.* **100,** 691–697.

37. Kakolewski, J. W., Deaux, E., Christensen, J., and Case, B. (1971). Diurnal patterns in water and food intake and body weight changes in rats with hypothalamic lesions. *Am. J. Physiol.* **221,** 711–718.

38. Cornich, S., and Cattene, C. (1978). Fatty acid synthesis in mice during the 24 hr cycle and during meal feeding. *Hormone Metab. Res.* **10,** 276–290.

39. Fuller, R. W., and Diller, E. R. (1970). Diurnal variation of liver glycogen and plasma free fatty acids in rats fed ad libitum or single daily meal. *Metabolism* **19,** 226–229.

40. Scott, G., and Potter, V. (1970). Metabolic oscillations in lipid metabolism in rats on controlled feeding schedule. *Fed. Proc.* **29,** 1553–1559.

41. Larue-Achagiotis, C., and Le Magnen, J. (1979). The different effects of continuous night and daytime insulin infusion on the meal pattern of normal rats: Comparison with the meal pattern of hyperphagic hypothalamic rats. *Physiol. Behav.* **22,** 435–440.

42. Larue-Achagiotis, C., and Le Magnen, J. (1983). Effects of a daytime infusion of exogenous insulin on subsequent nocturnal food intake and body weight. *Physiol. Behav.* **30,** 573–576.

43. Langhans, W., and Scharrer, H. (1987). Role of fatty acid oxidation in the control of meal pattern. *Behav. Neural Biol.* **47,** 7–16.

44. Balasse, E. O. (1971). Effect of free fatty acids and ketone bodies on glucose uptake and oxidation in the dog. *Horm. Metab. Res.* **3,** 403–409.

45. Balasse, E. O., and Neef, M. A. (1974). Operation of the "glucose-fatty acid cycle" during experimental elevations of PFFA in man. *Eur. J. Clin. Invest.* **4,** 247–252.

46. Pitts, G. S. (1941). A diurnal rhythm in blood sugar of the white rat. *Am. J. Physiol.* **139,** 109–112.

47. Pauly, J. E. (1967). Circadian rhythms in blood glucose. *Am. J. Anat.* **120,** 627–629.

48. Jolin, T., and Monthes, A. (1971). Daily rhythms of plasma glucose and insulin in rats. *Horm. Res.* **4,** 153–156.

49. Bellinger, L. L., Mendel, V. E., and Moberg, H. (1978). Circadian insulin, growth hormone, prolactin, corticosterone and glucose rhythms in fed and fasted rats. *Horm. Metabol. Res.* **7,** 134–135.

50. Hara, E., and Saito, M. (1980). Diurnal changes in plasma glucose and insulin responses to oral glucose load in rats. *Am. J. Physiol.* **238,** E463–E467.

51. Gagliardino, J. J., and Pessacq, M. T. (1974). Diurnal variations of the action of insulin on the muscle glycogen synthesis. *J. Endocrinol.* **63,** 171–177.

52. Gagliardino, J. J., Bellone, C. F., Angeletti, A., Pessacq, L., and Notti, H. P. (1975). Insulin-induced hypoglycemia: 24 hr variations. *Acta Diabet. Latina* **13,** 134–139.

53. Pessacq, M. T., and Gagliardino, J. J. (1972). Daytime dependence of the serum FFA, glucose and IRI levels. *Acta Physiol. Latino-Amer.* **22,** 87–92.

54. Petersen, S. (1978). Feeding, blood glucose and plasma insulin of mice at dusk. *Nature (London)* **275,** 647–649.

55. Mrosovsky, N., and Sherry, D. F. (1980). Animal anorexias. *Science* **207,** 783–842.

56. Odum, H. P., Rogers, D. T., and Hicks, D. (1974). Homeostasis of the fat components of migrating birds. *Science* **143,** 1037–1038.

57. Fabry, P. (1964). Metabolic consequences of the pattern of food intake. In *Handbook of physiology: Section 6: Alimentary canal,* Vol. 1 (Charles F. Code and W. Heidel) (pp. 31–49).

58. Hill, D. J., Anderson, J. C., Lin, D., and Yakubu, F. (1988). Effect of meal frequency on energy utilization in rats. *Am. J. Physiol.* **245,** R616–621.

59. Bitman, B., Layman, G. K., Glore, S., and Link, G. (1985). Effects of prolonged meal feeding on the body composition of adult rats. *Nutr. Res.* **4,** 476–483.

60. Lima, F., Hell, L., and Timo-Iaria, C. (1980). Carbohydrate metabolism and food intake in food restricted rats: Relationships between the metabolic events during the meal and the degree of food intake. *Physiol. Behav.* **29,** 445–449.

61. Curi, R., and Hell, L. (1986). Metabolic changes of 20 week-restriction schedule in rats. *Physiol. Behav.* **36,** 239–243.

62. Lima, F., Hell, L., Timo-Iaria, C., Dolinkov, K., and Pupo, C. (1980). Carbohydrate metabolism and food intake in weanling rats: Effects of an expected meal. *Physiol. Behav.* **29,** 633–647.

63. Peters, J. R., and Owen, D. R. (1979). Metabolic effects of altered meal frequency in man. *Horm. Metab. Res.* **11,** 524–525.

64. Sugano, Y. (1983). Heat balance of rats acclimated to diurnal 2 hr-feeding. *Physiol. Behav.* **30,** 289–293.

65. Woods, S. C. (1976). Conditioned hypoglycemia. *J. Comp. Physiol. Psychol.* **90,** 1164–1168.

66. Woods, S. C., Vasselli, J. R., Kaestner, E., Szakmari, G. A., Milburn, P., and Vitiello, M. V. (1977). Conditioned insulin secretion and meal feeding in rats. *J. Comp. Physiol. Psychol.* **91,** 128–133.

67. Valle, F. P. (1968). Effect of exposure to feeding-related stimuli on food consumption in rats. *J. Comp. Physiol. Psychol.* **66,** 773–776.

68. Balagura, S., and Harrell, L. E. (1975). Neuroendocrine conditioning: Conditioned feeding after alterations in glucose utilization. *Am. J. Physiol.* **228,** 392–396.

69. Balagura, S. (1968). Conditioned glycemic responses in the control of food intake. *J. Comp. Physiol. Psychol.* **65,** 30–32.

70. Weingarten, H. C. (1983). Conditioned cues elicit feeding in satiated rats. *Science* **120,** 431–433.

71. Weingarten, H. P., and Martin, G. M. (1989). Mechanisms of conditioned meal initiation. *Physiol. Behav.* **45,** 735–740.

72. Jouhaneau, J., and Le Magnen, J. (1980). Behavioral regulation of the blood glucose level in rats. *Neurosci. Biobehav. Rev.* **4** (Suppl. 1), 53–63.

73. Morgan, C. T., and Morgan, J. D. (1941). Studies in hunger: The effect of insulin upon the rat's rates of eating. *J. Genet. Psychol.* **67,** 153–156.

74. MacKay, E. M., Callaway, J. W., and Barnes, R. (1940). Hyperalimentation in normal animals produced by protamine insulin. *J. Nutr.* **20,** 59–66.

75. Porte, D., Jr., and Woods, S. C. (1981). Regulation of food intake and body weight by insulin. *Diabetologia* **20** (Suppl.), 274–280.

76. Booth, D. A., and Brookover, T. (1968). Hunger elicited in the rat by a single injection of bovine crystallin insulin. *Physiol. Behav.* **3,** 439–446.

77. Steffens, A. B. (1969). The influence of insulin injections and infusions on eating and blood glucose in the rat. *Physiol. Behav.* **4,** 823–828.

78. Larue-Achagiotis, C., and Le Magnen, J. (1983). Effects of a daytime infusion of exogenous insulin on subsequent nocturnal food intake and body weight in rats. *Physiol. Behav.* **30,** 573–576.

79. Larue-Achagiotis, C., and Le Magnen, J. (1979). The different effects of continuous night and daytime insulin infusion on the meal pattern of normal rats: Comparison with the meal pattern of hyperphagic hypothalamic rats. *Physiol. Behav.* **22,** 435–440.

80. Le Magnen, J., (ed.) (1986). *Hunger.* Cambridge: Cambridge University Press, 164 pp.

81. Towby, N., Bhacco, E., Geliebter, A., Stewart, J., and Hashim, S. (1985). Insulin increases body fats despite controlled food intake and physical activity. *Am. J. Physiol.* **21,** 1–4.

82. Ritter, R. C., Roelke, M., and Neville, M. (1978). Glucoprivic feeding behavior in absence of other signs of glucoprivation. *Am. J. Physiol.* **234,** E617–621.

83. Morrison, S. (1984). Synergistic stimulation of food intake by simultaneous insulin and cold. *J. Appl. Physiol.* **57,** 28–33.

84. VanderWeele, D. A., Haraczkiewicz, E., and DiConti, M. A. (1980). Pancreatic glucagon administration, feeding, glycemia and liver glycogen in rats. *Br. Res. Bull.* **5** (Suppl. 4), 17–22.

85. Balagura, S., Kanner, M., and Harrell, L. E. (1975). Modification of feeding patterns by glucodynamic hormones. *Behav. Biol.* **13,** 457–466.

86. Langhans, W., Scharrer, E., and Geary, M. (1986). Pancreatic glucagon effect on satiety and hepatic glucose production are independently affected by diet composition. *Physiol. Behav.* **36,** 483–488.

87. Langhans, W., Wiesenreiter, F., and Scharrer, E. (1983). Plasma metabolites and food intake reduction following heparinoid injection in rat. *Physiol. Behav.* **30,** 113–119.

88. Martin, J., Novin, D., and VanderWeele, D. A. (1978). Loss of glucagon suppression of feeding after vagotomy in the rat. *Am. J. Physiol.* **234,** E314–318.

89. MacLaughlin, C. L., Gingerich, R. L., and Baile, C. A. (1984). Decreased food

intakes and body weights in rats immunized against pancreatic glucagon. *Physiol. Behav.* **33,** 723–727.

90. Danguir, J., and Nicolaïdis, S. (1980). Circadian sleep and feeding patterns in the rat: Possible dependence on lipogenesis and lipolysis. *Am. J. Physiol.* **238,** E223–228.

91. Larue-Achagiotis, C., and Le Magnen, J. (1979). Dual effects of 2-deoxy-D-glucose on food intake in the rat: Inhibition at night and stimulation in the daytime. *Physiol. Behav.* **23,** 865–870.

92. Naito, C., Yoshitoshi, Y., Higo, K., and Ookawa, H. (1973). Effects of long-term administration of 2-deoxy-D-glucose on food intake and weight gain in rats. *J. Nutr.* **103,** 730–737.

93. Stricker, E. M., and Rowland, N. (1978). Hepatic versus central origin of stimulatus for feeding induced by 2-deoxy-D-glucose in rats. *J. Comp. Physiol. Psychol.* **92,** 126–132.

94. Smith, G. P., and Epstein, A. N. (1969). Increased feeding in response to decreased glucose utilization in the rat and monkey. *Am. J. Physiol.* **217,** 1083–1087.

95. Booth, D. A. (1972). Some characteristics of feeding during streptozotocin-induced diabetes in the rat. *J. Comp. Physiol. Psychol.* **80,** 238–249.

96. de Castro, J., and Balagura, S. (1975). Meal patterning in the streptozotocin-diabetic rats. *Physiol. Behav.* **15,** 259–264.

97. Panksepp, J., and Ritter, M. (1975). Mathematical analysis of energy regulatory patterns of normal and diabetic rats. *J. Comp. Physiol. Psychol.* **89,** 1019–1028.

98. Booth, D. A. (1972). Feeding inhibition by glucose loads, compared between normal and diabetic rats. *Physiol. Behav.* **8,** 801–805.

99. Carpenter, R. G., and Grossman, S. P. (1983). Early streptozotocin diabetes and hunger. *Physiol. Behav.* **31,** 175–178.

100. Friedman, M. (1978). Hyperphagia in rats with experimental diabetes mellitus: A response to a decreased supply of utilizable fuels. *J. Comp. Physiol. Psychol.* **92,** 109–117.

101. Friedman, M. I. (1977). Insulin-induced hyperphagia in alloxan-diabetic rats fed a high-fat diet. *Physiol. Behav.* **19,** 597–599.

102. Friedman, M. I., Emmerich, A. L., and Gil, K. M. (1980). Effects of insulin on food intake and plasma glucose level in fat-fed diabetic rats. *Physiol. Behav.* **24,** 319–325.

103. Friedman, M., and Ramirez, I. (1985). Food intake in diabetic rats: isolation of primary metabolic effects on fat feeding. *Am. J. Physiol.* **249,** R43–51.

104. Richter, C. K., and Schmidt, D. C. (1939). Behavioral and anatomical changes produced in rats by pancreatectomy. *Endocrinology* **25,** 698–700.

105. Kanarek, R. B., and Ho, L. (1984). Patterns of nutrient selection in rats with streptozotocin-induced diabetes. *Physiol. Behav.* **32,** 639–645.

106. Bartness, T. J., and Rowland, N. E. (1983). Diet selection and metabolic fuels in three models of diabetes mellitus. *Physiol. Behav.* **31,** 539–545.

107. Kakolewski, J. W., and Valenstein, E. S. (1969). Glucose and saccharine preference in alloxan diabetic rats. *J. Comp. Physiol. Psychol.* **68,** 31–37.

108. Van Itallie, T. B., Beaudoin, R., and Mayer, J. (1953). Arterioveinous glucose differences, metabolic hypoglycemia and food intake in man. *J. Clin. Nutr.* **1,** 208–216.

109. Van Itallie, T. B., and Hashim, P. (1960). Biochemical concomitants of hunger and satiety in man. *Am. J. Clin. Nutr.* **8,** 577–592.

110. Stunkard, A. J., and Wolff, H. G. (1954). Correlation of arteriovenous glucose differences, gastric hunger contractions and the experience of hunger in man. *Fed. Proc.* **13,** 147.

111. Bernstein, L. M., and Grossman, M. I. (1956). An experimental test of the glucostatic theory of regulation of food intake. *J. Clin. Invest.* **35,** 627–633.

112. Thornston, R., and Horvath, S. M. (1975). Blood glucose level after eating and after omitting breakfast: Studies on teenagers and young adults. *J. Am. Diet. Assoc.* **47,** 474–477.

113. Polak, C., Green, J., and Smith, G. (1989). Blood glucose prior to meal request in humans isolated from all temporal cues. *Physiol. Behav.* **49,** 529–534.

114. Lucas, F., Bellisle, F., and Di Maio, E. (1987). Spontaneous insulin fluctuations and the preabsorptive insulin response to food ingestion in humans. *Physiol. Behav.* **40,** 631–636.

115. Rodin, J., Bach, J., Peranini, E., and De Fronzo, F. (1985). Effect of insulin and glucose on feeding behavior. *Metabolism* **34,** 827–833.

116. de Castro, J. M., and Kreitzman, S. M. (1985). A microregulatory analysis of spontaneous human feeding pattern. *Physiol. Behav.* **35,** 329–336.

117. MacCutcheon, N. B., and Tennissen, A. M. (1989). Hunger and appetitive factors during total parenteral nutrition. *Appetite* **13,** 129–141.

118. Jordan, H. A. (1969). Voluntary intragastric feeding: Oral and gastric contributions to food intake and hunger in man. *J. Comp. Physiol. Psychol.* **68,** 498–506.

119. Cornel, C. A., Rodin, J., and Weingarten, H. (1989). Stimulus induced eating when satiated. *Physiol. Behav.* **45,** 695–704.

120. Birch, L. L., McPhee, L., Sullivan, S., and Johnson, S. (1989). Conditioned meal initiation in young children. *Appetite* **13,** 105–113.

121. Bellisle, F., Lucas, F., Amrani, R., and Le Magnen, J. (1984). Deprivation, palatability and the microstructure of meals in human subjects. *Appetite* **5,** 85–94.

122. Thornton, R. H., and Horvath, S. M. (1968). Blood glucose influenced by either one or two meals. *J. Am. Diet. Assoc.* **58,** 215–217.

123. Apfelbaum, M., Reinberg, A., Assan, R., and Lacatis, D. (1972). Hormonal and metabolic circadian rhythms before and during a protein diet. **8,** 867–873.

124. Schlierf, G., and Donoy, E. (1973). Diurnal patterns of triglycerides, FFA, blood glucose and insulin during carbohydrate ingestion in man and their modification by nocturnal suppression of lipolysis. *J. Clin. Invest.* **52,** 735–740.

125. Malherbe, C., de Gasparo, M., Hertogh, R., and Hoet, J. (1969). Circadian varia-

tion of blood sugar and plasma insulin levels in man. *Diabetologia* **5,** 397–404.

126. Aparicio, N. J., Puchulu, F. E., Gagliardino, J. J., Ruiz, M. D., Llorens, J. M., Ruiz, J., Lamas, A., and de Miguel, R. (1974). Circadian variation of the blood glucose, plasma insulin and human growth hormone levels in response to an oral glucose load in normal subjects. *Diabetes* **23,** 132–137.

127. Sensi, S., and Capani, F. (1976). Circadian rhythm of insulin induced hypoglycemia in man. *J. Clin. Endocrinol. Metabol.* **43,** 462–465.

128. Grossman, M. I., and Stein, I. F., Jr. (1948). Vagotomy and the hunger producing action of insulin in man. *J. Appl. Physiol.* **1,** 263–269.

129. Stunkard, A. J., Van Itallie, T. B., and Reis, B. B. (1955). Mechanism of satiety: Effect of glucagon on gastric hunger contractions in man. *Proc. Soc. Exp. Biol. Med.* **85,** 258–261.

130. Welle, S. L., Thompson, D. A., Campbell, R. G., and Lilivivathana, U. (1980). Increased hunger and thirst during glucoprivation in humans. *Physiol. Behav.* **25,** 397–403.

Chapter Three

Brain Mechanisms of Feeding: The Hunger–Satiety System

For a long time, classic neurophysiologists, by neglecting behaviors, and behaviorists, by neglecting the brain, forgot the oldest experiment: When an animal or a human is beheaded, all behaviors are immediately lost.

Brain processing of systemic information about the metabolic state will be examined here. Brain processing of sensory information, the other compound of the stimulation to eat, will be studied in Chapter 5.

What are the targets of the systemic stimulus to eat or not to eat, of the hunger and satiety studied in preceding chapters, and how do they subserve the behavioral output? During the last four decades, numerous works were designed to elucidate this major brain mechanism. A large variety of techniques and procedures were and are still used: local electrical stimulations or self-stimulations, lesions, electrophysiological recordings, and neurochemical and neuropharmacological approaches. These works merit a full book; they will only be overviewed here.

I. *Intracranial Administration*

A first and apparently easy approach to the problem of determining the brain target that induces the brain arousal of eating was to infuse various agents or drugs, suspected to be representative of the systemic stimulus of feeding or interfering with this stimulus, into the brain ventricles or locally into various structures. Hundreds of compounds were thus injected, and the results are reported in thousands of publications. These compounds were physiological agents involved in oxidative and/or intermediary metabolisms (e.g., glucose, insulin, and catecholamines; blockers of the activity of these agents; neurotransmitters; their agonists or antagonists). It is interesting to observe that almost all the compounds, intraventricularly or locally infused in either deprived or satiated animals, augmented or inhibited food intake. Hence, researchers claimed that the injected molecule or the neuronal system modified by the injection was physiologically involved in feeding. When they renewed intake in satiated rats, or blocked intake of hungry rats, the compound was claimed to be involved in satiety. When it enhanced or inhibited feeding in deprived rats, it was claimed to be involved in meal initiation. Thus, hundreds of compounds were and still are presented as involved in hunger and satiety or both. However, some of these experiments, when examined with caution, provided some valuable information.

The early findings of the dramatic effects of medial and lateral hypothalamic lesions led authors to localize microinfusion in these two areas. For the same reason, intraventricular administrations were generally made either in lateral or in the third ventricles adjacent to these areas. The interpretation of such intraventricular injections regarding the structure responsible for the recorded effect is, of course, hypothetical. The degree and space of diffusion of injected solutions are unknown, and this diffusion may lead the injected compound to act simultaneously on various sites. In both local and intraventricular infusions, an important cause of artifacts was recognized by many researchers. The placement of the cannula used to inject intracranially is sufficient to produce drastic changes in food intake.

Here the description of the effect of injections localized in various hypothalamic nuclei will not be dissociated. Later, and taking into account the very distinct functions in feeding of the ventromedial and lateral areas, attention will focus on the latter region, which will appear critically involved in the stimulation of feeding.

A. Anesthetics

The injection of anesthetics was a source of reliable information. Injecting procaine into the ventromedial region and saline into the lateral region caused a statiated rat to eat. Procaine into the lateral region and saline into the ventromedial region prevented a deprived rat from eating (1). This blockade of eating by procaine into the lateral region was obtained in the anterior and not in the dorsal portion of this area (2). Procaine into the ventromedial region produced hyperinsulinemia and hypoglycemia like that of the electrolytic lesion of the same nuclei (3). A chronic infusion of morphine in the third ventricle for 5 days provoked hypoglycemia, weight loss, and a reduction of food intake (4).

Morphine infused into the periventricular region stimulated feeding. Anesthesia in the ventromedial region by pentobarbital lasted for 10 min. During this time, only microinfusion started a meal (5). Pentobarbital injected into the third ventricle of normal rats also briefly stimulated eating in satiated rats. This intraventricular administration of pentobarbital immediately induced an insulin release like that of the ventromedial lesion. The increase of intake follows (6).

B. Glucose and Insulin

Because it was known that glucose was the substrate of brain oxidative metabolism in fed animals, the effects of intracranial administration of glucose, insulin, and glucose antimetabolites on both intakes and peripheral metabolic events were tested early. A 35% reduction of food intake was reported after an intraventricular administration of glucose plus insulin (7). According to Herberg (7), the intraventricular administration of glucose or insulin alone had no effects. Infused into the third ventricle, glucose produces a dose-dependent suppression of intake realized by short meals only (8). A microinfusion into the ventromedial region has any effect on intake (1). According to Panksepp and Nance (9), this ventromedial administration of glucose would have a long-term reducing effect, whereas the same administration in the lateral region would not change intake. On the contrary, according to Booth (10), this glucose administration in the lateral region blocks the feeding response to peripheral insulin administration. A more significant result was reported by Glick and Mayer (11). Intraventricular infusion of phloridzin, which competes with cellular glucose uptake, produces a strong hyperphagia that lasts for several days. The authors showed that phloridzin effectively interacted with the uptake of gold-thioglucose, leading to localized necrosis in the ventromedial region in mice. Based on this observation, they suggested that phloridzin produced hyperphagia by producing glucopenia in glucosensitive sites in the ventromedial region or elsewhere. This interpretation was later supported by other data.

Numerous researchers injected insulin either intraventricularly or locally mainly in the ventromedial region. Various important works on the effects of chronic intraventricular administrations will be examined elsewhere in relation to the potential role of insulinoreceptors in the lipostatic mechanism (see Chapter 9). A microinfusion of insulin into the ventromedial region reduced intake in normal as well as in diabetic rats. Hatfeld *et al.* (12) Iguchi *et al.* (13) compared the acute effects of insulin intraventricularly and locally injected on blood glucose level. Administration in the third ventricle did not change the blood glucose level. A control saline injection in the lateral region produced hyperglycemia: Insulin slightly reduced this hyperglycemia. In the ventromedial region, 90–100 μIV produced a dose-dependent hypoglycemia. Convergent results of other investigators support the conclusion of these authors—that the two hypothalamic sites belong to a central glucoregulatory system acting on blood glucose level via descending parasympathetic pathways from the lateral region and sympathetic pathways from the ventromedial region to the pancreas. Injected in suprachiasmatic nuclei, insulin reduced the nocturnal intake and augmented the diurnal intake, whereas the same injection induced a rise of plasma insulin at night and a fall during the day (14, 15). Finally, insulin antibody injected in the ventromedial region, thus blocking a possible action of endogenous insulin, produced a transitory hyperphagia when injected at night only (16).

Together, these dispersed results indeed do not give a clear picture of the role of circulating glucose and insulin and of their respective actions on the central nervous system (CNS) in the feeding mechanism.

C. Glucose Antimetabolites

How does the intraventricularly injected glucose antimetabolite 2-deoxy-D-glucose (2-DG) stimulate food intake at a dose 40 times lower than the dose active systemically (17)? This is a confirmation that the effect of the systemic administration is centrally mediated. Various studies (which cannot be detailed here) showed that the metabolic effects of the systemic administration [hyperglycemia and rise of plasma-free fatty acids (PFFA)] are obtained by intraventricular and local ventromedial infusions. The effect of local injections on food consumption is controversial. According to Kanner *et al.* (18), 2-DG injected into the lateral region increases by 79% the *ad libitum* intake but is inactive in 12-hr deprived rats. Injected into the ventromedial region, 2-DG reduces intake in both fed and deprived rats. Berthoud and Mogenson (19) and Miselis and Epstein (20), however, failed to replicate these results. Both studies found no response after the lateral injection. Other works suggested a dissociation of the 2-DG targets by producing the hyperglycemic–lipolytic pattern and the stimulation of intake. The latter is certainly not the ventromedial region. After the electrolytic destruction of this region, 2-DG is demonstrated to be still active in inhibiting intake, as it does at night in intact rats (21).

However, it is also clearly demonstrated that the main target of 2-DG is the lateral region. In rats recovered from aphagia induced by the lateral-electrolytic or 6-OH-DOPA lesions, 2-DG as well as insulin peripherally injected no longer stimulates eating. These responses to both insulin and 2-DG are said to be responses to a "glucoprivic challenge"; however, they can be dissociated. After a Zona Incerta lesion, the response to insulin is preserved and that to 2-DG is eliminated: After a hypothalamic depletion of noradrenaline induced by a mesencephalic lesion 2-DG also becomes ineffective, whereas insulin action is maintained (22, 23).

Finally, an intriguing result must be mentioned. The administration of 5-thioglucose (another glucose antimetabolite) stimulates food intake when infused either in the lateral or in the fourth ventricle. After an obstruction of the Monroe hole by which the lateral and third ventricles communicate with the fourth ventricle, 5-thioglucose in the latter still stimulates intake. It is no longer active when injected in the lateral ventricle (24). Without excluding a role for the lateral region, this experiment indicated for the first time a role, later confirmed, for glucosensitive sites in the hindbrain in the glucoregulation, including feeding. This was reminiscent of the famous experiment more than 100 years ago by Claude Bernard, who produced diabetes by a "piqure" in the floor of the fourth ventricle.

D. Neurochemicals

The most abundant and fashionable literature on brain function in feeding deals with neurotransmitters, neurohormones involved in synaptic transmissions, and with biochemical constituents of neuronal fibers. The pharmacological interest (i.e., the interest to find drugs stimulating or inhibiting feeding), is self-evident; however, progress of pharmacological knowledge of the basic brain mechanisms of feeding is still limited. To claim that agonists or antagonists of intraventricularly injected dopamine, or serotonin, are involved because they either stimulate or inhibit a feeding response is misleading. How and where are these neurotransmitter systems concerned in feeding mechanisms and dissociated from their role in many other brain functions? Local injections in various nuclei are more fruitful. Nevertheless, the fact that adrenergic or DOPAminergic systems are related to feeding in a particular structure is useful only if it is a tool to identify the trajectory of a specific neuronal circuitry subserving feeding, i.e., if the neurochemical is used as a "tracer." This is indirectly the case when neurotoxins (5-HTP, 6-OH-DOPA) are fruitfully used to destroy not a particular point, as in electrolytic lesions, but, rather, one or several identified pathways. Such works on brain lesions will be examined.

In the 1960s, the first extensive work was reported by Grossman (25). At this time, the dissociation of pathways respectively involved in feeding and drinking was a current problem. The authors found evidence of this dissociation by the difference of what they called the neurochemical coding of the two systems. Injecting either norepinephrine or carbachol (a stimulant of acetylcholinergic fibers) into various loci of the hypothalamus and limbic cortex showed a different mapping of points in which feeding was stimulated by norepinephrine and drinking by carbachol. In these two cases, the mapping was revised, but regarding feeding, it was the first demonstration, later confirmed, that noradrenergic pathways were, in some manner, indeed implicated in feeding mechanisms in these areas. The first study on a restricted area came from Slangen and Van der Gutten (26). Stimulants and blockers of various neurotransmitters were infused in the perifornical region. It was revealed that α- and not β-adrenergic agonists stimulated feeding in satiated rats. In this region, serotonin and dopamine were ineffective. The role of α- and particularly α_2-receptors was amply confirmed (27). A peripheral administration of clonidine (α_2-receptor agonist; 0.1–1 mg/kg) increased the 6-hr intake of an undeprived rat. The antagonist yohimbine shifted the dose–response curve to the right (28). According to Leibowitz and Brown (27), stimulating α-receptors elicits feeding and stimulating β-receptors induces satiety. Using local norepinephrine microinfusions, the more sensitive site was the periventricular nuclei. A threshold dose of 1–3.2 μg in these nuclei stimulates a feeding response. This threshold varies between night and day but without any relation to the day–night cycle of feeding. However, the stimulating effect of injecting norepinephrine into the periventricular region, always tested on one meal, is practically null when examined on the 24-hr meal pattern. The first meal is augmented, but the postmeal interval is proportionally prolonged. No effects on subsequent meal sizes and postmeal intervals are observed, and the 24-hr intake is unchanged. In the perifornical region, DOPA blockers reversed the effect of β-adrenergic agonists. Thus, a DOPA system would be associated with the β-adrenergic system (29). The fact that the 6-OH-DOPA lesion of lateral hypothalamic striatal pathways induce aphagia and the lesion of dopaminergic pathways to the ventromedial region induce hyperphagia will provide more convincing evidence of this involvement of dopaminergic fibers. But many other works also present results indicating a role for serotoninergic, neuropeptidergic, gabaergic, etc., pathways, in the same regions in which others claimed the specific involvement of adrenergic synapses or fibers (for a review on serotonin, see 30). The particular role of brain opiates and opioreceptors will be examined elsewhere (Chapter 5) in relation to their role in brain rewarding systems.

Neuropeptide Y injected peripherally (31) or in the periventricular region (32) elicits a strong feeding response. Hence, the unavoidable proposal was presented that neuropeptide Y was the "molecule" of feeding, acting for some as a meal initiator, for others as a satiety inhibitor. More reasonably it is thought that neuropeptide Y, like a considerable number of varied molecules, disturbs the normal process of feeding when injected into the brain.

Using a push–pull cannula makes it possible to follow the rate of nor-epinephrine or DOPA release during feeding (33, 34). This fascinating technique certainly will provide rich insights on the intrinsic neuronal functioning and perhaps will also contribute to a better understanding of the neurochemical background of feeding mechanisms.

II. *Electrical and Self-Stimulation of the Lateral Hypothalamus*

For a long time, it was thought and taught that the ventromedial hypothalamus (VMH) was a "satiety center" and the lateral hypothalamus (LH) a "feeding center." Moreover, it was admitted that the VMH was the active site responding to peripheral satiety signals. The VMH was claimed to inhibit the lateral region through hypothetical connections between the two sites, and it was assumed that feeding occurs when the LH is disinhibited. It was a longlasting and historical mistake. Its origin was in the early discovery of hyperphagia and obesity resulting from the electrolytic lesion of the VMH. This hyperphagia was interpreted as an effect of the loss of the animal capacity to be and to remain satiated by eating its foods. It was reinforced by the early finding by Mayer (35), using the gold-thioglucose lesioning of glucoreceptor sites in the VMH. This far-reaching finding, which was later fully confirmed, was misinterpreted inasmuch as the VMH was thought to be a satiety center. Hence, it was claimed that satiety systemic signals associated with high glucose tissue uptake acted upon these glucosensitive sites. Further studies ruled out these now-abandoned notions. The VMH-lesioned rat was shown to be unaltered on its capacity to be satiated. As will be seen elsewhere, the VMH and its glucosensitive sites were demonstrated to be critically involved, not in satiety, but in the control and regulation of the body fat mass. Finally, the reality of connections between the VMH and LH was dismissed by Sclafani *et al.* (36). Knife-cuts between the two regions did not change their respective functions. In addition, the combined VMH and LH lesions (the former inducing hyperphagia, the latter aphagia) gave evidence of their independent functions.

Thus, in the following sections dealing with the hunger–satiety system, attention will be restricted to the LH and other regions and to various demonstrations that they are indeed the critical neuronal system of the stimulation to eat or not to eat.

A. Electrical Stimulation of the Lateral Hypothalamus

Soon after the considerable discovery by Anand and Brobeck (37) that the LH lesion induced aphagia (see Chapter 3, Section III), Delgado and Anand (38) showed that the electrical stimulation of the same region caused hyperphagia. This electrically induced feeding was called "stimulus-bound feeding" (SBF). Found first in the cat, it was later confirmed in many other species. Steinbaum and Miller (39) reported that repeated stimulations inducing a sustained overeating lead to obesity in rats. Rats were stimulated twice per day for 1 hr for 11 days. In the two daily sessions, they consumed up to 2.5 times their control *ad libitum* intake and, after 11 days, became obese (Fig. 3.1). When they were

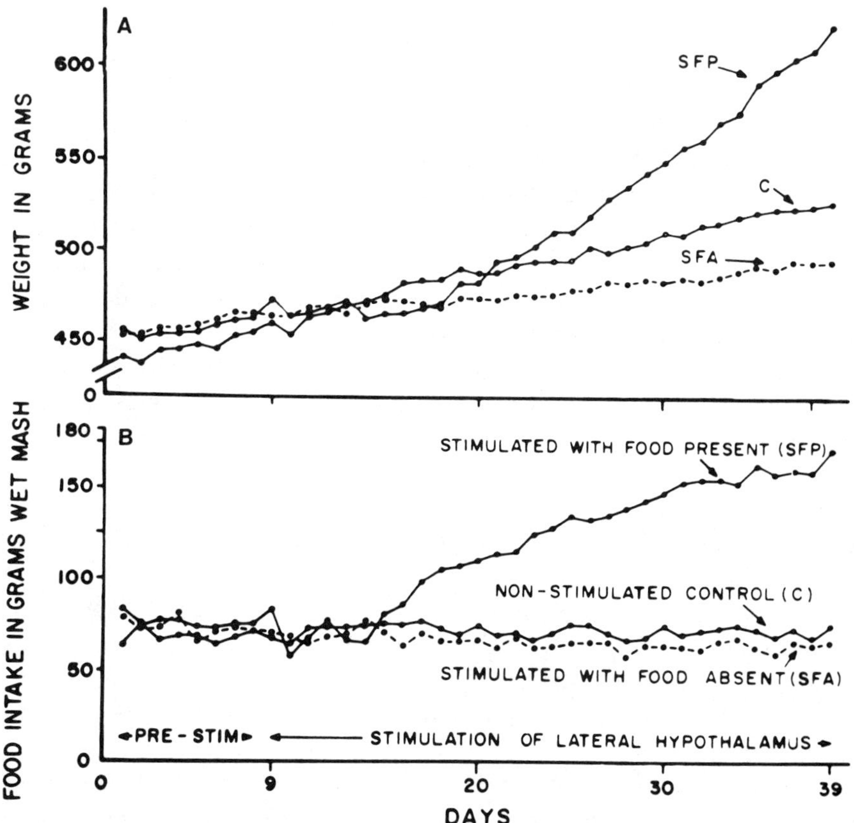

Figure 3.1 Effect of hypothalamic stimulation on (A) body weight and (B) food intake. Group means are shown [from Steinbaum and Miller (39)].

limited to 0–100% of their previous free intake during the stimulations, they compensated exactly for this restriction if allowed to eat freely between the sessions. In so doing, their daily intake did not exceed the control.

Difficulties and controversies arose from the fact that the electrical stimulation produced either feeding or drinking or both. Various studies regarding electrode placements, thresholds of current to obtain one or the other response, and histological controls ascertained that a specific neuronal system was stimulated when feeding and no drinking were exhibited (40, 41). Early on the question was raised as to whether or not the stimulation reproduced the state induced by food deprivation, i.e., hunger. Coons *et al.* (42) elegantly demonstrated that this was so. Rats were trained to press a lever to obtain their food when electrically stimulated. When the response was learned, it was maintained at the same level by food deprivation substituted for the LH stimulation. The same results were shown when concentration of quinine in the food, necessary to prevent intake, increased with the current intensity like it increases with food deprivation (43).

Unilateral or bilateral ablations of the cortex as well as a cortical-spreading depression transitorilly alter SBF. The frontal cortex seems primarily involved (44). A SBF cannot be obtained in any other brain regions except in the mesencephalon. In the vicinity of the descending pathway of the medial forebrain bundle, electrical stimulation induces feeding. However, contrary to that obtained in the LH, the feeding response is not immediately interrupted at the end of the stimulation and continues afterward (45).

Other works, examined later, demonstrate that the electrical stimulation of the LH in addition to an equivalent of the feeding stimulation by food deprivation also provides a reward equivalent to the rewarding effect by eating a food.

B. Metabolic Effects of Lateral Hypothalamic Electrical Stimulation

Peripheral changes (hyperglycemia or hypoglycemia, insulin, or adrenal catecholamine releases) brought about by the LH electrical stimulation by acting in turn on the LH or elsewhere could not be excluded as causes of the feeding responses. Thus, it was important to assess these peripheral consequences of LH stimulation.

Hyperglycemia and hyperinsulinemia are observed during the stimulation when food is present and eaten, and hyperglycemia without an elevation of plasma insulin between sessions when the food is absent (46). Hyperglycemia was generally confirmed (47). The recording of neural activity in various efferent pathways to the pancreas, the liver, and adrenals demonstrated that stimulation in the dorsal part reduced the firing rate in the pancreatic branch of the vagus nerve, whereas stimulation in the ventral LH augmented it. The dorsal stimulation increased activity in the splanchnic nerve (48). A medial stimulation increased and a dorsal and anterior LH stimulation decreased the neural activity in the adrenal nerve. This activity, augmented by intraventrally injected 2-DG was reduced during the LH dorsal stimulation (49). Thus, both sympathetic and parasympathetic descending pathways seem to be stimulated, and hyperglycemia may be the result of an adrenal catecholamine release.

Relations between the effects of the stimulation and those of insulin and 2-DG administrations on intake were of interest. A positive correlation was found between blood glucose level and the current threshold to obtain the feeding response. Insulin administration did not lower the threshold. On the contrary, 2-DG lowers this threshold, but only after a considerable latency of 20 min (50, 51). Thus, no evidence indicates that the LH electrical stimulation provokes hypoglycemia, which could be a cause of the feeding response.

C. Lateral Hypothalamic Self-Stimulation

Rats are easily trained to lever-press to self-stimulate LH sites in which SBF is previously observed. The study of this self-stimulation behavior and of its relations to feeding provided a series of striking results that proved the role of the LH in the hunger–satiety system, on the one hand, and of its normal stimulation by glucopenia on the other.

The two series of combined feeding stimuli (food deprivation and food sensory stimuli), separately or combined, increase the rate and lower the threshold of LH self-stimulation. Satiety induced either by gastric distension or by glucose or glucagon administrations interrupts self-stimulation. Margules and Olds (52), and later many others, showed that LH self-stimulation was augmented by food deprivation. This elevation in the rate of self-stimulation was shown better correlated to the weight loss than to the deficit of intake (53). Conversely, in rats rendered obese either by gavage (54) or by previous SBF (55), the rate of self-stimulation declines as a function of weight gain. When obesity is reached, rats no longer self-stimulate but press another lever to escape a programmed stimulation. In rats becoming obese following the VMH lesion, the self-stimulation also declines progressively. The self-stimulation is negatively correlated to the hyperphagia and the rate of weight gain (56).

Glucose solution intragastrically or intraveinously administered inhibits self-stimulation (57). It is augmented by insulin-induced hypoglycemia and inhibited by glucagon-induced euglycemia or hyperglycemia (58). Finally, inflating a balloon in the stomach interrupts self-stimulation and rats readily lever-press to escape a programmed stimulation (59).

A role for external food stimuli was demonstrated by Coons and Cruce (60). The current intensity for self-stimulation was that just below the threshold of SBF. At this level, rats lever-pressed only when a food was present in its vicinity. Before pressing, they sniffed the food and bit the lever. In another experiment (61), rats in a shuttle box were stimulated on one side and then escaped from the stimulation to the other side. The time spent on the stimulation side increased when a food was present and highly palatable.

The combination of food deprivation and various levels of food palatability allowed Coons and White (62) to assess the quantitative interrelationships between the two combined stimulations of feeding. Thus, they provided evidence of their synergistic interactions. We will later return to this essential point.

Several other studies provided convergent and convincing data indicating that the self-stimulation in the LH is related not only to the stimulation of feeding but also to the reinforcement by eating a food manifested in the learning of food palatability. The pairing of the presentation of a new food with self-stimulation induced a conditioned food preference for that food, proportional to the number of pairings (63). Moreover, it was indicated that the ability of rats to self-stimulate the LH was correlated to its ability to respond to a palatable sensory stimulus. Among two lines of rats selected as high and low self-stimulators, the first ones were shown to be the highest spontaneous drinkers of a saccharin solution (64).

By self-stimulating their LH, rats engage in a sort of "electrical meal": They are stimulated by the same stimuli that initiate an oral meal; however, satiation and satiety are not provided by eating a food, and this stimulation of the electrical meal persists indefinitely. It persists far more in this condition than it does in the sham-feeding experiments, in which the food passing the mouth gives some elements of satiation. This electrical meal terminates when the normal satiating process (stomach filling) and postingestive satiety factors are experimentally introduced.

In many works, relations between the self-stimulation of the LH and of other brain sites were studied. The correlations between self-stimulation of LH and of the tegmentum area, known to be involved in brain reward, were emphasized. One experiment among others will be cited here. Rats selected as high self-stimulators of the LH are also high self-stimulators of the ventral tegmentum. These rats also respond more to pain and receive a stronger analgesia by tegmentum stimulations (65). In the LH as well as in the tegmentum, self-stimulation was higher at night than during the day (66).

III. *The Lateral Hypothalamus and Other Brain Lesions*

Anand and Brobeck's discovery (37) of aphagia induced in rats by a LH electrolytic lesion was rapidly confirmed in other species (e.g., cat, monkey). Furthermore, the effects of electrical stimulations added to this basic finding indicate that the LH lesion destroyed a focus of blood-borne and neural signals required to bring about the behavioral output. Hundreds of works were carried out during the two subsequent decades (1951– to 1980) to assess and to extend this evidence. Scientists are rarely convinced by self-evidence, and, despite the early, simple conclusion, Anand and Brobeck's finding gave rise to various controversies. The abundant literature on the matter, now of declining interest, will only be overviewed in this chapter.

A. Neuroanatomy

Aphagia and its typical features (described below) are obtained by the bilateral electrolytic lesion of the extreme lateral border of the hypothalamus. The stereotaxic coordinates generally used are AP 6mm; 2mm lateral to the sagittal sinus; 0.9 mm up from the base of the brain. Lesions of the anterior and posterior parts of the area give two types of responses. Lesioning of the anterior region produces an active rejection of food and aphagia modified by a previous manipulation of body weight. Lesioning of the posterior region introduces a sensory-motor impairment leading to passive aphagia. In this case, food put in the mouth is swallowed and not rejected (67). For a long time, researchers thought that the LH lesion interrupted undefined fiber pathways in the region. Without excluding a role for such dysconnections, it was shown that the destruction of cell bodies in the site also played a role in the induced syndrome. Lesioning by kainic acid infusion, which specifically destroys cell bodies, produces the syndrome (68). The involved pathways were investigated in many experiments using unilateral or disparate bilateral lesions, by knife-cuts or electrolytical lesions. A major finding was that a lesion of the orbitofrontal cortex produces the same syndrome as that of the LH lesion. A unilateral LH or orbitofrontal lesion combined with the contralateral LH or orbitofrontal lesion has the same effect (69). Later, electrophysiological data confirmed these close links between the frontal cortex and the hypothalamus in the feeding control.

A more general role for the cortex is documented. In rats recovered from aphagia, a cortical spreading depression resumes aphagia (70, 71). An olfactory bulbectomy in recovered rats has the same effect (72). The implication of the medial forebrain bundle (MFB) was also indicated in several works. Another mode of lesioning was the use of 6-OH-DOPA and its depleting effect in dopaminergic pathways. In the LH, MFB, or nigrostriatum, 6-OH-DOPA produced aphagia. The interest of this finding was not that the involved pathways are dopaminergic but, rather, was the possibility that these pathways could be traced by histofluorescent techniques (73, 74); however, various other experiments demonstrated that the electrolytic LH and 6-OH-DOPA lesions of adjacent pathways did not have the same significance. According to Stricker *et al.* (73) the 6-OH-DOPA lesion would spare the functions of glucosensitive sites in the LH and would impair pathways needed for the sensorimotor expression of feeding.

B. Aphagia and Recovery from Aphagia in Lateral Hypothalamic-Lesioned Rats

Following the lesion, aphagia lasts from some days to months, according to the extent of the lesion and (as later seen) to the preoperative body weight of the animal. When aphagia persists, the rat must be nursed to survive. A progressive recovery of feeding occurs. Typical successive phases of this recovery were observed (75). Initially, the rat is stimulated to eat only high-palatability. When the previous daily intake of various foods is recovered, many irreversible sequels of the lesion are observed. As already mentioned, the most dramatic change is the alteration of the prandial pattern (Fig. 3.2). A rat becomes a nibbler, taking very prolonged meals at night made up of short bouts of eating separated by short pauses. The diurnal periodicity is maintained and even accentuated. As seen in Fig. 3.2, exactly the opposite occurs after the VMH lesion. Larue and Le Magnen (76) demonstrated that this disturbed meal pattern, exhibited by LH-recovered rats, is also produced by olfactory bulbectomy, amygdala, and stria terminalis lesions. This allows the assumption that the loss of the prandial pattern was essentially due to a definitive loss of sensory (particularly olfactory) projections to the LH. Therefore, it was also assumed that these sensory projections were required for a normal meal pattern.

Controversy stemmed from experimental discrepancies on the role of either a sensorimotor impairment or a loss of responding to peripheral metabolic stimuli as a cause of aphagia. Rats preoperatively trained to obtain their food by lever-pressing continued to press after the lesion but did not eat the obtained food. When lever-pressing to feed themselves intragastrically, they performed a metabolically adapted self-intragastric feeding. The authors concluded that aphagia was due to a sensorimotor oral deficit and not to what they called a "motivational" one (77). Opposite results and conclusions were reported by Rodgers *et al.* (78). Rats were found identically aphagic in an oral and self-intragastric

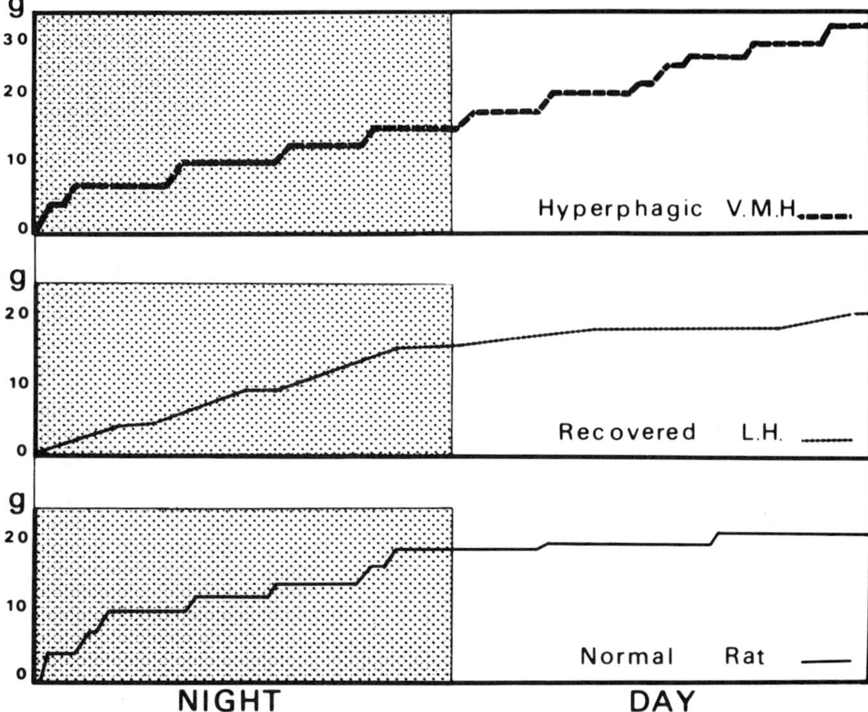

Figure 3.2 Daily feeding pattern compared among normal, LH-, and VMH-lesioned rats.

feeding. The conclusion was that aphagia was not the result of a loss of motor reflexes of eating but a loss of "hunger."

The recovered rat remains highly responsive to the palatability of foods. Its intake is more affected by quinine adulteration of the food than that of controls (79). However, LH-recovered rats regain a normal response to caloric dilution of the diet and to cold exposure.

A prominent fact is that recovered rats no longer respond to insulin (80); they also no longer respond to 2-DG (81). Blass and Kraly (82) argued that the maintained response to cold and other homeostatic challenges, despite the loss of response to 2-DG, proves that the glucoprivic condition that they consider to be induced by 2-DG is normally involved in a regulatory control of food intake. Stricker *et al.* (68) argued otherwise, by showing that the two series of responses are indeed correlated. Depending on the extent of lesions and on insulin doses, the responses to cold and insulin are either both maintained or eliminated.

C. Metabolic Effects of Laterial Hypothalamic Lesions

As in electrical stimulation, it was of importance to ask whether the lesion destroys the critical site of feeding stimulation or it acts directly or indirectly on feeding by modifying metabolic events such as insulin release and blood glucose level.

Glycemia and basal prandial insulin levels clearly are not modified in LH-recovered rats (83, 84). A slight glucose intolerance is due to dehydration in rats, which remain slightly adipsic after recovery of feeding. This fact is not consistent with the notion that the LH interrupts the parasympathetic vagal pathways modulating the pancreatic insulin release. A subdiaphragmatic vagotomy introduces typical metabolic and behavioral changes that do not at all resemble those caused by the hypothalamic lesion.

LH-lesioned rats recovering feeding regain their weight loss during aphagia, but their weight is limited to 87–90% of that of controls. This lowered body weight is "defended." Underfed or intragastrically overfed rats return to their low body weight at the cessation of treatments. Based on these observations, Grijalva *et al.* (85), Powley and Keesey (86), and Keaesey *et al.* (87) proposed the apparently provocative notion that the LH lesions lower the point of body weight equilibrium that results from the body energy balance. They observed that the duration and amplitude of aphagia were reduced, and aphagia was eventually eliminated after a preoperative lowering of body weight by restriction. They suggested that the rat is necessarily aphagic to establish the dictated lowered body weight as an effect of the lesion. This notion is not supported by the fact that after prolonged aphagia and a resulting drastic weight loss rats progressively increase their weight before reaching the chronically maintained underweight. Whatever the facts, the conclusion of these authors—that the LH is mainly involved in the regulation of body weight—is questionable. The "regulation" of body weight is not affected because the recovered rats "defend" their low weight like intact rats defend their normal weight. Why this weight is regulated at this slightly low level cannot be understood without understanding the overall mechanisms involved in the body energy balance.

Overall, it seems well established that the LH lesion impairs the brain responsiveness to neural and humoral inputs necessary for normal feeding. Why and how is this deficit recovered? The fact that a cortical spreading depression resumes aphagia suggests that feeding is then taken over by a whole brain diffuse response to glucopenia. The residual impairments, particularly the permanent loss of responsiveness to insulin and 2-DG, indicate that a specific and sensitive locus of the feeding control was irreversibly destroyed in the LH.

D. Other Brain Lesions

Of course, a large number of other brain lesions severely affect various aspects of feeding. This is the case of lesions or transections of all sensory pathways (e.g., the olfactory bulbectomy, cortical or subcortical lesions affecting motor performance). It is still the case of lesions of sympathetic and parasympathetic efferents and afferents involved in CNS modulation of the pancreas, adrenals, the liver, and the gastrointestinal tract. Out of the VMH and amygdala lesions (which will be studied later), various lesions in other hypothalamic nuclei and in the mesencephalon merit attention.

The periventricular nuclei are particularly interesting, mainly because these nuclei are the more sensitive sites in which norepinephrine infusions can stimulate feeding. However, it was seen that this effect is indeed very modest. Lesioning periventricular nuclei induces hyperphagia, whose similarities and differences with those induced by VMH lesions will be examined. The exact significance of this hyperphagia is presently unknown. The hypophagia induced by dorosomedial nuclei lesions was extensively studied by Bernardis and Bellinger (88). No change in insulin and blood glucose levels was observed as a possible cause. Hypophagia was demonstrated to be the primary cause of the lowest weight and linear growth. Lesions in area postrema and in the adjacent nucleus of the solitary tract also produce hypophagia (89). The role of suprachiasmatic nuclei in the control of metabolic cyclicity, at least in their synchronization with dark–light cycles, is well documented. The question was raised about their role in feeding periodicities. Effects of their lesion casts doubt on such a role. After the lesion, various responses (e.g., activity, water intake) associated to feeding continue to be entrained by the time of meal (90, 91). The conclusion of most researchers is that feeding cyclicities depend on an endogenous clock and are independent of a driving by the suprachiasmatic nucleus which is involved only in the synchronization of the free feeding with a dark–light cycle. The opposite opinion is supported by Stoinev *et al.* (92). They observed different responses to insulin administration at the beginning and the end of the day. These different responses, recorded during 1–4 hr, disappear after the suprachiasmatic nuclei lesion.

IV. *Brain Glucosensitive Neurons and Feeding*

Mayer and Marshal (93) were the first to discover that brain sites exhibit a specific high affinity for glucose and perhaps act as chemosensors of brain glucose availability. In mice, the injection of gold-thioglucose (GTG) induced a necrosis strictly located on the VMH. This localization, at the time denied by some, was fully ascertained by using the labeled compound. Characteristics of these VMH glucoreceptors and their insulin dependence will be examined elsewhere in relation to their involvement in the regulation of body fats (see Chapter 9, Section IV).

No necrosis by GTG was observed in the LH of mice. However, one unique experiment demonstrated (better than all other data) that such GTG-sensitive neurons indeed exist in the LH and, in addition, that they are indispensable for feeding. In rats, a local implant of GTG, like the electrolytic lesion, induced aphagia (94). This aphagic effect of the GTG lesion is blocked by 2-DG, like the hyperphagic effect of the VMH-GTG lesion is also blocked by the competitive action of 2-DG. The GTG-induced aphagia is obtained in diabetic and nondiabetic rats and, thus, is not insulin-dependent. This is contrary to the effect of the same lesion in the VMH, repetitively demonstrated to be insulin-dependent.

A. Hepatic Glucoreceptors

Epinephrine injection inhibits food intake. As shown earlier, this effect in relation to the meal pattern is easily explained by the hyperglycemic and lipolytic action of epinephrine. Epinephrine also produces liver glycogenolysis. It was the basis for the imaginative notion (95, 96) that hepatic glucoreceptors responding to glucose and glycogen loads of the liver were critically involved in the control of food intake. No experimental data supported this notion until the finding by Niijima (97) that the electrical activity in fibers of the hepatic branch of the vagus nerve varied inversely with concentrations of hexoses injected in the portal vein. Nothing in this experiment indicated a relation of this neural activation to food intake nor proved a role for hepatic glucoreceptors in its control. However, it was suggested that terminal fibers in the liver could be stimulated in an unknown manner by carbohydrates in the portal blood. As yet, the function of this response of hepatic vagal fibers is unclear, except that it is certainly not involved in the generation of hunger–satiety signals. Multiple and convergent results discarded the role of putative hepatic glucoreceptors in the control of intake. Liver total denervation, section limited to the hepatic branch of the vagus or the transplantation of the liver in rats and dogs did not substantially change either the cumulative food intake or the detailed meal pattern as well (98–103). Phentolamine (adrenergic blocker) blocks the hyperglycemic and glycogenolytic effects of epinephrine but does not prevent the reduction of intake by epinephrine (104). After a transection of the hepatic branch of the vagus or a total vagotomy, epinephrine and glucagon still reduce feeding. Thus, this reduction is independent of liver metabolism. Moreover, feeding suppressions by glucagon and epinephrine are not correlated to their respective hyperglycemic effects (105, 106). Comparisons of intraportal and intravenous administrations of glucose or of glucagon provide contradictory results. In cats, glucose infusions have an identical effect on insulin and glucagon releases by intraportal and femoral veins and in intact and liver-denervated animals. In rats, glucose via intraportal and intrajugular routes produces hyperglycemia, but intraportal glucose only suppresses intake (107).

B. Electrophoretic Localization of Glucosensitive Neurons in the Lateral Hypothalamus

The existence of LH neurons responding to blood glucose and responsible for the evident role of the LH in feeding was extensively studied by Oomura (108).

Electrophysiological responses of LH neurons to local applications of glucose, insulin, PFFAs, and their combinations were investigated. Using the multiple micropipette iontophoretic technique, 20–25% of explored neurons were found glucosensitive and localized in the ventral portion. The neuronal activity is suppressed by glucose application as an effect of the membrane hyperpolarization. This response to glucose is facilitated by insulin in a dose-dependent manner; on the contrary, glucagon local application suppresses the glucose responsiveness. Those neurons inhibited by glucose are activated by application of PFFAs, a surprising result because hyperglycemic high PFFAs are generally associated with feeding suppression.

Such studies, although of interest, were not conclusive. The electrophoretic application of glucose in the environment of nervous cell bodies is a specific experimental condition. The relationships between the glucose responsiveness of the neurons and their possible activation or inhibition by blood- or cerebrospinal fluid-borne metabolites and glucoregulatory hormones are unknown. Surprisingly, responses of the same neurons to a lowering of glucose in their environment or to 2-DG application were not investigated neither was their response to blood or cerebrospinal fluid contents. Finally, in such anesthetized animals, the effects of these local applications in the LH on the meal pattern could not and were not tested. Thus, complementary experiments in these various directions might strengthen the interest and significance of these studies.

C. Electrical Activity Related to Feeding in the Lateral Hypothalamus

The ideal experiment would be to measure simultaneously the LH neuronal activity in the glucosensitive LH neurons and these responses to fluctuations of circulating glucose before, during, and after a free intake in unanesthetized rats. This acrobatic experiment has not yet been carried out because, if not impossible it would technically be extremely difficult. Separate experiments in which the electrical activity of glucosensitive sites was examined in relation to either meal patterning or fluctuation of peripheral blood glucose levels should approximately provide the same sort of evidence.

One question is whether an electrical activity modulated by changes of blood glucose level acts to bring about the glucoregulatory mechanisms through descending pathways or to activate feeding. The assertion that this electrical activity governs feeding will be strongly supported if the same changes of this neural activity are demonstrated to be concomitant with feeding events.

The first study of single-unit firing during feeding was carried out by Hamburg (109). The resting firing rate in LH units was found to be stopped or reduced at the start of eating. No modification was seen at this time in other parts of the hypothalamus. When the food was withdrawn, the LH activity was immediately reestablished, even if the rat was still masticating. An electrical stimulation of the VMH (i.e., eating) stopped the LH activity. What are these neurons randomly picked up in the LH, and what is the significance of their inhibition by eating? This is still unknown (109).

Another study dealing with the circadian variation of LH activity only suggested relation to feeding. The firing rate of about 9/sec during the day increased to 40–51 at the beginning of the night (110) (Fig. 3.3). A long-term study on 23 neurons identified as glucosensitive on free-moving rats showed an elevation of firing at night in only 3 neurons; the other 20 neurons were inhibited when the rat ate like they are when locally stimulated by glucose (111). The authors (Ono *et al.*, 1981) interpret this and other data by suggesting that different LH neurons could be implicated in prandial events and in the diurnal periodicity of feeding.

More convincing data on the role of LH neurons in both hunger and satiation, and on their relationship to the palatability of foods in the monkey (112), will be described in relevant chapters.

D. Central Nervous System Responses to Circulating Glucose and Insulin

The first investigation on Central Nervous System (CNS) responses to circulating glucose and insulin was carried out by Anand *et al.* (113, 114). They looked for changes in multi- or single-unit activities in the VMH and LH in relation to variations of blood glucose production. In rats, cats, and dogs, they found that hyperglycemia following an intravenous glucose administration elevated activity in the VMH and reduced it in the LH. An insulin-induced hypo-

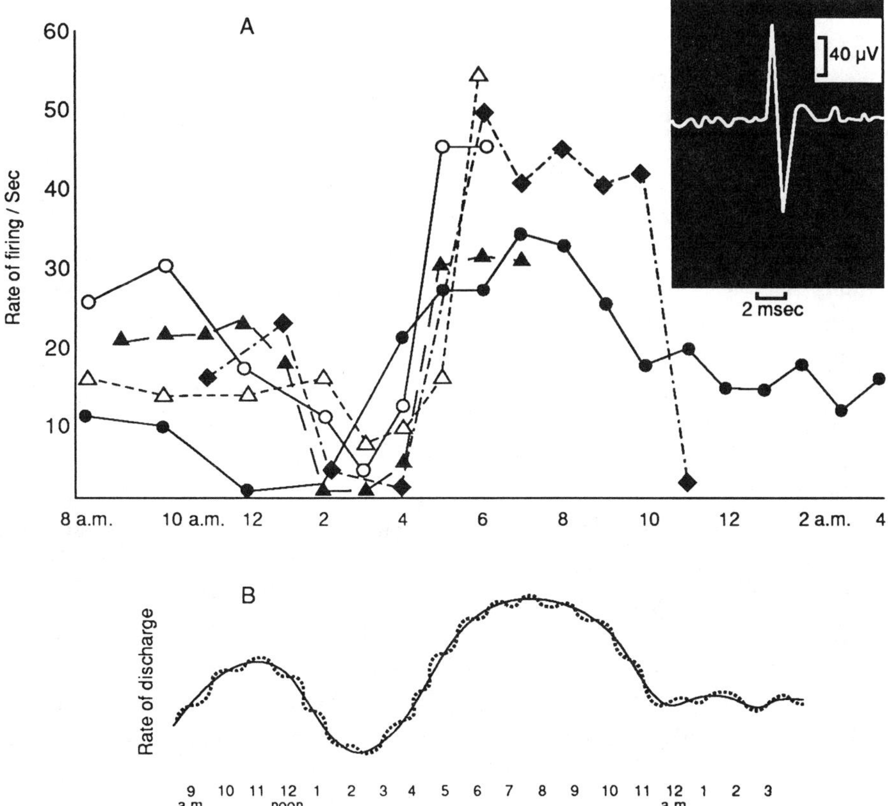

Figure 3.3 Spontaneous fluctuations in rate of firing of cells in hypothalamus. A. Rate of discharge of five lateral hypothalamic neurons responsive to portal vein influences. These cells were monitored for as much of the day as possible, with no injections given after the cell was originally located. B. Diagrammatic representation of overall rhythmic pattern of typical lateral hypothalamic cell, emphasizing super position of rhythms (doted line) on basic daily pattern (solid line). Insert shows a typical action potential configuration. [From Steinbaum and Miller (39).]

glycemia gave the inverse responses. Brown and Melzac (115) confirmed the suppression of LH neuron multiunit activity in cats. In addition, they showed that other hypothalamic regions were also responding (mammillary bodies, zona incerta, prefornical area). Miller and Rabin (116) extended this notion of a nonspecific LH or VMH responsiveness.

None of these works explored a quantitative relationship between spontaneous or induced changes of blood glucose levels and the firing rate of putative glucosensitive neurons in the LH. This was carried out successfully by Himmi *et al.* (117) (Fig. 3.4).

A chronic intracardiac catheter was implanted in anesthetized rats. A microflow of blood was drawn and blood glucose level determined every 2 min. Hyperglycemia and hypoglycemia were induced, respectively, by intravenous injection of glucose and phlorizine infusions. The firing rate of LH single units was simultaneously recorded in the same rats. The moderate induced hyperglycemia (30% of the basal level) decreased neuronal activity in 13 neurons among the 30 explored. In eight of these responding neurons, a significant correlation appeared between the increased or decreased activities and the hyperglycemic level. In other rats, 10 LH neurons responded to phloridzine-induced hypoglycemia by either decreasing or increasing activities. Four neurons responded in the opposite direction to a succession of hyperglycemia–hypoglycemia. In a last experiment, LH neuron responses to spontaneous fluctuations of blood glucose level during 2 hr were investigated. In six cases, a significant inverse relationship was found between changes of blood glucose level and neuronal activities.

Despite the small number of explored neurons, these results provide strong evidence for the activation of specific LH neurons by the circulating glucose level. These results associated with the previously described works on the relationship between the neuronal activity and feeding and, most of all, with aphagia induced by an LH-GTG implant finally clarify that glucosensitive neurons in the LH are targets of the systemic stimulus to eat.

But apparently they are not the only ones. Earlier, it was shown that in

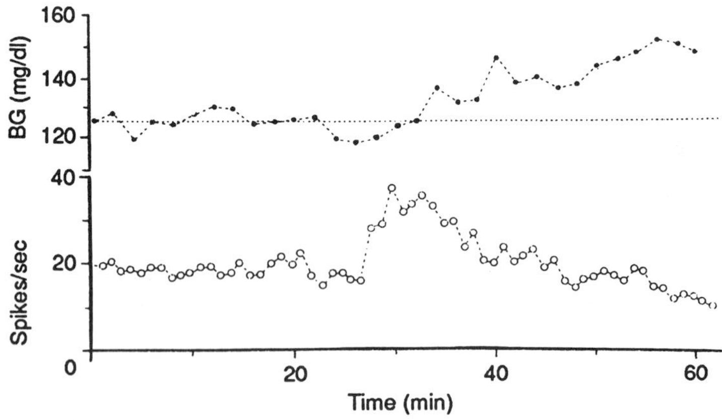

Figure 3.4 Recording of an inverse relationship between glycemia and neuron activity.

mice GTG labeled some sites in the hindbrain (area postrema, nucleus tractus solitarius) in addition to the VMH (118). In the nucleus tractus solitarius, neurons are found changing their activity after intraportal glucose infusions or in relation to blood glucose level (119, 120). The same neurons respond to local glucose (119). In area postrema, glucosensitive neurons are dissociated from other neurons responding to hypertonic NaCl, and from others presumably involved in nausea and learned taste aversion (119). Thus, consistent with the Ritter experiment (see Chapter 2, Section VI), it seems that the hindbrain is also glucose-responsive. However, the relation of this responsiveness to the elicitation of eating is not yet clearly established.

E. Brain Glucose Uptake and Feeding

A continuous cellular glucose uptake in the whole brain is a vital requirement. However, it was thought that different levels of uptake in various regions could reveal their specific role in feeding. The technique of ^{14}C-2-DG offered the possibility of a direct and regional evaluation of glucose uptake. Peripheral injection of ^{14}C-2-DG peripherally labels all areas of the brain as a function of their respective oxidative metabolism at the time of injection. ^{14}C-2-DG injected in rats during their bar-pressing for foods shows an augmentation of the labeling and therefore of the glucose uptake in most regions of the brain (e.g., cortex, thalamus, substantia nigra, fornix and medial forebrain bundle, nucleus tractus substains) (121). This is not surprising and confirms that almost all the brain is activated in the sensorimotor performance of eating. More significant is the difference of the overall glucose uptake in the cortex found between the nocturnal period of high intake and the diurnal period of low intake. A maximal glucose uptake was found at night (3 A.M.) in rats (0.98 μmol/min/g of tissue). A 30% reduction was measured at 3 P.M. (122). A comparable result was reported by Newman and Booth (123).

No attempt has yet been made to see a specific change of the glucose uptake in the critical sites of the hypothalamus, such as a reduced uptake in the LH glucosensitive neurons. Early O_2 consumption measurement following a ^{14}C-glucose tubing in the stomach showed a higher consumption in the medial LH hypothalamus than in fed rats. This difference disappeared under food deprivation. The ratio of VMH to LH O_2 consumption varied inversely with the level of intake (124, 125). The use of the modern technique of the hyperinsulinic-euglycemic clamp in unanesthetized animals opened a new possibility. Rats placed in this clamp condition during 3 hr received either ^{14}C-2-DG to determine the glucose uptake or labeled insulin to explore the insulin binding in various brain areas. In this condition of maintained hyperinsulinism and maintained basal blood glucose, a reduced glucose uptake was observed in the examined regions. The maximal reduction (-30%) was found in the LH. The insulin binding was elevated in the same region by the 3 hr of hyperinsulinemia (126). Further study associated with a concomitant measure of food intake should indicate the exact significance of this elaborated experiment.

The use of the ^{14}C-2-DG labeling of brain tissue to evaluate brain glucose uptake may be misleading. Young and Deutsch (127) injected ^{14}C-2-DG peripherally in hypoglycemic and hyperglycemic rats. They found a considerable elevation of labeling of the brain in hypoglycemia and its reduction in hyperglycemia. They concluded that brain glucose uptake was inversely correlated to blood glucose. However, it is well known that 2-DG competes with glucose in the glucose transport system across the blood–brain barrier. Thus, the entry of the injected ^{14}C-2-DG depends on the ratio of ^{14}C-2-DG to glucose level. It is increased in hypoglycemia and reduced in hyperglycemia. Consequently, the labeling of brain tissue by ^{14}C-2-DG does not indicate glucose uptake but, rather, the degree of substitution of this uptake for the glucose one.

A preliminary report by Meyerovitch *et al.* (128) is suggestive. Vanadate injection is known to increase glucose utilization in the brain. Added to the water drunk by rats or injected intraventricularly, the compound strongly reduces the intake and also the body weight of undeprived or deprived animals.

F. Effects of Alloxan Intraventricular Administration

A new insight on the nature of brain glucoreceptors and of their potential analogy with the β cell pancreatic glucoreceptors was provided by intraventricular administrations of alloxan. Alterations of either the normal glucoregulation or food intake by intraventricular administration of alloxan could differentiate various types and localization of such glucoreceptors. According to Woods and McKay (129), alloxan in lateral ventricles does not affect intake nor the normal regulation of blood glucose. However, treated rats show a lesser response to both 2-DG and 24-hr food deprivation. Interestingly, these effects of alloxan are suppressed by hyperglycemia, as they are at the level of pancreatic β cell to obtain alloxan-induced diabetes. Thus, brain glucoreceptors altered by alloxan would be membrane glucoreceptors analogous to those involved in the glucose-stimulated insulin release in the pancreas. The effects of alloxan injected either in the lateral or in the fourth ventricles could be suggestive (130–132). In both ventricles, an insignificant effect on *ad libitum* food intake was recorded. However, the stimulating effects of 2-DG as well as of insulin were strongly suppressed more in the fourth than in the lateral ventricles. In contrast, the hyperglycemic effect of 2-DG was preserved. Furthermore, alloxan in the fourth ventricle alters the feeding response both to intraventricular administration of GTG and to peripheral administration of 2-DG. This confirms the notion that peripherally injected 2-DG acts centrally. Hyperglycemia is maintained after administrations of glucose antimetabolites, suggesting again that brain sites involved in hyperglycemic and feeding responses to 2-DG are different. Other analogies of the effect of alloxan on brain glucoreceptors with its action on pancreatic β cells were added to the Woods' data. L-glucose inefficient to block the effect of alloxan on β cells, contrary to Δ-glucose, was also inefficient to prevent the intraventricular alloxan to suppress the 2-DG stimulation of feeding.

References

1. Epstein, A. N. (1960). Reciprocal changes in feeding behavior produced by intra-hypothalamic chemical injections. *Am. J. Physiol.* **199**, 969–974.
2. Mabel, J., Baile, C., and Mayer, J. (1966). Hyperphagia induced by ventricular pressure and pentobarbital in normal and hypothalamic obese rats. *The Lancet* **7761**, 472.
3. Duggan, J. P., Storlien, L. H., Kraegen, E. W., and Booth, D. A. (1988). Effect of procaine injection into VMH area of the rat on serum insulin, glucose, corticosterone and gastric emptying rate. *Physiol. Behav.* **43**, 29–34.
4. Dhatt, R. K., Rattan, A. K., and Mangat, H. K. (1988). Effects of chronic intra-cerebroventricular morphine to feeding responses in male rats. *Physiol. Behav.* **43**, 553–557.
5. Maes, H. (1980). Time course of feeding induced by pentobarbital injections into the rat's VMH. *Physiol. Behav.* **24**, 1107–1114.
6. Woods, S. C., and Porte, D. J. (1975). Effect of intracisternal insulin on plasma glucose and insulin in dog. *Diabetes* **24**, 905–909.
7. Herberg, L. J. (1960). Hunger reduction produced by injecting glucose into the lateral ventricle of the rat. *Nature (London)* **187**, 245–246.
8. Kurata, K., Fujimoto, K., Sakata, T., Etou, H., and Fukagawa, K. (1986). D-glucose suppression of eating after intra-3rd ventricle infusion in rat. *Physiol. Behav.* **37**, 615–620.
9. Panksepp, J., and Nance, D. M. (1972). Insulin, glucose and hypothalamic regulation of feeding. *Physiol. Behav.* **9**, 447–451.
10. Booth, D. A. (1968). Effects of intrahypothalamic glucose injection on eating and drinking elicited by insulin. *J. Comp. Physiol. Psychol.* **65**, 13–16.
11. Glick, Z., and Mayer, J. (1968). Hyperphagia caused by cerebral ventricular infusion of phloridzin. *Nature (London)* **219**, 1374.
12. Hatfield, J. S., Millard, W. J., and Smith, C. J. V. (1974). Short-term influence of

intraventromedial hypothalamic administration of insulin on feeding in normal and diabetic rats. *Pharm. Biochem. Behav.* **2**, 223–226.

13. Iguchi, A., Burleson, P. D., and Szabo, A. J. (1981). Decrease in plasma glucose concentration after microinjection of insulin into VMN. *Am. J. Physiol.* **240**, E95–100.

14. Nagai, K. (1982). Effect of intracranial injection of insulin on the circadian feeding rhythms of rats. *Brain Res.* 175–180.

15. Mori, S., Nagai, K., Hara, M., and Nakagawa, A. (1985). Time dependent effect of insulin in the central nervous system on blood glucose. *Am. J. Physiol.* **249**, R23–30.

16. Strubbe, J. H., and Mein, C. G. (1977). Increased feeding in response to bilateral injection of insulin antibodies in the VMH. *Physiol. Behav.* **19**, 309–314.

17. Muller, E., Cocchi, D., and Forni, A. (1971). A central site for the hyperglycemic action of 2-DG in mouse and rats. *Life Sci.* **10**, 1057–1068.

18. Balagura, S., and Kanner, M. (1971). Hypothalamic sensitivity to 2-deoxy-D-glucose and glucose: effects of feeding behavior. *Physiol. Behav.* **7**, 251–256.

19. Berthoud, H. R., and Mogenson, G. J. (1977). Ingestive behavior after intracerebral and intracerebroventricular infusions of glucose and 2-DG. *Am. J. Physiol.* **2**, R127–133.

20. Miselis, R. R., and Epstein, A. N. (1975). Feeding induced by intracerebroventricular 2-DG in the rat. *Am. J. Physiol.* **229**, 1438–1447.

21. Larue, C., and Le Magnen, J. (1971). 2-DG induced food intake in rats after VMH lesion. *Physiol. Behav.* **26**, 313–317.

22. Walsh, L. L., and Grossman, S. P. (1975). Loss of feeding following 2-DG but not insulin after zona incerta lesion in the rat. *Physiol. Behav.* **15**, 481–485.

23. Bellinger, L. L., Bernardis, L. L., and Dill, R. E. (1980). Ventral noradrenergic bundle "area" lesions producing reduced body weight and activity. *Physiol. Behav.* **25**, 783–789.

24. Ritter, R., Slusser, P. G., and Stone, S. (1981). Glucoreceptors controlling feeding and blood glucose: Location in the hindbrain. *Science* **213**, 451–452.

25. Grossman, S. P. (1969). A neuropharmacological analysis of hypothalamic and extrahypothalamic mechanisms concerned in the regulation of food and water intake. *Ann. N.Y. Acad. Sci.* **157** (Art. 2), 902–917.

26. Slangen, J., and Van der Gutten, J. (1979). The activity of central catecholamines neuron system and their role in normal feeding behavior. *Neurosci. Lett.* Suppl. 3.

27. Leibowitz, S., and Brown, L. (1978). Histochemical and pharmacological analysis of noradrenergic injection to the periventricular hypothalamus in relation to feeding stimulation. *Brain Res.* **141**, 119–128.

28. Singer, G., and Willis, G. L. (1977). Biochemical and pharmacological basis for the LH syndrome. *Brain Res. Bull.* **2**, 485–489.

29. Matthews, J., Booth, D. A., and Storlien, J. (1978). Factors influencing feeding elicited by intrahypothalamic noradrenaline in rats. *Brain Res.* **141**, 119–128.

30. Even, P., and Nicolaïdis, S. (1986). Metabolic mechanism of the anorectic and leptogenic effects of the serotonin agonist fenfluramine. *Appetite* **7** (Suppl.), 141–163.

31. Morley, J. E., and Levine, A. S. (1987). Effects of neuropeptide Y on ingestive behavior in the rat. *Am. J. Physiol.* **252**, R599–609.

32. Stanley, B. G., Chin, A. S., and Leibowitz, S. F. (1985). Feeding and drinking elicited by central injection of neuropeptide Y: Evidence for a hypothalamic site(s) of action. *Br. Res. Bull.* **14**, 521–524.

33. Myers, R. D., Peinado, J. M., and Minano, F. J. (1988). Monoamine transmitter activity in L.H. during its perfusion with insulin and 2-DG in sated and fasted rats. *Physiol. Behav.* **44**, 633–643.

34. Hoebel, B. G., Hernandez, L., Schwartz, D. H., Mark, G. P., and Hunter, G. A. (1989). Microdialysis studies of brain norepinephrine, serotonine and dopamine release during ingestive behavior. In *Psychobiology of eating disorders*. Volume 575 (L. H. Schneider, S. J. Cooper, and K. A. Halmi, eds.) *Ann. N.Y. Acad. Sci.* 171–193.

35. Mayer, J. (1953). Glucostatic mechanisms of regulation of food intake. *N. Engl. J. Med.* **249**, 13–16.

36. Sclafani, A., Berner, C. N., and Maul, G. (1975). Multiple knife cuts between the medial and lateral hypothalamus in the rat: A reevaluation of hypothalamic circuitry. *J. Comp. Physiol. Psychol.* **88**, 210–217.

37. Anand, B. K., and Brobeck, J. R. (1951). Localisation of a "feeding center" in the hypothalamus of the rat. *Proc. Soc. Exp. Biol. Med.* **77**, 323–324.

38. Delgado, J. M. R., and Anand, B. K. (1953). Increase of food intake induced by electrical stimulation of the lateral hypothalamus. *Am. J. Physiol.* **172**, 162–168.

39. Steinbaum, E. A., and Miller, N. E. (1965). Obesity from eating elicited by daily electrical stimulation of hypothalamus. *Am. J. Physiol.* **208**, 1–5.

40. Wise, R. (1974). Lateral hypothalamic electrical stimulation: Does it make the rat hungry? *Brain Res.* **67**, 187–210.

41. Huang, Y. H., and Mogenson, G. J. (1972). Neural pathways mediating drinking and feeding in rats. *Exp. Neurol.* **37**, 269–286.

42. Coons, E. D., Levak, M., and Miller, N. E. (1965). Lateral hypothalamus: Learning of food setting response motivated by electrical stimulation. *Science* **150**, 1320–1321.

43. Tenen, S. S., and Miller, N. E. (1964). Strength of electrical stimulation of lateral hypothalamus, food deprivation and tolerance for quinine in food. *J. Comp. Physiol. Psychol.* **58**, 55–62.

44. Rice, T. W., and Campbell, J. (1973). Effects of neocortical ablation on eating elicited by hypothalamic stimulation. *Exp. Neurol.* **39**, 347–370.

45. Waldbillig, R. J. (1975). Attack, eating, drinking and gnawing elicited by electrical stimulation of mesencephalon and pons. *J. Comp. Physiol. Psychol.* **89**, 200–212.

46. Steffens, A. B. (1965). Influence of reversibility on eating behavior, blood glucose and insulin in the rat. *Am. J. Physiol.* **227**, 142–173.

47. Booth, D. A., Coons, E. E., and Miller, N. E. (1969). Blood glucose responses to electrical stimulation of hypothalamic feeding area. *Physiol. Behav.* **4**, 991–1002.

48. Kita, H., Niijima, A., Oomura, Y., Ishizuka, S., Aou, S., Yamabe, K., and Yoshimatsu, H. (1980). Pancreatic nerve response induced by hypothalamic stimulation in rats. *Brain Res. Bull.* **5** (Suppl. 4), 163–168.

49. Yoshimatsu, J., Oomura, Y., Katagushi, K., and Niijima, A. (1987). Effects of hypothalamic stimulations and lesions on adrenal nerve activity. *Am. J. Physiol.* **253**, R418–424.

50. Berthoud, H. K., and Baettig, K. (1974). Effects of insulin and 2 DG on plasma glucose level and LH eating threshold in the rat. *Physiol. Behav.* **12**, 547–556.

51. Hernandez, L. L., and Briese, E. (1971). Insulin inhibition of hypothalamic self-stimulation. *Acta Latino Am.* **21**, 57–63.

52. Margules, D. L., and Olds, J. (1962). Identical "feeding" and "rewarding" systems in the lateral hypothalamus of rats. *Science* **135**, 374–375.

53. Blundell, J. E., and Herberg, L. J. (1968). Relative effects of deficit and deprivation period on rate of electrical self-stimulation of lateral hypothalamus. *Nature (London)* **219**, 627–628.

54. MacNeil, D. (1974). Lateral hypothalamic self stimulation: Excess of body weight. *Physiol. Psychol.* **2**, 51–52.

55. Hoebel, B. G., and Thompson, R. D. (1969). Aversion to lateral hypothalamic stimulation caused by intragastric feeding or obesity. *J. Comp. Physiol. Psychol.* **68**, 536–543.

56. Ferguson, N. B. L., and Keesey, R. E. (1971). Comparison of VMH lesion effects upon feeding and the lateral hypothalamic self-stimulation in the female rats. *J. Comp. Physiol. Psychol.* **74**, 263–271.

57. Balagura, S. (1968). Influence of osmotic and caloric loads on lateral hypothalamic self-stimulation. *J. Comp. Physiol. Psychol.* **66**, 325–328.

58. Balagura, S., and Hoebel, B. G. (1967). Self-stimulation of the lateral hypothalamus modified by insulin and glucagon. *Physiol. Behav.* **2**, 337–340.

59. Hoebel, B. G. (1968). Inhibition and disinhibition of self-stimulation and feeding: Hypothalamic control and post-ingestional factors. *J. Comp. Physiol. Psychol.* **66**, 89–100.

60. Coons, E. E., and Cruce, J. A. F. (1968). Lateral hypothalamus: Food current intensity and maintening self-stimulation of hunger. *Science* **159**, 1117–1119.

61. Mendelson, J. (1969). Lateral hypothalamic stimulation: Inhibition of aversive effects by feeding, drinking and gnawing. *Science* **166**, 1431–1433.

62. Coons, E. E., and White, N. (1969). Tonic properties of oral sensation and modulation of intracranial stimulation: The CNS waiting of internal and external factors governing reward. *Ann. N.Y. Acad. Sci.* **157**, 219–226.

63. Ettenberg, A., and White, N. (1978). Conditioned taste preferences in the rat induced by self-stimulation. *Physiol. Behav.* **21,** 363–368.

64. Ganchrow, J. R., Lieblich, I., and Cohen, E. (1981). Consummatory responses to taste stimuli in rats selected for high and low rates of self-stimulation. *Physiol. Behav.* **27,** 971–976.

65. Moreau, J. L., Cohen, E., and Lieblich, I. (1984). Ventral tegmental self-stimulation, sensory activity and pain reduction in rats selected for high and low rates of LH self stimulation. *Physiol. Behav.* **33,** 825–830.

66. Terman, M., and Terman, J. S. (1970). Circadian rhythm of brain self-stimulation behavior. *Science* **168,** 1242–1244.

67. Schallert, T., and Whishaw, I. Q. (1978). Two types of aphagia and two types of sensorimotor impairment after lateral hypothalamic lesions: Observations in normal weight, dieted and fattened rats. *J. Comp. Physiol. Psychol.* **92,** 720–741.

68. Stricker, E. M., Swerdloff, A. F., and Zigmond, M. J. (1978). Intrahypothalamic injections of kainic acid produce feeding and drinking deficits in rats. *Brain Res.* **158,** 470–473.

69. Braun, J. J. (1975). Neocortex and feeding behavior in the rat. *J. Comp. Physiol. Psychol.* **89,** 507–522.

70. Teitelbaum, P., and Epstein, A. N. (1962). The lateral hypothalamic syndrome recovery of feeding and drinking after lateral hypothalamic lesions. *Psychol. Rev.* **69,** 74–90.

71. Teitelbaum, P., and Cytawa, J. (1965). Spreading depression and recovery from lateral hypothalamic damage. *Science* **147,** 61–63.

72. Edwards, D. A., Bryan, B., and Florian, V. A. (1979). Olfactory system involvement in the recovery of feeding and drinking after LH damage in the rat. *Physiol. Behav.* **22,** 1163–1169.

73. Stricker, E. M., Friedman, M. I., and Zigmond, M. J. (1975). Glucoregulatory feeding in rat after 6-OH-DOPA or LH-lesions. *Science* **189,** 895–896.

74. Singer, G., and Willis, G. L. (1977). Biochemical pharmacological basis for the LH syndrome. *Brain Res. Bull.* **2,** 485–489.

75. Teitelbaum, P., and Stellar, E. (1954). Recovery from the failure to eat produced by hypothalamic lesions. *Science* **120,** 894–895.

76. Larue, C., and Le Magnen, J. (1970). Effects of the removal of the olfactory bulbs upon hyperphagia and obesity induced in rats by V.M.H. lesion. *Physiol. Behav.* **5,** 509–513.

77. Baillie, P., and Morrison, S. D. (1963). The nature of the suppression of food intake by LH lesions in rats. *J. Physiol. (London)* **165,** 227–245.

78. Rodgers, W. L., Epstein, A. N., and Teitelbaum, P. (1965). LH aphagia: Motor failure or motivational deficit? *Am. J. Physiol.* **208,** 334–342.

79. Loang, R. (1976). Dissociation of gustatory and weight regulatory responses to quinine following LH lesion. *J. Physiol. (Paris)* **69,** 978–985.

80. Epstein, A. N., and Teitelbaum, P. (1967). Specific loss of the hypoglycemic control of feeding in recovered lateral rats. *Am. J. Physiol.* **213,** 1159–1167.

81. Wayner, M. J., Yin, T. H., Barone, F. C., Lee, H. K., and Tsai, C. T. (1979). Effects of discrete destruction of functionally or identified chemosensitive hypothalamic neurons on ingestive behavior. *Physiol. Behav.* **23,** 385–390.

82. Blass, E. M., and Kraly, F. S. (1974). Medial forebrain bundle lesions: Specific loss of feeding to decrease of glucose utilization in rats. *J. Comp. Physiol. Psychol.* **86,** 679–692.

83. Steffens, A. B., Mogenson, G. J., and Stevenson, J. A. F. (1972). Blood glucose, insulin and free fatty acids after stimulation and lesions of the hypothalamus. *Am. J. Physiol.* **222,** 1446–1452.

84. Gray, R. H. (1971). The effect of bilateral destruction of the hypothalamus feeding centre on fasting blood sugar, glucose absorption, and glucose tolerance in the rat. *Aust. J. Exp. Biol. Med. Sci.* **49,** 225–232.

85. Grijalva, C. V., Novin, D., and Bray, G. A. (1980). Alterations in blood glucose, insulin and FFA LH lesions or parasagittal knife cuts. *Brain Res. Bull.* **5** (Suppl. 4), 109–118.

86. Powley, T. L., and Keesey, R. E. (1970). Relationship of body weight to the LH feeding syndrome. *J. Comp. Physiol. Psychol.* **70,** 25–36.

87. Keesey, R. E., Powley, T. L., and Kemnitz, J. W. (1976). Prolonging LH anorexia by tube-feeding. *Physiol. Behav.* **17,** 367–371.

88. Bernardis, L. L., and Bellinger, L. L. (1981). Dorsomedial hypothalamic hypophagia: Self-selection of diets and macronutrients, efficiency of food utilization, "stress eating", response to high-protein diet and circulating substrate concentrations. *Appetite* **2,** 103–113.

89. Contreras, R. J., Fox, E., and Drugovich, M. L. (1982). Area postrema lesions produce feeding deficits in the rat: Effect of preoperative dieting and 2-DG. *Physiol. Behav.* **29,** 875–884.

90. Clarke, J., and Coleman, N. (1986). Persistent meal associated rhythms in SCN-lesioned rats. *Physiol. Behav.* **36,** 105–113.

91. Nagai, K., Nishio, T., Nagakawa, H., Nakamura, S., and Fukuda, Y. (1978). Effect of bilateral lesions of the suprachiasmatic nuclei on the circadian rhythm of food intake. *Brain Res.* **142,** 384–389.

92. Stoinev, A. G., Ikonomov, O. C., Tarkolev, N., and Shisheva, A. (1980). Effects of the SCN lesions on the circadian variations in PRA and IRI levels in the rat. *Acta Physiol. Akad. Scientiarum Hungaricae* **56,** 431–435.

93. Mayer, J., and Marshall, N. B. (1956). Specificity of gold-thioglucose for ventromedial hypothalamic lesions and obesity. *Nature (London)* **178,** 1399–1400.

94. Smith, C. J. V. (1972). Hypothalamic glucoreceptors: The influence of GTG implants in VMH and LH areas of normal and diabetic rats. *Physiol. Behav.* **9,** 391–396.

95. Russek, M. (1971). Hepatic receptors and the neurophysiological mechanisms controlling feeding behavior. *Neurosci. Res.* **4**, 213–282.

96. Russek, M., Rodriguez-Zendejas, C., and Sina, T. (1968). Hypothetical liver receptors and anorexia caused by adrenaline and glucose. *Physiol. Behav.* **3**, 249–257.

97. Niijima, A. (1980). Glucose sensitive afferent nerve fibers in the liver and regulation of blood glucose. *Brain Res. Bull.* **5** (Suppl. 4), 175–179.

98. Louis-Sylvestre, J., Servant, J., Molimard, R., and Le Magnen, J. (1980). Effects of liver denervation on the feeding pattern of rats. *Am. J. Physiol.* **239**, R66–R70.

99. Louis-Sylvestre, J., Larue-Achagiotis, C., Michel, A., and Houssin, D. (1990). Feeding pattern of liver-transplanted rats. *Physiol. Behav.* **48**, 321–326.

100. Bellinger, L. L., Trietley, G. J., and Bernardis, L. (1976). Failure of portal glucose and adrenaline or liver denervation to affect food intake in dogs. *Physiol. Behav.* **16**, 299–304.

101. Bellinger, L. L., and Williams, F. E. (1983). Liver denervation does not modify feeding responses to metabolic challenges or hypertonic NaCl-induced water consumption. *Physiol. Behav.* **30**, 463–470.

102. Tordoff, M. G., Hopfenbeck, J., and Novin, D. (1982). Hepatic vagotomy (partial hepatic denervation) does not alter ingestive responses to metabolic challenges. *Physiol. Behav.* **28**, 417–424.

103. Egli, R., Langhans, T., and Scharrer, E. (1984). Selective vagotomy does not prevent compensatory feeding as a response to body weight changes. *J. Autonom. Nerv. Syst.* **10**, 159–170.

104. Langhans, T., and Scharrer, E. (1985). Dissociation of epinephrine hyperglycemia and anorexia effects. *Physiol. Behav.* **34**, 456–463.

105. Bellinger, L. L., and Williams, R. A. (1986). Glucagon and epinephrine suppression of food intake in liver denervated rats. *Am. J. Physiol.* **250**, R349–358.

106. MacIsaac, L., and Geary, M. (1985). Partial liver denervation dissociates inhibitory effects of pancreatic glucagon and epinephrine on feeding. *Physiol. Behav.* **35**, 233–237.

107. Tordoff, M., and Friedman, M. (1989). Effect of hepatic portal glucose concentrations on food intake and metabolism. *Am. J. Physiol.* **257**, 1474–1482.

108. Oomura, Y. (1984). Neural network of glucose monitoring system. *J. Autonom. Nerv. Syst.* **10**, 159–170.

109. Hamburg, M. D. (1971). Hypothalamic unit activity and eating behavior. *Am. J. Physiol.* **220**, 980–985.

110. Schmitt, M. (1973). Circadian rhythmicity in responses of cells in the lateral hypothalamus. *Am. J. Physiol.* **225**, 1096–1101.

111. Ono, T., Nishino, H., Sasaki, K., Fukuda, M., and Muramoto, K. I. (1981). Long-term lateral hypothalamic single unit analysis in feeding behavior of freely moving rats. *Neurosci. Lett.* **26**, 79–83.

112. Rolls, E. T., Burton, M. J., and Mora, F. (1976). Hypothalamic neuronal responses associated with the sight of food. *Brain Res.* **111**, 53–66.

113. Anand, B. K., China, G. S., and Singh, B. H. (1962). Effect of glucose on the activity of the hypothalamic "feeding centers." *Science* **138**, 597–598.

114. Anand, B. K., China, G., Sharma, K., Dua, S., and Singh, B. (1964). Activity of single neurons in hypothalamic feeding center: Effect of glucose. *Am. J. Physiol.* **207**, 1149–1154.

115. Brown, K. A., and Melzac, R. (1969). Effect of glucose on multiunit activity in the hypothalamus. *Exp. Neurol.* **24**, 363–373.

116. Miller, D. S., and Rabin, B. M. (1975). Effect of glucose on multiple activity in the hypothalamus of female rats. *Exp. Neurol.* **49**, 418–428.

117. Himmi, T., Boyer, A., and Orsini, J. C. (1988). Changes in lateral hypothalamic neuronal activity accompanying hyper- and hypoglycemias. *Physiol. Behav.* **54**, 347–354.

118. Powley, T. L., and Opsahl, C. A. (1974). Ventromedial hypothalamic obesity abolished by subdiaphragmatic vagotomy. *Am. J. Physiol.* **226**, 25–33.

119. Adachi, A., Kobashi, M., Miyoshi, N., and Takamoto, G. (1989). Chemosensitive neurones in area postrema of rats: Their possible functions. *Appetite* **12**, 195.

120. Laughton, W. B., and Campfield, L. A. (1989). Electrophysiological recording from abdominal vagal afferents sensitive to peripheral metabolic state. *Appetite* **12**, 221.

121. Morimoto, A., Suzumi, M., Sakata, Y., and Murakami, N. (1984). Activation of brain regions in rats during food-intake operant behavior. *Physiol. Behav.* **33**, 965–968.

122. Crane, P. D., Braun, E. D., and Kleindorf, W. (1980). Cerebral glucose utilization in conscious rats: Evidence for a circadian rhythm. *J. Neurochem.* **34**, 1792–1806.

123. Newman, J. C., and Booth, D. A. (1981). Gastrointestinal and metabolic consequences of a rat's meal on maintenance diet ad libitum. *Physiol. Behav.* **27**, 929–940.

124. Panksepp, J. (1972). Hypothalamic radioactivity after intragastric glucose ^{14}C in rats. *Am. J. Physiol.* **223**, 396–401.

125. Panksepp, J., and Heilly, P. (1975). Medial and lateral hypothalamic oxygen consumption as a function of age, starvation and glucose administration in rat. *Brain Res.* **94**, 131–140.

126. Marfaing-Jallat, P., Pénicaud, L., Broer, Y., Mraovitch, S., Calando, Y., and Picon, L. (1990). Effects of hyperinsulinemia on local cerebral insulin binding and glucose utilization in normoglycemic awake rats. *Neurosci. Lett.* **115**, 279–285.

127. Young, W. G., and Deutsch, J. A. (1986). Effects of blood glucose level on [14]C-2 DG uptake in rat brain tissue. *Neurosci. Lett.* **69**, 89–93.

128. Meyerovitch, J., Schechter, Y., and Amir, S. (1989). Vanadate stimulates in vivo glucose uptake in brain and arrests food intake and body gain in rats. *Physiol. Behav.* **45**, 1113–1116.

129. Woods, S. C., and McKay, L. D. (1978). Intraventricular alloxan eliminates feeding elicited by 2-deoxyglucose. *Science* **202**, 1209–1210.

130. Murname, J., Ritter, R., and Langenheim, E. (1986). Glucoprivic feeding is impaired by alloxan lateral and 4th ventricle injection. *Am. J. Physiol.* **243,** R312–318.

131. Murname, J. M., and Ritter, S. (1985). Intraventricular alloxan impairs feeding to central and systemic glucoprivation. *Physiol. Behav.* **34,** 309–313.

132. Murname, J. M., and Ritter, S. (1985). Alloxan-induced glucoprivic feeding deficits are blocked by D-glucose and amygdalin. *Pharm. Biochem. Behav.* **22,** 407–413.

Chapter Four

The Sensory Stimulation
to Eat and Not to Eat

In preceding chapters, it was repetitively mentioned that the stimulation to eat, i.e., to take in the mouth a solid or liquid material, called a food, to masticate, and to swallow it, is a response to two sources of stimuli. The first one, previously studied, is the systemic stimulus generated by an energy deficit, now identified and increasingly active with food deprivation. The second one, combined or not with the former, is the stimulation to eat as a response to the sensory stimulation of a cephalic sensory apparatus by the food. In the *ad libitum* and scheduled conditions of feeding, in an identical and constant systemic stimulation, animals as well as humans will accept or not a food, depending on their sensory properties. They will eat different foods, giving signs of different strength of the initial stimulation also depending on sensory properties. Finally, they will eat very different amounts of these different foods. When they are food-deprived, their initial stimulation will be (at various levels of the deprivation) as high as the response to the sensory activity is. This sensory contribution to the overall stimulation of feeding is called the "palatability of foods." The combination of the systemic stimulation generating "hunger" with the sensory stimulation is called "appetite." Inasmuch as food palatability is sensory-specific, the term appetite is generally and validly referred to a particular food or to categories of foods.

Before examining in Chapter 5 how internal and external, blood-borne and neural signals are integrated and processed in the brain to govern a regulatory feeding, numerous problems raised by the palatability of foods and its contribution to adapted feeding responses will be examined here.

I. *Alimentary Stimuli*

A. Terminology

To understand and to be understood, speaking the same language is useful. Many scientific controversies as well as many disputes outside of science stem from the use of one word to designate different things and more than one word to designate one thing. Regarding palatability of foods, a tremendous confusion was and is still due to the use of a number of terms of totally, or approximately, synonymous meaning. In addition, each of these various terms is used by different authors with different meanings. The first equivocal term is in the title of this chapter. Alimentary stimuli can mean, and thus be understood as, either foods as sensory stimuli (their odor, taste, etc.) or foods as sensory stimuli of eating. In this chapter (deliberately entitled "alimentary stimuli"), the two meanings, "sensory activity" and "palatability," will be dissociated and their relationships examined. With the same (or almost the same) meaning as palatability, the following terms are commonly used: preference or aversion; taste preference, taste aversion, hedonic value, or hedonic property of foods; pleasantness or unpleasantness; like or dislikes; and incentive property of the food, secondary reinforcer, or "food reward." It is a true Tower of Babel.

The term "preference for" or "prefered food" is the palatability level of a food referred to a standard or to another food of a lower level. Such preferences are tested either in choices or in compared single presentations. The difficulty is that the preference may be different in the two tests. For instance, the preference for sucrose solutions as a function of concentrations is different in the two tests. The standard is water in testing preferences for flavored solutions. Tested in a choice or in compared single presentations, a preferential threshold and curves of preference–aversion as a function of increasing concentrations of the solution are established.

The two terms preference and aversion are misleading and mistaken. The preferential threshold is often confounded with the sensory threshold, also called "absolute threshold." Below the preferential threshold, the solution is tasted, and above the absolute threshold, the solution is tasted as it is demonstrated by a conditioning of the response. The term preference–aversion curve is also inadequate. Concentrations higher than the maximally preferred ones are not aversive in the proper meaning (see Chapter 4, Section IV). They are of a lower palatability level, like concentrations below the maximally preferred one. Of course, such "preferences" vary with the standards. The preference for a sucrose solution versus water has nothing to do with the preference for this solution versus the solid food. They vary with the testing condition: deprivations, previous testing, and time of day. Most of all, they depend on the measured parameters. A measure of the initial rate of licking or eating, the cumulative intake during the first minutes of exposure (brief-exposure technique), or the cumulative intakes until satiation or during 24 hr give quite different and differently significant results. We will come back to this point.

The term "hedonic value," or "property," is commonly used by psychologists. It is often equivocal by implying that the palatability of a food, i.e., its capacity to stimulate eating, is an intrinsic and immutable property of a food and a dimension of its sensory activity, like its quality and intensity are. Pleasantness and unpleasantness and likes and dislikes are used in human studies. These terms are appropriate only to designate the perceived and tested subjective feeling of "pleasure." Their evaluation is badly dissociated by subjects from intensity judgments. Here again the subjective judgment by a brief tasting and by an objective measure of intake until satiation or on a longer time will differ. We will come back later to the terms "incentive" (now abandoned), "secondary reinforcer," and "food reward," all essentially questionable and difficult to delineate.

B. Behavioral and Sensory Specificity

Every behavior is a specific pattern of responses to a specific panel of external sensory stimuli. A hungry male rat is stimulated by food stimuli to eat the food and not to eat or copulate its female. A sexually aroused male rat is stimulated by the sight and odors of its female to copulate it and not to eat or copulate a food. Food stimuli are palatable like female stimuli are sexual.

This self-evident principle of behaviors is instructive when applied to the two ingestive behaviors: feeding and drinking. They are two distinct behaviors. Their internal stimuli and brain mechanisms subserving each of them are different. Sensory stimuli, to which hungry and thirsty animals respond, also are different and specific to each behavior. Rats, presented with a choice of four bottles containing either water, an isotonic NaCl solution, a saccharin sweet solution, or a bitter quinine, were injected with insulin 30 min before a 30-min drinking test. Compared to a saline injection as control, they augmented the intake of the saccharin solution and not that of the other fluids. In similar experiments, the optimal palatability or preference to NaCl solutions shifted to the left or to the right after hypertonic or hypotonic loads, respectively. The same manipulation did not significantly affect the response to sweet solutions (1, 2) Figs. 4.1 and 4.2). Rats, deprived of foods and not of water or moderately

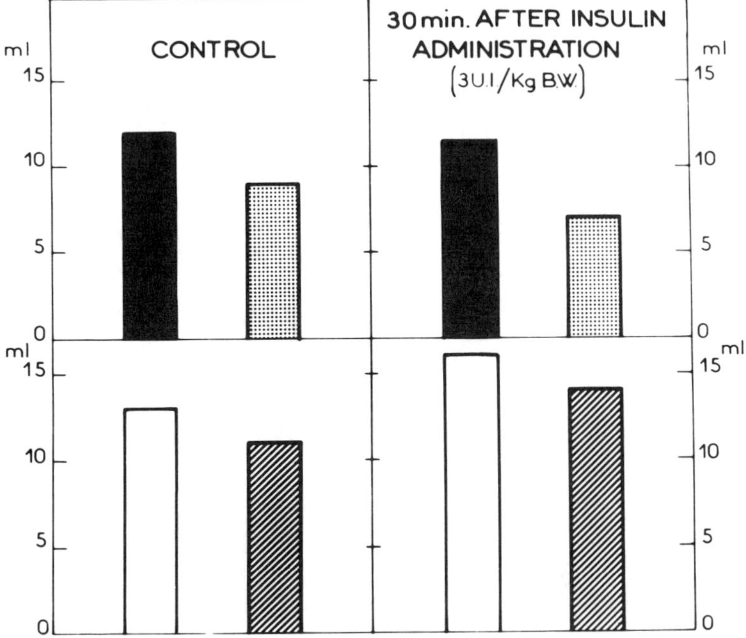

Figure 4.1 Acute and specific effects of insulin injection on the intake of a sweet solution. ■, 1% NaCl; ⊡, H_2O; □, 5% sucrose; ▨, 0.037% saccharin.

deprived of both, preferred saccharin in a choice of the sweet solution or water. Deprived of water and not of food, they preferred water, which they drank in large amounts (3). Thirsty rats offered a longer presentation of water versus saccharin initially preferred water and, after drinking a copious amount, shifted their preference to the sweet solution (3). We will return to the important fact that these behaviorally specific responses are taste-specific. In food-deprived rats, the initial response is identical to sugar and saccharin solutions and, thus, is a response to sweetness. As shown by Schulkin (4), the response of sodium-deprived rats to salt solution is a response to saltiness and not to the sodium content of the salt. We will see why this is the case of these gustatory stimuli and how such taste-specific responses will become, through conditioning, responses either to calories or to the hydroosmotic properties of the solution.

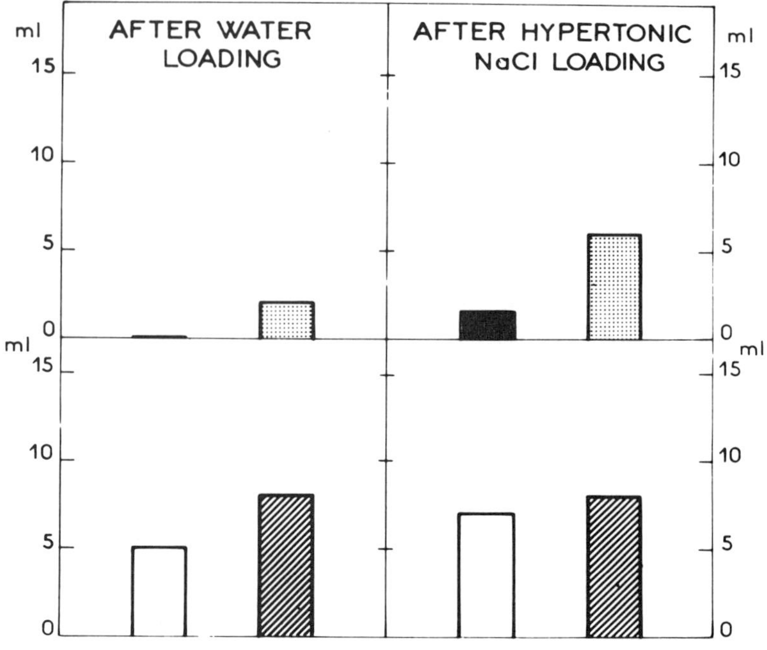

Figure 4.2 Specific effect of water and hypertonic NaCl loads on the intake of salty solution. ■, 18% NaCl; ▨, H$_2$O; □, 22% sucrose; ▨, 1% saccharin.

C. Sensory Quality, Intensity, and Palatability

In each sensory modality, the response of peripheral sensory receptors and the specific brain processing of this sensory information give rise to the two dimensions of the sensory perception: its quality and its intensity. In each of these sensory modalities (e.g., vision, audition, olfaction), the relationships between physical or chemical characteristics of the stimuli and the qualitative discrimination are studied by sensory physiologists. These characteristics of stimuli responsible for the quality (e.g., sweetness) are generally independent of the stimulus–intensity relationships. In other words and, for instance, at all levels of intensity, a sweet stimulus is sweet. Regarding intensity and in all sensory modalities, it is now well documented that the perceived intensity increases as a function of increased stimulations according to a power law (straight line in log–log coordinates) with a specific slope for each quality of the stimulus.

The important point here is that these responses of sensory apparatus and of sensory brain processing are not at all behaviorally dependent, whereas palatability is. Many experiments in animal models and humans demonstrate that the quality of the stimulus (e.g., sweetness or a particular odor), on which the ingestive response will be based, is not modified by changes of this response, i.e., of the palatability of the stimulus. Palatable or not, sweet is sweet. Sweetness becoming aversive by conditioning does not become bitter. A menthol odor is a menthol odor whatever the ingestive response it promotes. The same is true for the intensity–palatability relationships. Changing the palatability for a given stimulus, i.e., the preference, does not change the threshold of its sensory activity. Early on, for example, Campbell (5) showed that the preferential threshold for glucose solution in a choice versus water was considerably shifted by food deprivation, whereas the absolute threshold as determined by conditioning was unchanged. The same fact will be ascertained by measurements of electrophysiological thresholds in the gustatory chorda tympani nerves, which are not changed under treatment altering behavioral responses to the taste stimulus (see Chapter 5).

Finally, and decisively, the evolution of palatability and of sensory intensities as a function of increasing stimulus intensity is not parallel nor correlated to each other. A considerable number of experiments, particularly in humans, showed a monotonous increase of perceived intensity with the stimulus strength opposed to a typical curve of the palatability–intensity relationship with a maximal level at a given concentration of the stimulus. In other cases, such as food saltiness or olfactory flavors, one concentration of the stimulus only is palatable, both lower and higher concentrations being unpalatable. In all these cases, the versatility of palatability is observed, in contrasting to the stability of sensory responsiveness *per se*.

D. Measurements

The measurement of palatability as a component of the initial stimulation to eat by a measure of intake is difficult and may be misleading. it is due to the complex role of this initial strength of stimulation in the satiating process. We will see later that, from this initial level of stimulation, the amount eaten until satiation is both the effect of a specific satiating sensory capacity of the food not related to the initial palatability and to the sensory-independent effect of stomach filling. The amount eaten at the onset of satiety will be the result of a balance between the initial strength of stimulation and the two uncorrelated oral and gastric negative feedbacks. Thus, the amount eaten cannot be a measure of the initial stimulation. The only valid measure is that of the initial rate of eating. With liquid foods or sweet solutions, the recording of the licking rate as a reliable measure of palatability was finely studied by Davis (6). Rats lick at constant rate of seven to eight lappings per second. Durations of bursts of licking and/or interburst pauses are increased or decreased with concentrations of various sweet solutions. The integrated initial rate is also changed for each concentration with the state of the animal (food deprivation or a previous caloric load). The recording of the microstructure of the meal of a solid food also provides an identical measure of the initial rate of eating and of its changes with manipulations altering the palatability of foods.

E. Visual Stimuli

The visual appearance of food (shape, size, and color) is a component of the sensory stimulation to eat before it is taken in the mouth and, thus, is involved in the initial seeking, recognition, and selection of foods in the environment. The response is specific to the sensory characteristics. In monkeys, a false peanut as a lure is mistaken like rats are mistaken by saccharin prior to extinction (e.g., 7). In mice, rats, and monkeys, the piece sizes of the offered solid food affects intake (8, 9). However, this role of the sight of food in the overall palatability response is different among species. It is spontaneously predominant in birds and in primates whose feeding is said to be a mainly visual behavior. In the same species, visual cues, more than others, will be the basis for learned preferences or learned aversions. The opposite is true in rodents, in which chemical senses are the main basis for the oral analysis of foods.

Of course the visual appearance *per se* gives no information about nutritive properties of foods. Thus, the question of whether or not and how unlearned or learned responses to these visual cues will permit an anticipatory response adapted to nutrition efficiencies will be raised.

F. Olfactory Stimuli

Food odors through the nasal route provide a very important telesignal in the short- and long-distance detection and recognition of foods (e.g., of prey by predators). Four different properties of olfactory sensitivity are exploited in this function. The gazeous nature of olfactory stimuli permits air-borne diffusion and, partly, a localization of sources. The property of odorous chemicals to be adsorbed on the ground or other solids are the base for additional spatial repairs and trails. Moreover, thresholds of odorant activity in air dilutions are extremely low, up to 10^{-10}. Thus, only extreme minute traces of the odorant in a food allow its recognition. Finally, this recognition is highly specific. The peculiar property of the olfactory system, compared to taste, is to individualize practically all active molecules by a discriminable odor. In addition, mixtures of pure odorants are similarly discriminated from each other as a function of their respective composition. Such unlimited possibilities of food discriminations and food responses are offered by their smells. The sniffing of foods, particularly of novel foods, belongs to the exploratory feeding behavior.

With food in the mouth during mastication and salivation, olfactory cues play a role in the overall palatability responses via the pharyngeal route. It is assessed by the study of rats rendered transitorily anosmic. There, the specific properties of the olfactory stimulation also are operative. Thresholds of activity largely depend, via this route, on the volatility of odorous compounds in solid and liquid foods in which they are contained. But in the mouth, as outside the mouth, unlimited capacity of identification of foods to generate as much palatability responses is provided. Again, the question will be raised of whether or not and how adapted responses to the flavored foods will be possible since no correlation exists between olfactory qualities and intensities in a food and the nutritive properties of that food.

G. Taste Stimuli

The word "taste," like "goût" in French, is used in various ways. It is often used to designate, not a gustatory stimulus, but the sum of sensations provoked by a food in the oral cavity instead of the correct term "flavor." Surprisingly, a diet labeled by the addition of an odorous compound is called a "flavored diet" instead of "odorized diet." Taste for a food means palatability of a food, strengthening the trivial confusion between sensory stimulation *per se* and sensory stimulation to eat. Finally, the meaning is extended to "taste for women," "taste for music," and so on.

The gustatory apparatus is the only sensory modality exclusively differentiated to subserve feeding behavior. Its differentiation from olfaction is recognized from echinoderms in invertebrates and is present in all the vertebrate fillum. Taste receptors on the tongue surface and also in the palate and pharynx are stimulated by a number of chemicals dissolved in liquids or in saliva. Their threshold of activity is considerably higher than that of odorants. Most of them, except acids, are not mixed odorous and taste stimuli. At the level of peripheral receptors, each of them elaborates a specific cross-fiber pattern of discharges, which could be (like in olfaction) a basis for the individualization of stimuli.

However, the electrophysiological exploration of responses in the nucleus of the solitary tract, the primary gustatory projection, indicates four different patterns of neuronal responses corresponding to sweet, salty, acidic, and bitter substances, known as such in humans, and specified in animals by different and characteristic behavioral responses. Stimulating substances are not discriminated within each group. A conditioned aversion for a glucose or a saccharin solution is generalized to other sugar solutions. A conditioned enhanced aversion for quinine is not generalized to sweet, salt, or acid responses (10).

The responses to acid and bitter compounds will be treated in a chapter devoted to aversive responses (Chapter 4, Section IV). Responses to sweet and salty substances are very peculiar inasmuch as the sensory coding is a biochemical one. Most sweet compounds are carbohydrates and supply calories. Apparently, the pure salty taste is only stimulated by NaCl and LiCl. For this reason, unlearned responses preadjusted to nutritive properties of sweetened and salty solid or liquid foods are observed. But the correlation between taste and nutritive properties is only approximate. Saccharin and other sweeteners also are sweet. LiCl is highly toxic. In a couple of two-sugar solutions, the sweetest may be the less calorically dense solution. Thus, the universally manifested unlearned preferences or palatability of sweet solutions and sweetened foods (examined in the next chapter) must be and are modulated by conditioning, as other sensory responses are.

H. Mechanical Stimuli

The consistency and texture of foods play an important role in feeding responses. The origin of the difference, often neglected, is between liquid and solid foods. The masticatory pattern before swallowing is excluded for liquid foods. This pattern provides complex tactile and proprioceptive sensations on solid food, which vary of course with the hardness and physical texture of the food (see Chapter 7, Section II). They can represent very aversive stimuli. An equivalent of saccharin is the greasy texture of mineral oil such as vaseline, which is mistaken by rats as fat and strongly stimulates intake when added to a food like a true fat does (11). We will see that the fatness of cream and other dairy products is also perceived by humans and gives rise to specific responses (12).

I. Other Oral Stimuli

The chemical sensations of pungent, astringent, etc. are components of food flavors that are not always aversive. Finally, food temperature is a decisive factor of palatability, whatever the role played by other components of the flavor. This is true only in humans for the simple reason that humans are the only animal that cooks food and eats it when cooked within a narrow range of temperature.

II. *Synergistic Combination of Systemic and Sensory Stimuli in the Initial Stimulation of Eating*

As shown previously, the meal size of a constant and familiar food is not affected by the internal stimulus that initiates the meal in the *ad libitum* condition. On the contrary, the size of that meal is augmented as a function of the graded systemic stimulus at the time of onset after food deprivation. In addition, the rate of intake at the start of the meal, indicating the strength of the initial stimulation, is constant in the *ad libitum* condition. It increases with time of previous deprivations.

In the two conditions, a presentation of foods (e.g., sweetened or greasy food) other than the habitual one increases the meal size. Due to these effects, the new food is considered to be more palatable than the habitual lab-chow. This situation, among many others, confirms the dual internal and external stimulation to eat and solves the difficult problem of their combination and interrelationships.

A. Hunger Dependency of Palatability

The question to know whether or not eating can be elicited by sensory stimuli alone in a satiated animal is now raised. In humans, our common experience indicates that indeed we can eat a highly palatable item when satiated. Is this true in animal models? Rats presented glucose or saccharin solutions between spontaneous or scheduled meals drink these highly stimulating sweet solutions. Rats offered permanent sucrose solution in addition to *ad libitum* intake of the lab-chow drink large amounts of this solution. This sucrose intake, poorly or not compensated for by a reduced intake of the solid one, increases the total caloric intake and can produce obesity (13) or sucrose-induced obesity (14). In the *ad libitum* condition, the substitution of the habitual food by foods known as highly palatable for the rat augments the initial rate of eating and the meal size in a constant state of the systemic stimulation. This effect of food palatability is then identical to the effect of a previous deprivation on the intake of the lab-chow. It is a first indication of the identical and substitutive effects of the two stimulations of feeding.

Earlier the effect of deprivation on responses to glucose or saccharin solutions was demonstrated. In a 1-hr test, the choice between six concentrations of saccharin versus water was assessed after various times of food deprivation compared to satiety. The total intake of saccharin and this intake relative to water were increased by deprivation. The difference between the deprivation and satiety condition was as high as the saccharin concentration was (15). The preferential threshold in a choice between a sucrose solution and water was considerably lower in hungry rats (5). These relationships between palatability and hunger levels were examined in the following experiment. Rats were offered three versions of the same food made differently palatable by manipulation of their sensory properties in a state of 0, 24, and 48 hr of food deprivation. The eating rate from the beginning to the end of the meal was recorded and expressed in 0.01 g/min (Fig. 4.3). In the nine conditions, an initial increase of the rate of intake was observed followed by a negatively accelerated intake. The initial rate of eating of each palatable food was augmented as a function of time of food deprivation. At each time of deprivation, the initial rate was augmented as a function of the palatability level of the food. In other words, the substitutive and additive effects of food deprivation, or hunger, and of palatability on the initial strength of stimulation to eat were so demonstrated (16). Comparable results will be obtained later in humans (see Fig. 6.6 in Chapter 6, Section III).

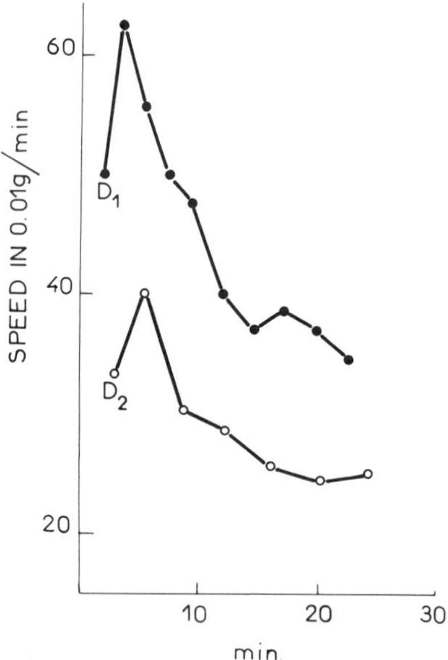

Figure 4.3 Eating rate as a function of the combined effect of previous deprivation and palatability of food.

Another experiment demonstrated that the different palatabilities of two foods and the determining effect of this differential palatability on their respective intakes were reduced by an increased hunger (17). The same food was made differently palatable. As a result of the learned preference, adding the odorant X to the food induced a higher intake of this version than that flavored by the other odorant Y, either in a choice or in separate presentations. Rats were tested in a 2-hr choice after 4 or 18 hr of previous food removals. In the first condition of mild deprivation, rats consumed 11 times more of food X than food Y (only 3.5 times more in the state of acute hunger). Thus, a high level of the internal stimulus overcomes the determining effect of palatability, which, on the contrary, predominates at a low level of hunger (17). The result is reminiscent of the common human experience showing that, when very hungry, humans are less finicky and less discriminating of the difference of palatability of foods.

A historical study (by far more elaborated) was carried out by Coons and White (18), who took advantage of the additive effect (already mentioned) of the rate of lateral hypothalamic (LH) self-stimulation and of the palatability of the offered food. They investigated the variation of current necessary to compensate for inverse variation of the concentration of the sucrose solution to maintain the same level of self-stimulation. A sophisticated mathematical analysis of the results allowed the authors to establish the laws of these relationships. Hunger (here, the intensity of LH stimulation) and palatability (here, sucrose concentration) have a multiplicative reciprocal effect on the response. But this reciprocity is not symmetric. An increase of hunger augments by the response to a sucrose concentration far more than the increases of these concentrations augment the response at a constant level of hunger. In a log–log coordinates, a 3-log increase of sucrose concentration was needed to compensate for a 1-log reduction of hunger.

This work confirmed the synergistic action of the internal and external stimulations. Under this effect, the initial palatability, inasmuch as it is hunger-dependent, results from the summation of the external and systemic stimuli to eat the food.

III. *Unlearned and Learned Palatability*

The term "unconditioned stimulus" (UCS) will be used here in its true meaning. The UCS of a response is the stimulus that elicits this response without a previous learning. The UCS is the acid eliciting salivation in the dog of Pavlov. The expression is erroneously and now commonly used to designate what is actually the "primary reinforcer." Here, the postingestive activity of a food to induce a persistent satiety will be demonstrated as the reinforcer that makes sensory stimuli associated to this effect "conditioned stimuli" (CS) of a learned response of eating. This "reinforcer" is not *per se* a stimulus of the response and therefore must not be called an UCS.

A. Unconditioned Stimuli to Eat

Sweet and salty tastes are UCS to eat or drink in all animal species studied. They are exhibited at birth. In rats, the absolute threshold of various sugars and their order of sapidity were established early. The threshold of neural discharges in the chorda tympani nerve corresponds to thresholds determined by intakes in a brief exposure to the solution. The increasing intensity with increasing concentrations showed a monotonous linear increase on log–log coordinates. The exponent of the straight line is different among various sugars. By contrast, the palatability determined by various methods varies with concentrations according to an inversed U-shaped curve. The palatability increases up to a point of inflection and then declines with the highest concentrations. This is observed in comparison of various concentrations of the sweet solution in a choice versus water or in single, short presentations. In a choice between two sweet solutions, the most concentrated is preferred (19). A curve of preference–aversion is known for saccharin solutions with the highest palatability at 0.20%. Analogous facts are observed with NaCl solutions. The so-called preference–aversion curve shows a maximal intake at the isotonic level, in a choice with water or in single presentations, i.e., at the level of normal blood osmolarity (0.9%). This coincidence is striking and has been, surprisingly, rarely pointed out.

The existence of an olfactory unlearned stimulus to eat is suspected but not clearly ascertained, at least in rats. However, Chapter 12, which is devoted to the ontogeny of behavior will show that suckling rats on the first day of life achieve the nipple attachment due to olfactory stimuli. Washing the nipples of the dam or performing an olfactory bulbectomy of the young prevents the attachment and leads to death of pups. Various odors added to the habitual foods initially elicit a reduction of intake, rapidly eliminated later on. No odorant seems to act like food sweetening does.

By contrast, as already said, greasy texture is clearly an UCS to eat, comparable to saccharin when this greasy texture is that of a calorically inert mineral oil (11).

B. Extinction of Unreinforced Unconditioned Stimuli

Saccharin solutions, sweet but providing no calories, were extensively studied to demonstrate the persistence or the disappearance of the response unreinforced by a postingestive nutritive effect. The intake pattern of a saccharin solution in a choice with water is different from the pattern in a choice between sucrose and water. Bursts of licking are less frequent but more prolonged with the sucrose solution (20, 21). Initially, as mentioned above, deprivation enhances saccharin like sucrose intake does (15). But two groups of fasting rats—the first one maintained on a sucrose solution, the second one on a saccharin solution, demonstrated the rapid extinction of the saccharin responsiveness. From day to day, the intake of sucrose solution increased in the first group; in the second group, the saccharin intake, after increasing on the first day, fell rapidly to 0 (22) (Fig. 4.4). In another study, two groups of rats received a 30-min daily presentation of a saccharin solution. In one group, the free intake was followed by a stomach-tubing of a glucose solution, in the other group by a stomach load of saline. After 15 days, after the associated tubing was inversed, the first group

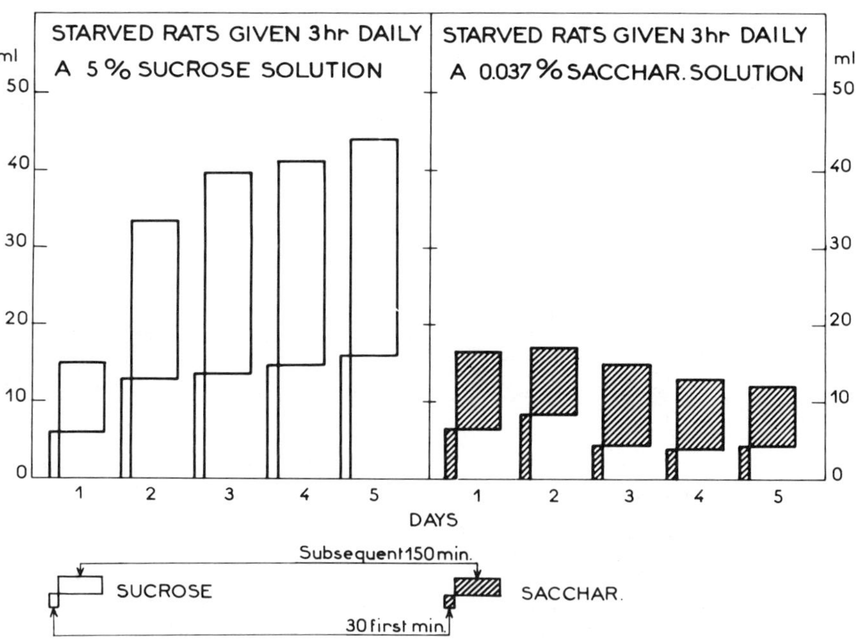

Figure 4.4 Extinction of saccharin preference in fasted rats.

exhibited a maintained high intake of the sweet solution. It was strongly reduced in the other group (23). A striking finding was reported in mice by Warren and Warren (24). The pure saccharin was compared to the Na-saccharinate, universally used in experimental studies. In a choice, the initial palatability of the two solutions was identical. In a choice with water, the preference for the Na-saccharinate persisted and for the pure saccharin disappeared as soon as the second presentation. The authors interpret the result as an effect of a reinforcement by the postingestive sodium with the saccharin salt.

The spontaneous response to greasy texture also leads to an extinction of the response. In a choice between two diets containing either 30% fat or 30% petrolatum, the preference for the fat-rich diet only was maintained. In a choice between the lab-chow and three diets with different petrolatum contents, the caloric intake became progressively identical on these three diets, like on diets diluted by addition of other inert material (11).

C. Postingestive Conditioning of Palatability

In the absence of other data, the preceding results would have been sufficient to conclude that an unlearned and apparently genetically determined palatability is maintained only if it is reinforced by the postingestive nutritive activity of the food or solution. A considerable number of results were accumulated to confirm this fact by demonstrating that the palatability of foods is learned. In this learning process, caloric and specific properties of the ingested food act as the primary reinforcer. Like in other forms of learning, this postingestive effect of the food makes the associated sensory stimuli "conditioned stimuli to eat." Sensory stimulations by the food (e.g., odor, taste), generally not correlated to nutritive properties, become the basis for an anticipatory quantitative and qualitative response to the nutritive properties and their adequacy to cover the present metabolic requirement.

The initial demonstration came from an experiment (today forgotten) by Harris *et al.* (25). Rats were made vitamin B-deficient through vitamin-deficient diets. Then they received a choice between the deficient diet and the same diet supplemented with vitamin B. The two versions were sensorily differentiated by the addition of a trace of odorant X in the deficient diet and Y in the supplemented one. Rats progressively developed a preference for the Y-supplemented diet. Then the inversion of flavors led them to follow the Y flavor and to prefer the deficient diet. After some time, they relearned the correct response and again preferred the supplemented diet.

This simple yet effective experiment inspired this author (J. L. M.) to perform a series of experiments by which he demonstrated, for the first time, the palatability learning of calorie-based responses (26–28). The procedure of flavored diet (today generalized in such studies) was used. Two odor-labeled versions of the same complete synthetic diet were alternatively presented to rats. One of them, flavored X in one group, flavored Y in another group, was presented in a 30-min meal in the morning, the other version in an afternoon meal. After 2 wk of these alternate presentations, rats were offered either a 4-hr choice between the two flavored diets or separate presentations of them during six days. When the alternate presentations of the two flavored versions were followed by a control saline injection, the final choice or separate meals showed an identical intake of the two versions. Then the injection of an agent susceptible to modify postingestive satiety was given following the repeated intake of version X in one group and of version Y in the other. Insulin administered immediately after the 30-min intake of X or Y induces a strong preference for the flavor previously associated to the saline control injection. In another experiment, a diluted diet (1.6 cal/g) was used. At the end of the intake of one of the two flavored diets, a glucose injection was given and, after each meal, adjusted to 25% of the calories just eaten freely in the meal. This repeated postprandial supplementation of the oral intake also induced a preference for the flavor previously associated to the nonsupplemented version. In a third experiment, insulin was injected following the intake of one flavor, glucose after the other one. No preference was then manifested during the final choice.

The interpretation of these results was that both postprandial insulin and glucose increased the postprandial duration of satiety and that this effect reinforced a reduced intake based on associated sensory stimuli. This diet would be treated as a calorically denser diet than the other one. However, it was proposed that this enhanced satiating effect of the food was not a result of a conditioned reduced initial palatability but, rather, of a conditioned–enhanced orally satiating capacity of the food. (Figs. 4.5 and 4.6).

The hypothesis that the reinforcing effect of postprandial glucose or insulin was due to a prolongation of the postprandial satiety was tested by using amphetamine added to the diet. Ten mg/100 g added to one of the flavored versions also induced a reduced preference for the version. In one experiment, four groups were compared with the same procedure: no flavors and no amphetamine, flavors and no amphetamine, amphetamine and no flavors, and flavors and amphetamine. The result showed that the conditioning required both the sensory discrimination of the two diets and the differential postingestive reinforcement.

Using a similar procedure, Booth (29) reexamined and elegantly extended

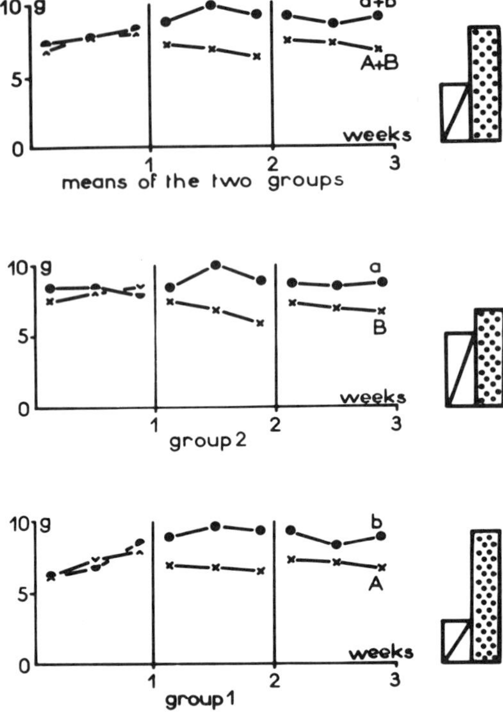

Figure 4.5 Postprandial reinforcement by glucose of the sensory control of food intake. ◨ previously supplemented; ⊡ not supplemented.

these results. These wonderful experiments will be detailed and commented upon in a later chapter devoted to the role of learning in the satiation process (Chapter 6, Section I).

Many other studies confirmed this palatability conditioning. The evolution of preferences between two sweet solutions due to their respective caloric contents, initially shown by Jacobs (30), was among these results. Fructose and glucose solutions were offered to rats. Initially, they preferred the first one, previously tested as the sweetest one. After 1 or 2 days, the preference was reversed in favor of the more caloric glucose solution. Sham- compared to real-feeding was suitable to complete these demonstrations of the postingestive reinforcement. Rats were sham-fed or fed freely by two differently flavored or differently placed diets. A flavor or place preference developed in the real-

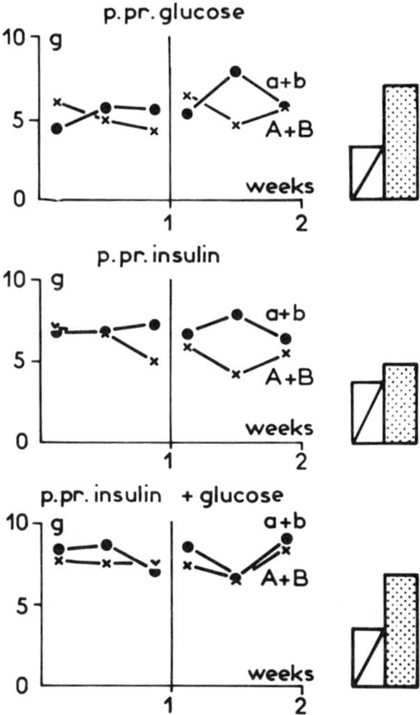

Figure 4.6 Comparison of the postprandial (p.pr.) reinforcement of the sensory control of intake by glucose, insulin, and insulin + glucose. ▨ previously supplemented; ▨ not supplemented.

feeding condition only, in which the postingestive activity of the food can induce the sensory-based response (31).

The last experiment cited here, by Sclafani and Mistenbaum (32), is particularly convincing. Fed rats were offered a permanent choice between two bottles, each containing differently flavored water. The intake of one of the two flavored waters was associated to an automatically monitored intragastric administration of starch, and intake of the other to an intragastric administration of water. After 4 days, a 92% preference for the flavored water previously associated to starch in the stomach was manifested. Both of the two flavors were then associated to water in the stomach and extinction of preference occurred in more than 4 days.

This palatability conditioning may occur very early in the life of the animal, even *in utero*. Apple juice injected in the amniotic fluid at day 20 of gestation induced a preference for an apple-flavored solution when tested after weaning of the young (33–36). The lactating mother rat injected daily with a minute amount of an odorant (citral) fed her young with a citral-flavored milk. After weaning, rats exhibited a preference for a citral-flavored diet (37). It was also demonstrated that the young rats acquired a preference for the food eaten by their mother during lactation through the flavored milk in the mother's diet (38).

D. The Reinforcer

Together the results suggest that the ability of a food to relieve hunger and to induce a persisting satiety is the reinforcer. However, satiation by filling the stomach alone seems to provide a reinforcement. Rats with a linked pylorus were offered the choice between two flavored nonnutritive solutions. The ingestion of one of them was associated to the stomach loading of a nutritive solution, the other to a water load. The preference for the first one was readily acquired (39).

Other researchers support the opposite notion—that the reinforcement is only a postabsorptive event. Recovery from malaise is shown to be a positive reinforcer. This supports the notion that the relief from hunger *per se* contributes to the learning of palatability. Rats injected with LiCl and made sick by this injection were offered a saccharin solution either 10, 60, or 90 min after the injection. Compared to the 10- and 60-min delay, the presentation at the time of recovery from the sickness induced the saccharin preference (40). In a choice between two flavored items, one ingested after an apomorphine injection, the other one before this injection, the second one (consumed at the time of recovery from the apomorphine effects) becomes preferred (41) (Fig. 4.7).

In one series of experiments, a role for the duration of postprandial satiety in the reinforcement of palatability was tested in the specific condition of scheduled meals (42, 43). Two flavored diets were presented alternatively, one of them followed by 2 hr until the subsequent scheduled meal, the other one followed by 6 or more hr until the presentation of the next meal. The induced palatability was critically affected by the postprandial time without foods. Initially, a long fast after a meal tends to be a negative reinforcer. Later (like in the conditioning of provisional appetite, described earlier), the long delay after a meal induced higher palatability of the flavor associated to that meal.

Such demonstrated learning of the sensory stimulation to eat, i.e., of palatability, explains many of the phenomena described earlier: the adjustment of intake to calorically diluted diet, the meal intake in the *ad libitum* condition anticipating postprandial expenditures, the adjustment to a single daily meal, etc. It is the key mechanism of feeding behavior and of its contribution to body energy balance.

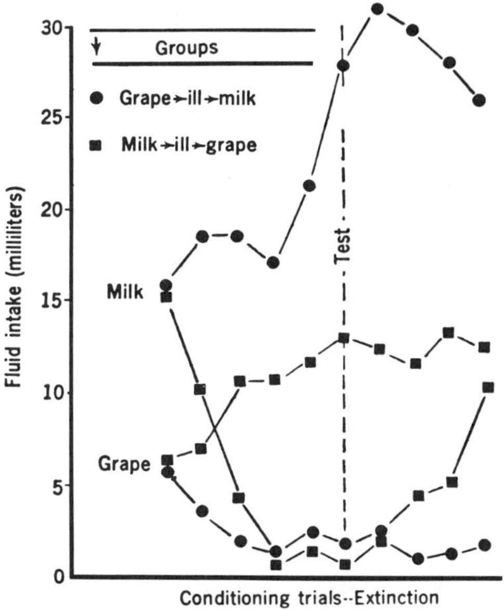

Figure 4.7 Recovery from illness as a positive reinforcer of palatability [From Green and Garcia (41).]

E. Reinforcement of Instrumental Learning or Maze-Learning by Foods

Learning a bar-pressing response or a maze to obtain and eat food, which
reinforces this learning, may provide some confirmation about palatability reinforcement. Favored by psychologists, such learning is viewed as a means to
study food reinforcement, i.e., the reinforcement of direct feeding and free-
feeding. However, the training for this active response to obtain food is not at all
the equivalent of free-feeding and of measurement of its parameters. Instrumental learning is largely (in the Pavlovian language) a "second-order conditioning."
In this conditioning, a CS associated to another stimulus makes it a CS. The first
CS (here, the palatability of the food) is said to be a "secondary reinforcer" or
food reward. If this is true, instrumental learning and maze-learnings can provide
some information about this secondary reinforcer but not necessarily about the
functional postingestive reinforcement.

In the old and classic experiment by Hull *et al.* (44), oesophagotomized
dogs were tested for training in a maze in three feeding conditions: oral sham-
feeding, sham-feeding plus a gastric load, and a gastric load alone. The reinforcing effect of the two latter conditions was higher than that of the former one, but
sham-feeding alone is reinforcement compared to nothing passing the mouth or

entering the stomach. Rats learn a maze when rewarded on one side by the sight of an inaccessible food (45). The response is better learned by hungry than by satiated rats. Rats also learn a maze with saccharin as a reinforcer. The strongest reinforcement comes from the largest volume and highest concentration of saccharin solution as a reward for the correct response (46). Saccharin was also shown to be strong reinforcer of an operant response, its effectiveness being augmented by previous deprivation (47).

The key question of the respective role of palatability and of postingestive events in such learning was examined by Smith and Dufy (48). Four groups of rats learned a maze: hungry or satiated, rewarded with 4 ml of a sucrose solution or only with a drop of this solution. The hungry group rewarded with 4 ml sucrose learned faster than the three other groups, but hungry or satiated rats rewarded with only the taste of the solution also learned.

Thus, such instrumental learning and maze-learning are apparently mainly reinforced by food palatability and, thus, are second-order conditionings. However, the postingestive action of the obtained food also contributes to the reinforcement. Nevertheless, these procedures are not a reliable source of information on the nature of this postingestive reinforcement.

F. Social Learning of Palatability

After weaning, a young rat is attracted to a food or to a place where that food was eaten by its mother (49). More striking is the fact that an adult rat exposed to another rat eating a given food acquires a preference for that food. This preference is not induced by the exposure to the food alone. An unpalatable and even aversive food (piquant) becomes palatable for a rat exposed to another rat eating that food. Thus, parental education, social pressure, and imitation are not absent as external reinforcers of food palatability in animal species. Rats differ less from humans than it appears.

IV. *Unconditioned and Conditioned Aversions*

Some unknown elements of genetic coding are expressed in innate responses of food preferences preadapted to species survival. In addition, inherited neurobehavioral mechanisms allow individuals to learn and relearn permanent responses to foods in accordance with their nutritional needs. Thus, animals and humans escape the danger of being underfed, or not fed at all. Natural selection also developed another ingestive mechanism by which animals and humans escape a more immediate danger for their survival by eating toxic materials. It is a defensive mechanism of inhibition to eat or aversively respond to food. This system of responses belongs to both ingestive and defensive behaviors. These stimulations of "not to eat" or aversive responses must not be confounded with nonpalatability. They are characterized by an active rejection: spitting (an orofacial typical expression of disgust) or nausea, which manifest a sort of inverse ingestive behavior.

A. Unconditioned Aversive Responses

Natural selection has differentiated two taste sensitivities—acid and bitter—generating aversive responses. They are present and operative at birth and mature later. As for sweet and salt sensitivities, electrophysiological exploration in the nucleus of the solitary tract showed characteristics of the two substance groups across neuronal patterns. Tested by behavioral responses or by recording discharges in peripheral nerves, the thresholds of responses to bitter compounds are by far the lowest compared to sweet, salty, and acidic ones. The threshold of acid rejection varies among acids according to pH. But beyond a given level of pH, acids are aggressive, primarily toward the oral mucosas. Salivation induced by sour stimulation, which dilutes and buffers acidity, seems to be a protective mechanism.

All organic and inorganic acids are acids. All bitter compounds, being initially rejected as such, are not toxic. Thus, the rejection of some nontoxic bitter liquids or solids will be susceptible to extinction based on experience, like the preference for saccharin is also extinguished. But compounds that are neither acidic nor bitter but rather salty or tasteless may be highly toxic, as is the case for LiCl, for example, which is not distinguished from NaCl. Here, instead of learning the safety of the substance, the safeguard will be achieved by a learned aversion.

The telesensitivity of odors via the nasal route, as, for instance, in other defensive behaviors against predators, is used to detect poisonous and inedible materials. Inherited responses correlated to toxicity of plants, from which the odors emanate, are suspected to be defensive signals. In the wild, animal species avoid poisoning by not eating poisonous plants present in their biotope (50). Many animal species reject the flesh of their own species (51). Finally, mechanical stimuli in the mouth can make many materials unacceptable and aversive; this is the case, of course, for nonmasticable materials. In mammals, all solid foods are aversive during the suckling period for this simple reason.

B. Neophobia and Familiarization

An aversive initial response to a novel food is exhibited in all animal species studied. This neophobia is manifested in a repeated sampling of minute amounts of the ingested food, as well described by various authors (52). Then a "familiarization," which is the suppression of neophobia, occurs rapidly. This appears as a learning of the food safety. This familiarization depends on the number and duration of exposures to novel foods, not on the intervals separating these presentations. A familiarization acquired in a brief initial presentation is manifested 90 days later (53). Neophobia favors the conditioned taste aversion (see below). Thus, familiarization may be considered as an extinction of the aversive neophobic response and, on the contrary, conditioned taste aversion as an experimental confirmation of the suspected danger of a novel food.

C. Extinction or Reinforcement of Aversive Responses

Saccharin intake disappears when not associated to the postingestive reinforcement by a food. Similarly aversive responses to nontoxic substances are eliminated when associated to a food reinforcement. On the contrary, they persist when they are reinforced by their postingestive deleterious consequences.

The habitual food of rats being adulterated by addition of a dose of a bitter compound higher than the gustatory threshold but lower than the toxicity threshold or by an innocuous bitter substance, is at first aversive and its intake is reduced. After 2 or 3 days, a normal intake is reestablished (54). These effects were studied by using two bitter compounds—sucrose–octo-acetate (SOA) and quinine-HCl, the first one bitter and safe, the second one bitter and slightly toxic. The intake of a food adulterated by SOA was reduced for 4 hr only. The food adulterated by quinine was accepted after some days but a reduced *ad libitum* intake leading to weight loss persisted. Surprisingly, the lower body weight thus induced was maintained after the removal of quinine from the food and rapid normal intake resumed (55).

In another study (56), an initial choice between SOA- and quinine-adulterated foods showed a preference for the latter, which was less bitter than the former. Following an *ad libitum* intake of the two diets, rats exhibited a higher intake on the SOA diet, i.e., the more bitter and initially more aversive but safe diet. This reversibility of the aversive response is not always achieved. In guinea pigs maintained on SOA-adulterated water for a long time, aversion to the bitter fluid is slowly attenuated, but the preference for pure water persists in a choice

(57). In rats, the strong aversion for food adulterated by the piquant chili pepper cannot be reversed (58).

The reinforcement of sensory responses by postingestive poisoning which makes food aversive, is called the conditioned, or learned, taste aversion (CTA). Such an acquired aversion is dramatic when the toxic food is tasteless or initially preferred when sweet or salty. Therefore, if the animal survives the first ingestion, the food palatability becomes aversive.

D. Aversive Responses Counteracted by Hunger

The hunger–aversion relationship results from the conflict between an ongoing behavior and a defensive one identical to that of rats obliged to pass an electric grill to reach their food. The CTA paradigm of an aversion induced by pairing the ingestion of a food with the toxic injection of LiCl (see below) gives a clear demonstration of this hunger–aversion relationship (59). Aversion induced by a dose of LiCl is attenuated or blocked by a pairing with a LH self-stimulation at low and high intensity of the current, respectively. Conversely, a flavor preference associated to the ICSS was attenuated in a dose-dependent manner by LiCl. Thus, an algebraic summation was manifested between induced preferences and aversions. As already mentioned, a blockade of intake induced by increasing times of food deprivation is achieved by increasing concentrations of quinine in the offered diet (60). Similarly, electoral stimulation of the LH blocks (as a function of its intensity) the aversive response to quinine-adulterated foods (61).

E. Conditioned Taste Aversions

That animals could learn an aversive response to the sensory aspect of a food to avoid poisoning by this food was more than suspected by observations of animals in their natural environment. It was first experimentally studied by scientists dealing with rat poisoning (50, 62, 63). These researchers, using dicoumarol (an anticoagulant) as poison, noted that, when they were not poisoned by the first ingestion, rats refused to eat again the bait in which the toxic compound was contained. To renew the eating and to kill the rat, the bait need to be changed, not the toxin. From these pioneering experiments, the term "bait shyness" was coined and is still used to designate the conditioned taste aversion.

As early as 1951, this author (J. L. M.) remarked, on the basis of these early works, that the two surprising characteristics of such learning were (1) the fact that the conditioned response was acquired after only one trial (one pairing of the sensory characteristics of the bait with the postingestive illness) and (2) that this associate learning tolerated a relatively long delay between the sensory and the reinforcing event. It was contrary to the prevailing dogma of Pavlov and could lead, he said, to a reexamination of the notions of Pavlovian conditioning (64).

This reexamination of one-trial, long-delay learning was undertaken only 20 years later. At this time, it suddenly gave rise to a rush of work (more than 1500 publications from 1970 to 1990). Most of these works deal with learning *per se* and its debated theories rather than with the physiological occurrence of this learning in the defensive ingestive responses. Thus, only works dealing with the last aspect will be briefly examined here.

The most dramatic and convincing experiment was due to Nachman, who was the first to call attention to this field of investigation. Rats were offered an LiCl solution to drink. The salty taste of LiCl not distinguished from that of NaCl, rats drank a substantial amount of the highly toxic solution. The day after, they refused to drink an isotonic NaCl solution. Then the same NaCl solution was repetitively offered during subsequent days. Rats in this extinction phase progressively resumed their drinking of the NaCl appetitive solution. The safety of this solution being learned, a new presentation of the LiCl solution mistook and killed rats (65).

Later, hundreds of works were designed with the same paradigm and its variations. A saccharin solution was used as a CS and the postingestive intraperitoneal injection of LiCl used as reinforcer, the USC. One-trial learning with the taste solution was also confirmed as well as the possible long delay between the CS and UCS. Increasing the delay reduced the induced aversion, but it was still obtained after a 24-hr delay (66). A preconditioning exposure to the saccharin solution impairs the induced aversion and, thus, neophobia was thought to be a requisite of the learning. The extinction of the learned aversion was also widely studied.

From this classical paradigm, various CS and various UCS were compared. *X*-irradiations particularly of the bully of the animal acts like the LiCl poisoning does. An electric shock substituted for LiCl acts also as UCS but eliminates the long delay. A 30-sec delay between the CS and the shock prevented the learning. The most interesting comparison was made between various toxic compounds used as UCS (67); SO_4Cu, Na-fluoro acetate are weak UCS, as is LiCl. But highly toxic compounds such as thallium salts, cyanure, and strychnine are weak UCS. Based on this surprising fact, the author attempted to identify the symptoms of toxicity and the delay of their apparition necessary for the reinforcing activity. Morphine and alcohol induce aversion in naive rats; they induce a preference in morphine- or ethanol-dependent rats.

Regarding sensory stimuli as CS, an aversion to odors is obtained in rats but requires repeated pairings. The association of the odor cue with a taste potentiates the induced aversion against the olfactory stimulus (68). In rats, the visual stimulus is a poor CS and is also readily obtained only when associated to a taste (69). LiCl induces a taste aversion better than a place aversion, in which visual cues are involved. But other UCS such as gualamine induce a place better than a taste aversion. Again the various symptoms of toxicities seem to be concerned in these differences (70). However, in monkeys, as in birds, the visual system is the best sensory channel to become the basis for an acquired aversion. With LiCl as the UCS, the aversion is obtained in one trial, as with the taste cue in rats (71). These species differences will be commented on with respect to central mechanisms of such preference and aversion learnings (Chapter 5).

Finally, a fascinating finding was that, like preferences, aversions can be conditioned *in utero*. Fetuses were exposed *in utero* to an odorant or a taste injected in the amniotic fluid while an external aversive stimulus was applied. They exhibited the acquired aversion 2 wk after birth (36). In another study, menthol was injected in the amniotic fluid of the mother associated or not with LiCl injection. Fetuses were exteriorized and their acquired aversive responses to menthol tested. Such learning was obtained from day 17 of pregnancy and then was exhibited by fetuses. This is an important result. Such prenatally learned responses exhibited at birth or later may be mistaken as innate. On the other hand, the result permits suspicion that such prenatal responses could be acquired to odorous or taste substances present in the mother's blood. Rats injected with an odorant acquired an aversion to the odorant acting both as CS and UCS (72, 73). Do the young of a diabetic mother rat acquire a glucose or ketone body odor aversion?

V. *The Palatability of Foods in Human Feeding*

"Dis-moi ce que tu manges; je te dirai qui tu es"
(Tell me what you eat: I shall tell you who you are) (74).

Human feeding habits are, among many others, a specific attribute of humankind compared to those of the animal kingdom. In addition, feeding behavior, like and perhaps more than other behaviors, specifies human communities and individuals and contributes to their cultural identity.

Humans invented fire about 500 thousand years ago. Simultaneously, they invented cooking food. This was certainly their first step away from animality. Much later, they invented "labourage and pâturage," the "production" of their food which no other animal species (except some ants) does. Finally, they invented "la cuisine," which was and still is one of their most exquisite creations. The main feature of humans is to develop ethical and esthetic rules. They developed an art from their sensory modalities and from all behaviors. Feeding in some manner becomes an art in humans.

Unfortunately, for some humans, eating is still an unsatisfied need for survival in undeveloped populations or in many tragic circumstances. By contrast, the intake of foods exceeds the physiological requirements in most developed countries. Everywhere, eating is a social and cultural activity, a source of refined and organized pleasure.

A. Unlimited Variety of Food Palatabilities in Human Feeding

The result and, partly, the purpose of food production, trade, and transport are to increase the variety of available sources of the three macronutrients: carbohydrates, fats, and proteins. These crude products represent an extreme variety of specific flavors. In addition, these basic food products are cooked and "prepared." More and more food industries are developing a market of prepared foods and look for an infinite differentiation of these products by the variety of their flavors and for a correspondence of these flavors with their palatability for the targeted population. Meanwhile, in most developed countries, "the cuisine"—the domestic preparation of highly refined dishes—is predominantly an activity of women. In France, for instance, hundreds of recipe books are published every year, and many successful magazines are devoted to cuisine. The prestige and fashion of "high cuisine" is increasing. Some "grands cuisiniers," known around the world, are true stars.

This is not just anecdotal, because the result of this enormous variety of crude and prepared dishes also introduces an unlimited variety of human responses to sensory activities of foods, which are the main feature of human feeding. Hundreds of palatable or unpalatable sources of different carbohydrates, fats, and proteins and of their mixtures are available or prepared. How are these unlimited palatability responses acquired? How are these responses to this multitude of sensory activities of foods adapted to their nutritive contents? The last and important question is: To what extent do palatability and individual appetites determine the choice and, thus, the purchase of one food over another? To what extent does palatability, whatever its origin, determine the amount eaten?

Regarding the choice and selection of foods, individuals shopping for and preparing their foods are submitted to at least three or four different pressures: price and personal income; knowledge of nutritional requirements; and, finally, likes and dislikes. This is a matter of marketing techniques in food industries. Competitions and choices are not among foods but, rather, categories of foods. A choice is made between meat and fish, among fruits, and among desserts, not between fish and cakes. In one study (75), among many others, foods were tested in an attempt to dissociate the role of four categories of factors: sensory preference, hunger and satiety, nutritional content, and environment in the choices of a group of subjects. According to this study, 66% of the variance was accounted for by likes and dislikes.

Regarding amounts eaten, we will return to the fact that the unitary intake of a food depends on its initial palatability and on its modulation by hunger and satiety; rather, it mostly depends on the sensory satiating capacity of each food. However, as a whole in a given population, the total amount of food eaten by this population is determined by the hunger–satiety system, despite the variety of available foods and likes and dislikes. The food market is unique because it is saturated at a permanent level. The total food consumption cannot be under or over this level determined by hunger and satiety and, therefore, by the energy and nutrient physiological requirements. Roughly, the total food consumption in the United States is 250 million multiplied by 2500 kcal/day. With a lower intake, Americans would be permanently hungry, with a higher intake they would become obese. Inasmuch as humans physiologically resist obesity (we will see how and how much), it is theoretically impossible to overcome this limit despite advertising and improvements of sensory qualities of foods. Thus, the choice and role of palatability in this choice will be within this frame. The promotion of a particular food will be necessarily compensated for by the rejection of another one.

B. Palatability Measurements in Humans

The palatability measurement of food in humans, also called pleasantness or like, is easier than it is in animal models; however, it is not without difficulties and confusion.

Various psychophysical classical techniques are used to score the pleasantness of food, postulated to be the subjective correlates of the palatability (category scaling or visual analogue). Various studies showed that this evaluation prior to intake does not predict the amount eaten. Yoghurts sweetened with various sucrose concentrations were at first tasted and their pleasantness evaluated; they were then offered *ad libitum*. A discrepancy was observed between pleasantness ratings and intakes. Subjects prefering the sweetest samples ate a maximal intake of a less sweet yoghurt (76). A similar comparison of tasting and intakes of food sweetened by either aspartam or sucrose indicated differences in the two measures (77).

Comparable results are reported by Pangborn and Pecore (78), who used more or less sweetened fruit juice. In another elaborated study, subjects were at first tested by category scaling for their likes and dislikes of a flavored syrup adulterated or not by quinine. Subjects who gave two points of divergence were retained for an intake test. A difference of 100 g of intake was observed for each point of the scale. Sixty percent of interindividual variances were due to palatability. The initial rate of eating was found to be augmented as a function of the initial rating of pleasantness. The conclusion of the authors, partly opposite to that of above studies, is that the category scaling of pleasantness is a good individual predictor of intake but a bad predictor of interindividual differences (79). We will return to these discrepancies between the subjective evaluation of pleasantness and the intake until satiety and the objective measure of palatability (see Chapter 6, Section III).

C. Sweetness and Saltiness

Sweetness elicits or potentiates ingestive responses in human infants as soon as the first day of life, saltiness at 2 or 3 mo (see Chapter 12). This provides evidence indicating that gustatory mechanisms are at work in newborns and that sweetness and saltiness are, in humans like in animals, UCS of feeding responses. Other evidence of the strong unlearned palatability of sweetness and saltiness is provided by the fact that humans, unsatisfied by sweetened or salty products, organize the yearly production of 10 million tons of sugars and salts.

However, a typical feature of human feeding is that these responses to sweetness and saltiness are rapidly modulated according to foods to which sweet or salt taste is associated. As early as 2 or 3 yr of age, children, like adults, accept and prefer sweetened milk, fruit, cakes, etc., and reject sweet vegetables or sweet meat. They like salty soup, bread, etc., and reject the same foods when sweetened. It is the first indication of the versatility of palatability responses and of their early modulation by learning. Here, external reinforcers (sociocultural pressure, eating experiences imposed by parents) establish what is called "food or eating habits." They are conditioned taste preferences or aversions.

Regarding sweetness, sensory thresholds and the supraliminar relationship between concentrations and perceived intensities of various sugars are well known. Simultaneous rating of intensity and pleasantness shows, like in rats, an inverted U-shaped curve of increase and decrease of pleasantness with increasing concentrations. The inflexion point decreases with a previous gastric glucose load and increases toward the highest concentration by privation. Both manipulations do not affect detection thresholds. With low concentrations of sugars, the pleasantness increases more or less rapidly than perceived intensity up to the inflexion point. This point of maximal pleasantness is at a level of identical sweetness for glucose and sucrose solutions corresponding to different concentrations of the two sugars. The same is observed in a comparison among four solid foods (80–83).

The perceived intensity is modified by other sensory aspects of the item. For instance, fruit juices identically sweetened were found to have different sweet intensities when presented with different colors. The most yellow juice was the sweetest (84). Most of all, pleasantness as a function of concentration is extremely variable from food to food. A very high concentration preferred in cakes, candies, etc. is not tolerated in a beverage. In addition it is well known that these likes and dislikes are individually highly variable, changing with age and affected by sex. Contrary to common belief, males showed the highest preference for sweets as compared to females (85).

The preference–aversion curve for NaCl tested on salty water also is of limited interest because the consumption of salty waters is not, of course, a very common eating habit. Response to food saltiness is very peculiar. An optimal concentration may be determined, which may vary from food to food. The peculiar fact is that, at this optimal concentration, which makes the food "good," the saltiness is not perceived. Either not enough or too much saltiness is perceived. The preference for a more or less salty soup was tested in subjects previously classified as high or low current consumers of salty foods. As expected, the optimal concentration of salt in the soup was found to be higher in the high consumers, while the rating of intensity was identical in the two groups. Many data indicated that a considerably excessive NaCl intake compared to the physiological need is a result of stress and of sensitivity to stress. An inquiry by Shepherd and Farleigh (86) and Shepherd *et al.* (87) seemed to validate this notion. High consumption of salty foods and hyperanxiety were found to correlate.

D. Food Consistencies

While Section VI of this chapter is devoted to the importance of olfactory stimuli (aromas) in human feeding behavior, another sensory component of the orosensory pattern is also of importance but, nevertheless, is often neglected: the consistency of tactile and proprioceptive stimuli. The palatability of meat and many other foods is mainly determined by their hardness or tenderness. The palatability of bread (French bread), biscuits, and various pastries greatly depends on sensations such as crusty, cracking, pasty, etc. The palatability of mature versus immature fruit is linked to softness and water content, which gives a melting sensation in the mouth. In high cuisine, the aim of using cream, butter, oil, or other items is to obtain the palatable creaminess and succulence of the dish. The response to fluidity and consistency is the basis for a delicate blending of liquid and solid foods in various feeding cultures. In many of these cultures, liquid food, or soup, is consumed at the start of a meal. It is significant that the more elegant word of *potage* comes from Latin *potere,* which means to drink. This initial intake of this aqueous dish could be linked to the feeding–drinking relationships mentioned earlier and to a disinhibition of feeding by an initial fluid intake. Soups are the subject of hundreds of recipes, and they are often mixtures of liquid and solid foods. Often of unrelatively low caloric value, they are prepared as "holders" of an unlimited variety of food aromas: vegetable, spices, meat, fish, crustacean flavors, etc.

E. Hunger and Palatability in Humans

The combination of the systemic and sensory stimulations of feeding seen in animals was also clearly demonstrated in humans. The recording of the chewing–swallowing pattern during test meals was used to compare the effect of previous deprivation and palatability of the food and of their combination on the microstructure of the meal. Subjects deprived of foods either since breakfast or overnight were tested at lunchtime. In the two conditions, they were offered either a highly or moderately palatable food to eat. Both the increase of previous deprivation and of the food palatability induced a larger and longer meal. The two factors were additive: Increasing deprivation as well as palatability similarly augmented the eating rate in the first quarter of the meal only (88) (Fig. 4.8).

F. Conditioned Taste Preferences in Humans

The conditioning of the palatability of foods by their postingestive nutritive activities, clearly demonstrated in animal species, was hardly investigated in humans. Inasmuch as the spectrum of food preferences is not inherited as determined in homozygous compared to heterozygous twins (see Chapter 12), it is self-evident that these preferences are acquired (89, 90). The fact that they are acquired for nutrients and not for noncaloric materials proves a general role for

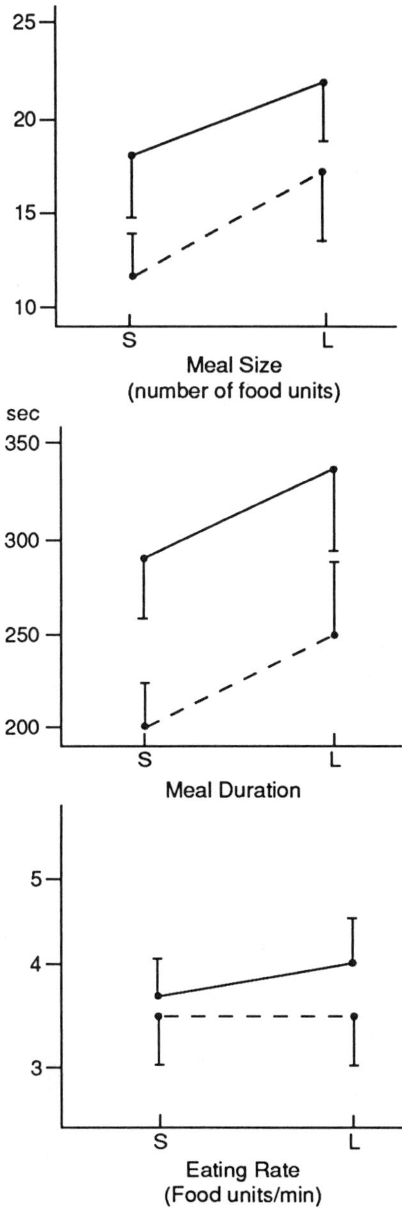

Figure 4.8 Meal size (number of cocktail-size sandwiches ingested), meal duration and eating rate observed in high (——) or low (---) preference meals, under low (L) or high (H) deprivation conditions [From Bellisle *et al.* (88).]

postingestive activities as reinforcers in taste learning. However, the role of external reinforcers provided by the sociocultural context is also self-evident. As will be detailed later, children learn to eat and to like or dislike foods as they learn to speak. However, this social learning is obviously reinforced or extinguished by the postingestive caloric and nutritive properties of foods. This is the general notion; however, this does not explain the individually specific spectra of food preferences in adulthood and their changes. Even within the same family and the same population, the hierarchy of preferences for fruit, meats, dairy products, and flavored beverages is a personal trait. The feeding history of each individual does not clearly explain these individual preferences.

Elegant experiments by Booth *et al.* (91) demonstrated the postingestive conditioning of flavor preferences. One of the experiments was the following: Subjects were offered sandwiches for lunch followed by yoghurt for dessert. Alternatively and prior to the meal, they received a starch solution containing either 65 or 5% starch. The yoghurt was differently flavored after the initial intake of the two solutions. After some days, subjects took more sandwich and more flavored yoghurt following the 5% starch solution. Then they received a 35% starch solution prior to the two alternated meals. They continued to prefer the yoghurt with the flavor previously associated to the 5% starch solution. This yoghurt repetitively eaten in the greater state of initial hunger had apparently acquired a slower satiating capacity, if not a higher initial palatability.

Other convincing evidence was provided by an extensive study on the changes of food preferences reported by prisoners in deportation camps associated with their severe inanition. This inquiry carried out on 60 exdeported (published in French in 1960) was totally ignored (92). The considerable body of facts reported cannot be detailed here. They strikingly demonstrated that in states of semiinanition, severe weight loss, and multiple nutritional deficiencies, the spectrum of food preferences and aversions was totally changed. During the deportation, terribly aversive spoiled foods or inert materials were considered delicious. After Liberation and during their refeeding, significant changes of their previous food preferences for sources of calories (mainly fats and proteins), for highly sweetened beverages, for sources of vitamins, and so on were noted. A craving for salt was also reported.

G. Unconditioned and Conditioned Aversions in Humans

Aversive responses to bitter and acid and to hard and otherwise orally aggressive materials are apparently unlearned responses; however, some of them are reversed by learning when they are associated to a positive reinforcer and not to a postingestive deleterious effect. This is the case for many slightly bitter foods or beverages, which at first disgust children, but later become palatable. The opposite inversion of preference to aversion by postingestive toxic effects, i.e., the conditioned taste aversion, is convincingly exhibited by humans (93). It is a common experience that food intolerance, a gastrointestinal pain, or other postingestive malaise after eating a particular food may lead to a subsequent and durable aversion to that food. Experimentally, it is demonstrated that the most potent reinforcer inducing aversion is nausea and vomiting more than diarrhea or oppression (94). We will see later that area postrema involved in nausea and vomiting is also critically involved in rats' conditioned taste aversions. Various authors reported that eating a food in a state of disease induced an aversion to that food. One subject receiving *per os* an antibiotic flavored with banana aroma during childhood acquired a true phobia for banana, which persisted 20 years later. Other examples of acquired odor aversions will be given in the next chapter.

Here again such conditioning does not explain the individual spectrum of food aversion and their versatilities. Many aversions are exhibited for foods that have never been tasted. Rozin (90) classified aversions studied in groups of subjects into three categories. Aversions would be determined either by disgust, by the presumed and reputed danger to eat the food, or even by ritual or cultural prohibition. In the latter case, it is not sure that a socially imposed prohibition for some meats, alcohols, and so on is associated to a true sensory aversion. The rejection and spitting of meat occasionally tasted by a vegetarian seems to be, rather, a neophobic response.

VI. *Odors, Aromas, and Perfumes*

The role played by natural aromas of foods and by aromas added to foods and beverages is of utmost importance in the day-to-day human feeding behavior. Although rarely examined as such and rarely investigated, some attention to the matter is justified here.

The field concerns food scientists and the food industry. It also belongs to anthropology of feeding inasmuch as the use of and likes and dislikes for aromas are highly specific to each population. They are a feature, not only of food habits, but also of cultural identity.

A. Ambient Odors and Aromas

We live in a world of ambient odors. Many of them are unpleasant at various levels, up to extremely repulsive, nauseating odors. The fight against odor pollution, at home and in cities, is the subject of particular technologies. Other odors are spontaneously pleasant: those of fruit, flowers, smokes, etc. This bipolar affective dimension of odors is extremely marked. No noise or sight is as aggressive and repulsive as a putrid or fecal odor. Except classical music and painting, no noise or sight is as pleasant as some perfumes. This extreme dichotomy is also quasi-unanimous in humankind. Many inquiries and scorings of odor pleasantness and unpleasantness showed identical judgments for the extremes in different populations such as Europeans and the Japanese. The variance increased for odors approaching indifference but also for odors relevant to specific foods and food habits of these two populations.

Some of these pleasant ambient odors of fruits and plants are also natural aromas of foods, i.e., the natural odor of food perceived from the mouth via the pharyngeal route. In addition to this natural flavor of the raw or cooked food, a considerable number of various aromas (spices, condiments, herbs, or synthetic odorants) are added to food products.

An intriguing case is that of the extremely disgusting smells of some foods when they are smelled in the environment but are nevertheless liked and even judged delicious as aromas. This is the case of the putrid and even fecal odors of some cheese, which are, nevertheless, highly appreciated. This is also the case of the putrid fish eaten for Christmas by Swedish people: Its odor is so horrible that the dish is served in separate and closed rooms in restaurants.

Generally, fragrant odors used as aromas are those of fruit, of some specific aromatic plants such as onion and garlic, and of those forming the family of spices. Odors used to make perfumes are those of flowers. But the border between them is very flexible and varies largely among populations. In addition, the separation between aroma and perfume odors and within aromas is different for solid foods and beverages. Except for some rose-tasting liquors, flower odors for perfumes are not used as aromas. Conversely, aromas of fruit or of other added aromas are not commonly used to make perfumes. It is inconceivable to prepare a perfume with a garlic or onion odor. However, many national differences are known by everyone. For example, soaps perfumed with aromas of fruit (raspberry, strawberry), like those in the United Kingdom, are not acceptable in France.

The likes or dislikes for some aromas and their differences in two populations may be easily explained, in some cases, by external conditioning. Anecdotal examples are that of the "venergren" (methylacetate) and of the eugenol flavor, both used in the United States. A common trick played by young Europeans coming back from the United States is to offer these cakes or candies to their friends and relatives. These victims are obliged to spit the cakes and "bonbons" with horrible faces, provoking laughs. Why? Methylacetate is, at least in France, the prevailing ambient odor of pharmaceutical offices. Thus, it is a strong odor of medicine and, as such, is aversive. Eugenol is the common antiseptic used by dentists in Europe. A similar example is the use of menthol or anise odors in beverages. In France, a peppermint flavor is rarely successful as flavor in liquors. In Hungary, the anise flavor and absinthe are unsuccessful. The reason is probably because, in France, toothpaste is almost always flavored with menthol and in Hungary with anethol. A peppermint liquor in France or an anise spirit in Hungary tastes like toothpaste.

B. Aromas with Some Foods, Not with Others

The association of various flavors in each particular dish and variations of these associations in various dishes in high cuisine obey unwritten and complicated laws. The multiple aromas associated in the same preparation compose a temporal pattern of various predominant flavors of mixed foods and sauces, from the mouth-taking to the swallowing. Again, these rules or inventions of great "cuisiniers" are very specific to the "taste" (in the large meaning of the word) in each country and are an aspect of their cultural identity. The association of jam and fruity aromas to meats is a sacrilege in the French cuisine, but not in others. Some aromas are reserved for foods eaten cold. Others are exclusively used as flavors of desserts or ice creams, such as fruity flavors and most of all, sweetness. No one has risked preparing a chocolate-flavored chicken.

Surprisingly, this field is "terra incognita," and has never been seriously investigated by psychologists and anthropologists. In fact, it belongs to a science that is not even recognized–food esthetics.

C. Aromas in Foods and Beverages

Another unexplained peculiarity is that some aromas in foods are liked, whereas the same aromas in beverages are disliked, and vice versa. Entire books are devoted to the taste of wines. In France, the tasting of wines and the ability to recognize their qualities are learned as basic education. The inability to distinguish a Bordeaux from a Burgundy wine is considered as severe as an inability to master the alphabet. The success of wine-tasting parties all around the world is a sort of artistic play. The wine flavors are associated with some prepared solid foods, rarely with desserts. Many spices are impossible as beverage flavors.

Coffee is a separate case. Its bitter taste and typical aroma do not favor likeness *a priori*. In fact, children are disgusted by it. It provides no calories and, thus, it is not a food. Nevertheless, the consumption of this bitter "nonfood" is such that, in the rank order of raw materials, income from the production and trade of coffee is only below that of oil. Why? Very likely this is due to the psychotonic reinforcement by caffeine, which also perhaps has contributed to the success of cola beverages.

D. Sensory and Chemical Analysis of Food

A technology called "sensory analysis of foods" was developed and is now extensively utilized by food industries. The techniques are subjective or objective. In the former, discriminative performances in intensity and qualities of the complete food flavors and pleasantness are tested by a panel of selected subjects. These techniques are described in specialized books.

A complementary technique is the chemical analysis of natural aromas. Gas chromatography coupled with mass spectrography allowed analysis of most of these natural aromas; and these analyses are still in progress for some aromas. Despite the very sophisticated techniques of chemical quantitative and qualitative analysis, two difficulties are encountered. The first one is the considerable complexity of mixtures of odor traces present in some natural aromas. Only two or three chemicals in the appropriate proportion reproduce the aroma of cream or of many fruits. For some aromas, hundreds of compounds must be separated by gas chromatography and identified. More than 500 compounds are found in the smell of coffee. The second difficulty is to know the respective role of each peak of the "aromagramme" in the intensity and quality of the mixture.

A funny example of these difficulties was given 20 years ago by a team of chemists attempting for the first time the chemical analysis of the aroma of fresh strawberry. After months of work, they had separated and hardly identified about 50 different chemicals. Happy with their success, they tried the final test of the mixture of the synthetic 50 compounds in the adequate proportion. Unfortunately, the mixture smelled strongly of rubber. Since these pioneering works, results are now far better, and, after analysis, it is now possible to obtain a considerable number of synthetic aromas identical to the natural ones or mimicking them perfectly. Many of these preparations of synthetic aromas patented by their producer are secret.

The compilation of Maarse and Wisscher (95) includes about 6000 analyzed volatile compounds concerning 374 products of the 12 following product groups:

alcoholic beverages
bread
dairy products and eggs
meat and poultry
seafood and seaweed
fruits

spices and condiments
vegetables
nuts
tea, coffee, and cocoa
cereals
miscellaneous

Volatile compounds do not equally contribute to the flavor. For example, according to Maarse and Visscher, the aroma of vanilla is essentially due to vanillin and ethyl-vanillin, 2 of the 182 identified volatiles in this product.

The oldest way to estimate the relative contribution of each volatile to the overall product aroma is to express them in terms of odor units, i.e., the ratio of measured concentrations to threshold concentrations (96). Furthermore, to weigh the respective concentrations of each volatile, the estimation of suprathreshold odorous intensity is often preferred (97, 98)

Table 4.1

Some Foods Whose Aroma Resides Largely in One Compound

Food	Character-impact compound
Banana	Isopentylacetate
Grape, "Concord"	Methylanthranilate
Grapefruit	Nootkatone
Lemon	Citral
Pear	*trans*-2,*cis*-4-Decadienoates
Cucumber	*trans*-2,*cis*-6-Nonadienal
Green pepper	2-Isobutyl-3-methoxypyrazine
Potato, raw	2-Isopropyl-3-methoxypyrazine
Mushroom, boiled	1-Octen-3-one
Shiitake	Lenthionine
Beetroot	Geosmin
Garlic	Di-2-propenyl disulphide
Watercress	2-Phenethyl isothiocyanate
Almonds	Benzaldehyde
Cheese, blue	2-Heptanone and 2-nonanone

[From H. E. Nursten, 1979, in D. G. Land and H. E. Nursten (eds.), *Progress in flavour research* (pp. 337–356). London: Applied Science Publishers.]

Table 4.2

Some Foods Whose Aroma is Due to a Mixture of a Small Number of Compounds

Foods	Character-impact compound	Contributory flavor compounds
Apple, "Delicious"	Ethyl-2-methyl butyrate	Hexanal *trans*-2-Hexenal
Bilberry		Ethyl-2-methyl butyrate Ethyl-3-methyl butyrate *trans*-2-Hexenal
Raspberry	1-*p*-Hydroxyphenyl-3-butanone	*cis*-3-Hexen-1-ol Damascenone α- and β-Ionone
Tangerine		Methyl-*N*-methyl anthranilate Thymol
Tomato		Hexanal *trans*-2-Hexenal *cis*-3-Hexenal *cis*-3-Hexen-1-ol 2-Isobutylthiazole Some high-boilers
Celery	3-Isobutylidene-3a, 4-dihydrophthalide 3-Isovalidene-3a, 4-dihydrophthalide	*cis*-3-Hexen-1-yl pyruvate 2,3-Butanedion
Onion, raw	Thiopropanal *S*-oxide (lachrymator)	Thiosulphinates Thiosulphonates
Onion, boiled	Propyl and 1-propenyl disulphides	
Cabbage, boiled	Dimethyl disulphide	2-Propenyl isothio-cyanate Dimethyl trisulphide
Potato, boiled	2-Ethyl-3-methoxy pyrazine	Methional
Butter	2,3-Butanedione	Ethanal Dimethyl sulphide

[From H. E. Nursten, 1979, in D. G. Land and H. E. Nursten (eds.), *Progress in flavour research* (pp. 337–356). London: Applied Science Publishers.]

In terms of current knowledge of flavor volatiles, foods are usually classified into four groups. (99).

Group 1. Foods whose aroma resides largely in one compound, a so-called character-impact compound. The other components have little or no importance.

Group 2. Foods whose aroma is essentially due not to a single compound, but to a mixture of a small number of compounds, called contributory flavor compounds.

Group 3. Foods whose aroma can only be reproduced accurately with quite a large number of compounds. A character-impact compound is unlikely to be present.

Group 4. Foods whose aroma cannot be reasonably reproduced, even by a complex mixture of specific compounds. No character-impact compound has been discovered.

Tables 4.1 and 4.2 provide some examples of Group 1 and 2 compounds, according to Nursten (99).

Typical examples of Group 3 are roast beef, bread, coffee, passion fruit, and some white wines and of Group 4 red wines and distilled beverages. In a recent compilation, Morton and MacLeod (100) quote newly identified flavors, principally those of several fruits: quinces, cherries, apricots, peaches, strawberries, plums, black currant berries, and melons.

References

1. Le Magnen, J. (1953). Activité de l'insuline sur la consommation spontanée de solutions sapides. *C. R. Soc. Biol. (Paris)* **147,** 1753–1757.
2. Le Magnen, J. (1953). Régulation immédiate de la prise spontanée d'eau et de sel chez le rat blanc dans des états imposés de déséquilibre hydrominéral. *C. R. Soc. Biol. (Paris)* **147,** 619–623.
3. Strouthes, A. (1971). Thirst and saccharine preference. *Physiol. Behav.* **6,** 289–294.
4. Schulkin, J. (1982). Behavior of sodium-deficient rats: The search for the salty taste. *J. Comp. Physiol. Psychol.* **96,** 628–634.
5. Campbell, B. A. (1958). Absolute and relative sucrose preference threshold for hungry and satiated rats. *J. Comp. Physiol. Psychol.* **51,** 795–800.
6. Davis, J. D. (1973). The effectiveness of some sugars in stimulating licking behavior in the rat. *Physiol. Behav.* **11,** 39–45.
7. Mora, F., Rolls, E. T., and Burton, M. J. (1976). Modulation during learning of the responses of neurons in the lateral hypothalamus to the sight of food. *Exp. Neurol.* **53,** 508–519.
8. Mensel, E. W., and Draper, W. A. (1965). Primate selection of food by size: Visible versus invisible reward. *J. Comp. Physiol. Psychol.* **59,** 231–239.
9. Balagura, S., and Harell, F. (1973). Effect of size of food on food consumption. *J. Comp. Physiol. Psychol.* **56,** 67–73.
10. Stewart, C., and Frasczek, S. (1988). The taste characteristics of Na-saccharine in the rat: A reexamination of the dual taste hypothesis. *Chem. Senses* **13,** 205–210.
11. Hamilton, C. L. (1964). Rat's preference for high fat diets. *J. Comp. Physiol. Psychol.* **58,** 459–460.

12. Drewnowski, A., Shrager, E. E., Lipsky, C., Stellar, E., and Greenwood, M. R. C. (1989). Sugar and fat: Sensory and hedonic evaluation of liquid and solid foods. *Physiol. Behav.* **45,** 177–183.

13. Kanarek, R. B. and Marks-Kaufman, R. (1979). Developmental aspects of sucrose-induced obesity in the rat. *Physiol. Behav.* **23,** 881–886.

14. Kanarek, R. B., Aprille, J. R., and Hirsch, L. (1987). Sucrose-induced obesity: Effect of diet on obesity and body weight. *Am. J. Physiol.* **253,** R158–166.

15. Bacon, W. E., Snyder, H. L., and Hulse, S. H. (1962). Saccharine preference in satiated and deprived rats. *J. Comp. Physiol. Psychol.* **55,** 112–114.

16. Le Magnen, J. (1968). The eating rate as related to deprivation and palatability in normal and hyperphagic rats. Third International Conference on the Regulation of Food and Water Intake, Haverford.

17. Le Magnen, J. (1969). Influence de l'état de faim sur la manifestation d'un appétit discriminatif par le rat blanc. *C. R. Soc. Biol. (Paris)* **154,** 80–83.

18. Coons, E. E., and White, Z. (1969). Tonic properties of oral sensations and the modulation of intra cranial stimulation. *Ann. N. Y. Acad. Sci.* **157,** 219–226.

19. Pfaffmann, C. (1955). Factors influencing taste sensitivity to sugars. *Am. J. Physiol.* **183,** 651–656.

20. Smith, M., and Michael, D. (1957). Consumption of sucrose and saccharine by hungry and satiated rats. *J. Comp. Physiol. Psychol.* **50,** 65–69.

21. Mook, D. G., and Wagner, S. (1987). Some determinants of the time course of the saccharin ingestion in hungry rats. *Appetite* **9,** 39–56.

22. Le Magnen, J. (1954). Le processus de discrimination par le rat blanc des stimuli sucrés alimentaires et non alimentaires. *J. Physiol. (Paris)* **46,** 414–418.

23. Capretta, P. J. (1964). Saccharine consumption and the reinforcement issue. *J. Comp. Physiol. Psychol.* **57,** 448–450.

24. Warren, T., and Warren, F. M. (1966). Soluble saccharine preference: A putative basis for persistency. *Nature (London)* **210,** 5103–5110.

25. Harris, L. J., Clay, J., Hargreaves, F. J., and Ward, A. (1933). Appetite and choice of diet: The ability of the vitamin B deficient rat to discriminate between diets containing and lacking the vitamin. *Proc. R. Soc. London B.* **113,** 161–190.

26. Le Magnen, J. (1956). Effet sur la prise alimentaire du rat blanc des administrations postprandiales d'insuline et le mécanisme des appétits caloriques. *J. Physiol. (Paris)* **48,** 789–802.

27. Le Magnen, J. (1959). Effets des administrations post-prandiales de glucose sur l'établissement des appétits. *C. R. Soc. Biol.* **153,** 212–215.

28. Le Magnen, J. (1969). Peripheral and systemic actions of food in the caloric regulation of intake. *Ann. N. Y. Acad. Sci.* **268,** 3107–3110.

29. Booth, D. A. (1972). Conditioned satiety in rats. *J. Comp. Physiol. Psychol.* **81,** 457–471.

30. Jacobs, H. L. (1974). Some physiological, metabolic and sensory components in the appetite for glucose. *Am. J. Physiol.* **213,** 1043–1054.

31. Van Vort, W., and Smith, G. P. (1983). The relationships between the positive reinforcing and satiating effect of meal in the rat. *Physiol. Behav.* **30,** 279–284.
32. Sclafani, A., and Mistenbaum, G. (1988). Robust conditioned hyperphagia induced by intragastric starch infusions in rats. *Am. J. Physiol.* **255,** R672–675.
33. Smotherman, W. P. (1982). Odor aversion learning by the rat fetus. *Physiol. Behav.* **29,** 769–771.
34. Smotherman, W. P. (1982). In utero chemosensory experience alters taste preference and corticosterone responsiveness. *Behav. Neural Biol.* **38,** 61–67.
35. Smotherman, W., and Robinson, T. (1985). The rat fetus and its environment: Behavioral adjustment to familiar, aversive and conditioned stimuli presented in utero. *Behav. Neurosci.* **99,** 521–530.
36. Stickrod, G., Kimble, D. P., and Smotherman, W. P. (1982). In utero taste odor aversion conditioning in the rat. *Physiol. Behav.* **28,** 5–8.
37. Le Magnen, J., and Tallon, S. (1968). Préférence alimentaire du jeune rat induite par l'allaitement maternel. *C. R. Soc. Biol. (Paris)* **162,** 387–390.
38. Galef, B. G., Jr., and Henderson, P. W. (1972). Mother's milk: A determinant of the feeding preferences of weanling rat pups. *J. Comp. Physiol. Psychol.* **78,** 213–219.
39. Deutsch, J. A., Puerto, A., and Wang, M. (1970). The pyloric sphincter and differential food preference. *Behav. Biol.* **19,** 81–589.
40. Hasegawa, Y. (1981). Recuperation from lithium-induced illness: Flavor enhancement for rats. *Behav. Neural Biol.* **33,** 252–255.
41. Green, T., and Garcia, J. (1971). Recuperation from illness: Flavor enhancement for rat. *Science* **173,** 749–751.
42. Le Magnen, J. (1957). Effet de la durée du jeûne post-prandial sur l'établissement des appétits chez le rat blanc. *C. R. Soc. Biol. (Paris)* **151,** 229–233.
43. Le Magnen, J. (1957). Etude d'un facteur post-ingestif de l'établissement des appétits chez le rat blanc. *Arch. Sci. Physiol.* **11,** 237–254.
44. Hull, C. L., Livingstone, J. R., Rouse, R. A., and Barker, A. N. (1951). True sham and oesophageal feeding as reinforcements. *J. Comp. Physiol. Psychol.* **44,** 236–245.
45. Schlosberg, H., and Pratt, C. (1956). The secondary reward value of inaccessible food for hungry and satiated rats. *J. Comp. Physiol. Psychol.* **49,** 149–152.
46. Hughes, L. H. (1957). Saccharine reinforcement in a rat maze. *J. Comp. Physiol. Psychol.* **50,** 431–435.
47. Sheffield, Y. D., Roby, T. B., and Campbell, B. A. (1954). Drive reduction versus consummatory behavior as determinants of reinforcement. *J. Comp. Physiol. Psychol.* **57,** 349–354.
48. Smith, M., and Dufy, M. (1957). Evidence for a dual reinforcing effect of sugar. *J. Comp. Physiol. Psychol.* **50,** 242–247.
49. Galef, B. G. (1989). Enduring social enhancement of rat's preferences for the palatable and the piquant. *Appetite* **13,** 81–92.

50. Franke, L. W. (1936). The ability of rats to discriminate between diet of various degrees of toxicity. *Science* **83,** 130–135.

51. Carr, W. J., Dissinger, M. L., and ScannaPieco, M. R. (1982). The stimulus basis of nature food aversion in Norway rats. *Physiol. Behav.* **28,** 281–287.

52. Barnett, S., Dickser, R., Martles, T., and Radar, E. (1988). Sequences of sampling and exploration in wild and laboratory rat. *Behav. Proc.* **3,** 29–43.

53. Bett, M., Domjan, M., and Haskins, T. (1978). Long-term retention of flavor familiarization: Effects of number and amount of prior exposure. *Behav. Biol.* **25,** 95–99.

54. Naim, M., and Kare, M. (1978). Diet palatability and growth efficiency: Evidence for inter-relationships in rats. *Life Sci.* **23,** 212–234.

55. Kratz, C. M., Levitsky, D. A., and Lustick, S. (1978). Differential effects of quinine and sucrose octoacetate on food intake in the rat. *Physiol. Behav.* **20,** 665–668.

56. Aravich, P. F., and Sclafani, A. (1980). Dietary preference behavior in rats fed bitter tasting quinine and sucrose octo-acetate adulterated diets. *Physiol. Behav.* **25,** 157–160.

57. Stark, K. D. (1963). Effects of early and prolonged experience with bitter water on its preferableness to guinea pigs. *Dissert. Abstr.* **24,** 859–860.

58. Rozin, P., Gruss, L., and Berk, G. (1979). Reversal of innate aversions: Attempts to induce a preference for Chili pepper in rats. *J. Comp. Physiol. Psychol.* **93,** 1001–1014.

59. Ettenberg, A., Sgro, S., and White, N. (1982). Algebric summation of the affective properties of rewarding and an aversive stimulus in the rat. *Physiol. Behav.* **28,** 873–877.

60. Miller, N. E. (1955). Shortcomings of food consumption as a measure of hunger: Results from other behavioral techniques. *Ann. N. Y. Acad. Sci.* **63,** 141–143.

61. Tenen, S. S., and Miller, N. E. (1964). Strength of electrical stimulation of lateral hypothalamus, food deprivation and tolerance for quinine in food. *J. Comp. Physiol. Psychol.* **58,** 55–62.

62. Armour, E. J., and Barnett, S. A. (1950). The action of dicoumarol on laboratory and wild rats and its effects on feeding behavior. *J. Hyg. Cambs* **48,** 158–172.

63. Rozka, T. (1953). Bait-shyness: A study of rat behavior. *Brit. J. Animal Behav.* 127–137.

64. Le Magnen, J. (1952). Quelques aspects des liens entre sensibilité chimique et appétit. Dans: "Le Comportement alimentaire et l'Appétit," H. Piéron (ed.), Réunion du CNERNA, Editions du CNRS, Paris, pp. 11–25.

65. Nachman, M. (1963). Learned aversion to the taste of lithium chloride and generalization to other salts. *J. Comp. Physiol. Psychol.* **56,** 343–349.

66. Nachman, M. (1970). Learned taste and temperature aversion due to lithium chloride sickness after temporal delays. *J. Comp. Physiol. Psychol.* **73,** 22–30.

67. Nachman, M., and Hartley, P. M. (1975). Role of illness in reproducing learned taste aversion in rats. *J. Comp. Physiol. Psychol.* **89,** 1010–1018.

68. Rusinyak, K., Hankins, R., and Garcia, J. (1979). Flavor illness-induced aversion: Potentiation of odor by taste in rats. *Behav. Neural Biol.* **25,** 1–17.

69. Galef, B. (1978). Novel taste facilitation of the association of visual cues to toxicity in rats. *J. Comp. Physiol. Psychol.* **94,** 907–916.

70. Bett, P., Bett, M., and Nickley, A. (1973). Conditioned aversion to distinct environmental stimulation resulting from gastrointestinal distress. *J. Comp. Physiol. Psychol.* **85,** 250–257.

71. Johnson, J., Beaton, R., and Hzall, K. (1975). Poison-based avoidance learning in non-human primates. *Physiol. Behav.* **14,** 493–497.

72. Micelli, D., Marfaing-Jallat, P., and Le Magnen, J. (1980). Ethanol aversion by induced parenteral administered ethanol acting both as CS and UCS. *Physiol. Psychol.* **5,** 433–436.

73. Maruniak, J. A., Mason, J. R., and Kostelc, ·J. G. (1983). Conditioned aversion to an intravascular odorant. *Physiol. Behav.* **30,** 617–620.

74. Brillat-Savarin. (1828). *La physiologie du goût.* Paris.

75. Pilgrim, F., and Kamen, J. (1963). Prediction of human food consumptions. *Science,* **139,** 501–502.

76. Lucas, F., and Bellisle, F. (1987). The measurement of food preference in humans: Do taste and spit tests predict consumption? *Physiol. Behav.* **39,** 793–743.

77. Monneuse, M. O., Bellisle, F., and Louis-Sylvestre, J. (1991). Responses to an intense sweetener in humans: Immediate preference and delayed effects on intake. *Physiol. Behav.* **49,** 325–330.

78. Pangborn, R. M., and Pecore, S. (1982). Taste perception of NaCl in relation to dietary intake of salt. *Am. J. Clin. Nutr.* **85,** 510–520.

79. Bobroff, E. M., and Kissileff, H. R. (1986). Effects of changes in palatability on food intake and the cumulative food intake curve in man. *Appetite* **7,** 85–96.

80. Moskowitz, H. R., Kunrich, V., Sharma, K., Jacobs, H., and Sharma, S. (1978). Effect of hunger, satiety and glucose load on taste intensity and taste hedonics. *Physiol. Behav.* **19,** 471–475.

81. Moskovitz, H. R., Kluter, R. C., Terling, J., and Jacobs, H. L. (1974). Sugar sweetness and pleasantness: Evidence for different psychological laws. *Science* **274,** 503.

82. Meyer, G. D., Mrig, R., and Engen, T. (1977). Influence of internal factors in the perceived intensity and pleasantness of gustatory and olfactory stimuli. In M. R. Kare and O. Maller (eds.), *Chemical Senses and Nutrition* pp. 104–123. New York: Academic Press.

83. Meyer, D. R. (1952). The stability of human gustatory sensititivy during changes in time of food deprivation. *J. Comp. Physiol. Psychol.* **45,** 303–326.

84. Pangborn, R. M., and Hansen, B. (1963). The influence of color on discrimination of sweetness in peer nectar. *Am. J. Physiol.* **76,** 315–317.

85. Weizenbaum, F. E., and Bensen, B. (1980). Relationship among reproductive variables, sucrose taste reactivity and feeding behavior in humans. *Physiol. Behav.* **24,** 1053–1056.

86. Shepherd, R. and Farleigh, C. A. (1989). Attitudes and personality related to salt intake. *Appetite* **7,** 343–354.

87. Shepherd, R., Farleigh, C. A., and Land, D. G. (1985). Estimation of salt intake by questionnaire. *Appetite* **6,** 219–233.

88. Bellisle, F., Lucas, F., Amrani, R., and Le Magnen, J. (1984). Deprivation, palatability and the microstructure of meals in human subjects. *Appetite* **5,** 85–94.

89. Rozin, P., and Millman, L. (1987). Family environment, not heredity accounts for family resemblances in food preferences and attitudes: A twin study. *Appetite* **2, 8,** 125–134.

90. Rozin, P. (1980). The psychological categorization of food and non foods: A preliminary taxonomy of food rejections. *Appetite* **1,** 193–201.

91. Booth, D. A., Mather, P., and Fuller, J. (1982). Starch content of associatively conditioned human appetite and satiation, indexed by intake and eating pleasantness of starch-paired flavors. *Appetite* **3,** 163–184.

92. Berger, P., and Le Magnen, J. (1960). La faim et les appétits chez l'homme en état de semi-inanition. *Ann. Nutr. Alim.* **14,** 101–133.

93. Bernstein, I., and Webster, M. (1980). Learned taste aversion in humans. *Physiol. Behav.* **25,** 363–366.

94. Pelchat, N., and Rozin, P. (1982). The specific role of nausea in the acquisition of food dislike in humans. *Appetite* **3,** 342–353.

95. Maarse, H., and Wisscher, C. A. (1989). Volatile compounds in food. Qualitative and quantitative data. TNO-CIVO Food Analysis Institute, Zeist, The Netherlands, 1377 pp.

96. Guadagni, D. G., Okano, S., Buttery, R. G., and Burr, H. K. (1966). Correlation and gas-liquid chromatographic measurements of apple volatiles. *Food Technol. (Chicago)* **20,** 518–521.

97. Dravnieks, A., and O'Donnell, A. (1971). Principles and some techniques of high-resolution headspace analysis. *J. Agr. Food Chem.,* **19,** 1049–1056.

98. Casimir, D. J., and Whitfield, F. B. (1978). Flavour impact values: A new concept for assigning numerical values for the potency of individual flavour components and their contribution to the overall flavor profile. In *Flavour of fruits and fruit juices symposium.* International Federation of Fruit Juice Producers, pp. 325–347. Bern.

99. Nursten, H. E. (1979). Why flavour research? How far have we come since 1975 and where now? In *Progress in flavour research* D. G. Land and H. E. Nursten (eds.), (pp. 337–356). London: Applied Science Publishers.

100. Morton, I., and MacLeod, A. (1990). Food flavours. Part C—The flavour of fruits. Amsterdam: Elsevier, 360 pp.

Chapter Five

Brain Mechanisms of Palatability

I. *Central Orosensory Projections and Responses*

A. Neuroanatomy

The general organization of sensory projection in the central nervous system and the brain processing of sensory information are described in all textbooks on sensory physiology.

In the various sensory modalities, afferent neurons from peripheral receptors terminate in their specific primary projection. From there, the sensory information is conveyed to secondary projections through both cortical and subcortical pathways. The first one is thought to be concerned with elaborated and mainly learned responses; the second one in basic "instinctive" behavioral outputs. Modern histological techniques of retrograde and orthograde labeling only disentangled their complicated network since the last two decades. In the olfactory system, from the first relay of the olfactory nerve in the olfactory bulb, mono- and polysynaptic projections reach almost all brain regions (1). Primary and secondary (orbitofrontal) projections are well recognized. The subcortical pathways include two series of pathways from olfactory bulbs to the hypothalamus. From the prepiriform cortex (primary projection), a direct monosynaptic pathway conveys the olfactory input to the lateral hypothalamus (LH) through the medial forebrain bundle. Another polysynaptic connection with the hypothalamus is achieved from the amygdaloid complex. From the corticomedial nuclei, pathways through the stria terminalis and the anterior commissure reach both the ventromedial and posterior hypothalamus (2). From basolateral and lateral amygdaloid nuclei, a direct amygdofugal pathway terminates in the LH (3). Another pathway links these amygdaloid nuclei to the nucleus of the solitary tract (NTS), the primary gustatory and vagal projection (4, 5)

191

The NTS is the primary projection of afferent gustatory neurons. It is connected to the adjacent area postrema and dorsomotor nucleus of the vagal nerve. From the NTS, two gustatory pathways, both cortical and subcortical, were identified by Norgren (6, 7) Hamilton and Norgren (8). From the NTS, in addition to the thalamocortical pathway, terminal fibers, through a pontine relay in the suprabrachial nuclei, reach the forebrain: the bed nuclei of substantia innominata and the lateral hypothalamus. From these limbic and diencephalic sites, both olfactory and gustatory information reach the frontal cortex.

In both olfactory and gustatory systems, a network of centrifugal pathways feeds back to the primary projections. Centrifugal systems reaching the olfactory bulbs are well documented (9–11). From the periventricular nucleus, a bidirectional monosynaptic pathway connects the hypothalamus and the NTS (12).

B. Electrophysiological Responses

Electrophysiological responses and their latencies complete histological studies. Stimulating the olfactory bulb or the peripheral organ by an odorant gives rise to single-unit responses in the ventrolateral branch of the medial forebrain bundle (MFB) (13) and in other parts of the LH (14, 15). The response is of short latency (200 msec) in the monkey (16). Using various odorants, the single-unit responses in the MFB are partly specific to the odorant; however, they are less specific than those in the olfactory bulb (13).

Gustatory responses in the NTS and parabrachial nuclei to various taste stimuli were extensively studied as key sites of quantitative and qualitative taste discriminations. In the hypothalamus, gustatory responses were found in the MFB and in the LH. In the MFB, each electrode placement gave responses partly specific of taste qualities: sweet, bitter, and NaCl. Interestingly, only in electrode placement where responses to sucrose applied to the tongue are recorded, an electrical self-stimulation is likely obtained through the same electrode (17). Such specific responses are found in LH units. Two groups of neurons are identified responding to aversive and "rewarding" stimuli, respectively. In the rostral LH region, neurons are activated by aversive stimuli, inhibited or not modified by rewarding stimuli (18).

Responses of amygdaloid nuclei and of amygdalohypothalamic pathways were investigated. Separate or combined electrical stimulations of corticomedial and basolateral nuclei of amygdala showed that two different pathways are involved and converge on single units recorded in the ventromedial hypothalamus (VMH). Responses are transmitted via the stria terminalis from the corticomedial nuclei and via the ventral amygdalofugal pathway from the basolateral nuclei

(19). Interestingly, electrical stimulation in the LH producing the stimulus-bound feeding activates neurons in basolateral amygdala and in the prepiriform cortex (20). Exploration of sensory responses in central, corticomedial, and lateral nuclei of amygdala revealed responses to visual, olfactory, and gustatory stimulations, some units often responding to two of these sensory modalities.

Thus, electrophysiological as well as morphological data demonstrate the convergence of sensory inputs in diencephalic sites shown to be responsive to hunger and food deprivation and to blood-borne signals. These convergences provide the first indication about the brain substrate subserving the behaviorally observed combination of external and internal stimuli.

C. Effects of Lesions

Effects of electrolytic and neurotoxic lesions of various brain regions on palatability-dependent feeding responses have been widely investigated. Here, only palatability responses, i.e. unconditioned or conditioned preferences, will be examined. Effects of lesions on spontaneous or learned aversions will be overviewed in a subsequent chapter (Section II of this chapter).

Most hypothalamic and amygdaloid lesions affect the manifestation and/or the acquisition of preferences. They suppress or eliminate, irreversibly or not, the discrimination of foods from nonfoods in a choice as well as in single presentations or in instrumental responses (21, 22). As mentioned earlier, the LH lesion, in addition to induced aphagia, impairs palatability response manifested during and after recovery. Rats who later resume eating sometimes accept high-palatability foods only. Fully recovered, they exhibit an irreversible alteration of their meal pattern in which unrecovered sensory projections to the region seem to be involved. Olfactory bulb ablations, transection in medial amygdala and of stria terminalis in LH-intact rats induce the same nibbling pattern (23). The LH, not the VMH, is concerned. A pre- or postoperative olfactory bulb ablation did not prevent the induction and the persistency of hyperphagia following the VMH lesion, concurrent with the nibbling pattern (24).

Based on studies of the VMH lesion, the hyperreactivity of hyperphagic rats to food-related stimuli, called their finickiness, will now be examined. The transection of stria terminalis projecting in the VMH also induced this finickiness in intact rats and added its effect in VMH-lesioned rats. This addition limited hyperphagia and weight gain. The conclusion of the author was that finickiness introduced by stria terminalis transection is independent of the VMH and that the VMH-lesion-induced finickiness is independent of amygdaloid connections (25).

A number of convergent results emphasized a critical role of basolateral and lateral nuclei of amygdala in the manifestation and/or acquisition by experience of food preferences and aversions. After corticomedial nuclei lesions, rats lose their neophobic response to a novel food (21). Lateral nuclei of amygdala are terminal of two dopaminergic and noradrenergic distinct pathways originating in the nucleus ceruleus. Lesion of the norepinephrine pathway by 6-OH-DOPA provokes hyperphagia and weight gain. A 6-OH-DOPA desipramine lesion depleting DOPA produces hypophagia and weight loss (26).

D. Hunger and Palatability Dependency of Neuronal Responses

Fiber discharges in peripheral gustatory nerves do not depend on hunger and satiety nor on palatability, i.e., on the capacity of the stimulus to drive or not drive a feeding response. In rats sodium-depleted by adrenalectomy, threshold responses to NaCl are not modified (27). In a state of hypoglycemia and hunger induced by insulin injection, thresholds to sweet stimuli are unchanged (28). Changes of activity in the olfactory nerve as a function of a motivational state have never been demonstrated.

By contrast, single and multiunit activities recorded in the NTS of rats (not of monkeys) are hunger- and palatability-dependent (29–31). This modulation depends on centrifugal pathways coming from the LH. In one study, gustatory responsive neurons in the NTS were identified by stimulation of the chorda tympani nerve. The neuronal responses of these neurons were altered by a LH electrical stimulation (29). Extracellular recordings were carried out on NTS neurons classified as chemical, thermal, or mechanical neurons according to their responses to lingual stimulations. Sixty-four percent of neurons were found activated by LH stimulation. A prestimulation of the LH facilitated by 20–80% the response to lingual stimulation for 20–120 msec. Only chemical gustatory neurons were so facilitated. On the contrary, thermal- and mechanical-responsive neurons were inhibited (30). Rats exhibiting the LH stimulation-bound feeding were selected by multielectrode placements. In the anesthetized rats, stimulation of the LH through the electrode, which provided the stimulus-bound feeding, elicited responses in gustatory neurons in the NTS and the pontine relay. The responses, either inhibitory or excitatory, were identical to those elicited by the lingual stimulation (31). Thus, it seems that through centrifugal pathways from hypothalamic sites (in which hunger and palatability responses are generated) a feedback system governs the palatability-dependent responses recorded in the NTS, in which they are independent of the primary processing of the peripheral message in terms of quality and intensity.

An opposite point of view was presented in a series of works. According to the reported results, the modulation by hunger and palatability of NTS responses to gustatory stimuli in rats would be a direct effect of the blood glucose level on the structure. This action would lead not only to changes of neuronal responses in relation to hunger and palatability but also to changes of the intensity and quality coding at this level. Under hyperglycemia following intravenous glucose injection, unit responses in the NTS to glucose applied to the tongue are reduced. Insulin injection, as an effect (according to the authors) of the increased glucose utilization, has the same effect. Sodium depletion reduces responses of neurons best responding to salts in the NTS and considerably augments this response to salts in neurons best responding to sweetness. An analysis of the cross-fiber pattern induced in the NTS by saccharin shows a shift of this pattern in the direction of that recorded for bitter aversive substances when saccharin was made aversive by conditioning. A behavioral test shows that hyperglycemia induces a fall of perceived sucrose intensities. The licking rate of rats was recorded for some seconds on 16 concentrations of glucose solutions. Under hyperglycemia induced by intravenous glucose, the licking rate of 1.2 M glucose by hyperglycemic rats is the same as the rate of licking a 0.6 M solution in normal rats (32–34).

The significance of these data in rats is questionable. The same change of intensity responses under glucose and insulin injections is surprising. However, its specificity to sweet stimulation of the tongue and to NTS sweet responding neurons is convincing. Instead, the change of the cross-neuronal pattern to salt in Na-depletion and to saccharin made aversive is intriguing and misleading. It is not at all evidence that NaCl becoming "good" in Na-depletion becomes sweet and that saccharin becoming bad becomes bitter. It is only strong evidence indicating that the electrophysiological exploration of single units in the NTS cannot separate the cross-neuronal pattern responsible for the quality coding from changes of the NTS responses to peripheral stimulation associated to hunger and satiety and to the palatability level of stimuli. As mentioned earlier, these changes are likely governed from hypothalamic and limbic sites through centrifugal pathways.

Results comparable to those recorded in rats in the NTS, the primary projection of the gustatory system, are found in the olfactory bulb, the primary projection of the olfactory system. The multi-unit activity in the mitral cell layer is elevated by food deprivation and reduced by an induced satiety, but only as a response to food odor (Fig. 5.1). In the rat, hunger and satiety do not change responses to a neutral odor that is unrelated to food. Like in the NTS, centrifugal pathways to the olfactory bulb seem to be responsible for this effect (35, 36).

In rats, 63% of explored neurons in the LH responded to three olfactory

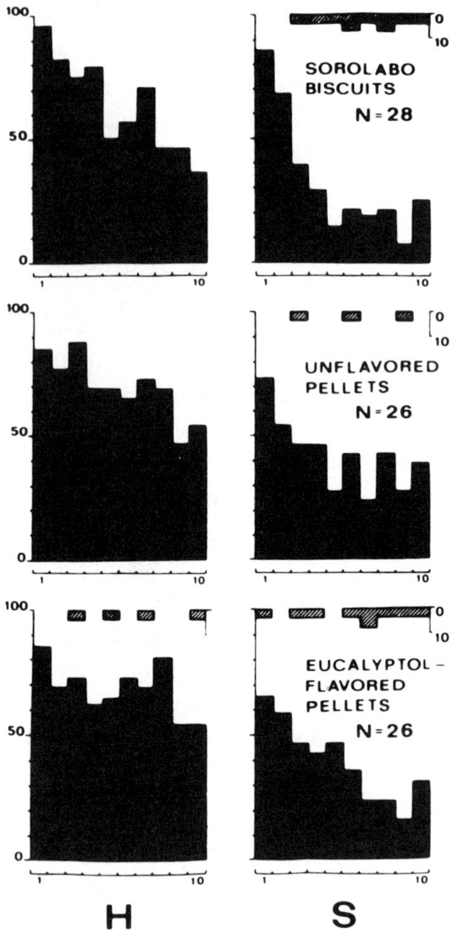

Figure 5.1 Hunger-dependent centrifugal modulation of electrophysiological olfactory bulb responses to food odors in hungry (H) and satiated (S) rats. Abscissa: rank order of the stimulus; ordinate: percent positive and negative response. [From Pager (11).]

stimuli by three odorants. These neurons responded to one or two odorants, not to sound or light. Their firing rate was augmented by food deprivation and was reduced as a function of the degree of satiation. A fascinating result was that the odor responsiveness was more frequent in neurons otherwise identified as glucosensitive neurons than in nonglucosensitive units and that the firing rate of the glucosensitive sites was more elevated by food deprivation than that of the nonglucosensitive sites (37).

In a series of experiments, monkeys showed hunger- and palatability-dependent responses in the LH. In LH single units and also in substantia innominata, the sight or taste of an object known and accepted as food by the monkey activates neurons in the hungry animal only. The response progressively disappears during the ingestion of a glucose solution until the rejection of this solution by the satiated monkey. This modulation by hunger and satiety is absent in the globus pallidus involved in the motor response and is also absent in the NTS, the insular gustatory and the occipital visual cortex (38) (Fig. 5.2). The neuronal response in the LH and substantia innominata appears after a latency of 200 msec, and the ingestive response after 411 msec, suggesting that the sensory stimulation indeed initiates the motor output (16). The sight of a glucose-containing syringe, made aversive by its repeated presentation with a mixture of glucose and hypertonic saline, no longer activates LH neurons in the hungry monkey (39, 40). In the orbitofrontal cortex (secondary taste projection), responses are hunger- and palatability-dependent, contrary to those recorded in the NTS primary projection. In the caudolateral portion of the orbitofrontal cortex, 47% of explored neurons respond to taste, 16% to olfactory, and 10% to visual stimuli. Some of them respond in a multimodal manner. All these sensory responses are

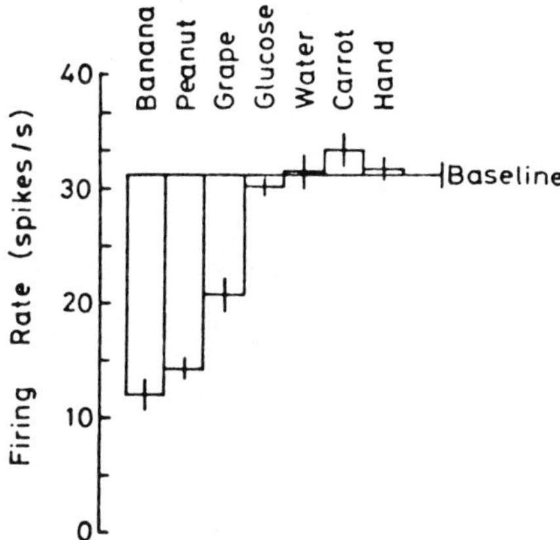

Figure 5.2 Hunger-dependent activation of single units in the lateral hypothalamus of the monkey by the sight of its familiar food. s, sec. [From Burton *et al.* (38).]

modulated by hunger (41, 42). The same discriminative response of amygdaloid neurons to foods and no foods and to palatability levels is found (43, 44).

E. The Learning of Palatability

Contrasting with the abundant works dealing with brain mechanisms of conditioned taste aversion (Section II of this chapter), few works are devoted to the learning of palatability, i.e., of conditioned taste preference.

In monkeys, the firing rate of LH neurons manifests the learning. These neurons, which respond only to a food stimulus in a hungry animal, progressively respond to an object when it is paired with the ingestion of a food. A black bottle containing glucose solution and a white one containing saline were offered to a monkey. After a repeated and alternate tasting of the two solutions, the recorded neuron was activated by the presentation of the black bottle to the hungry monkey. A false peanut visually identical to a true one initially activated the neuron. After tasting the false and the true peanuts, the latter only activated the neuron (45) (Fig. 5.3).

In rats, the firing rate of a single unit in the tegmentum was recorded. Rats were trained to press a lever. The delivery of food by pressing was associated to a sound and no delivery to another sound. Progressively, the recorded neuron

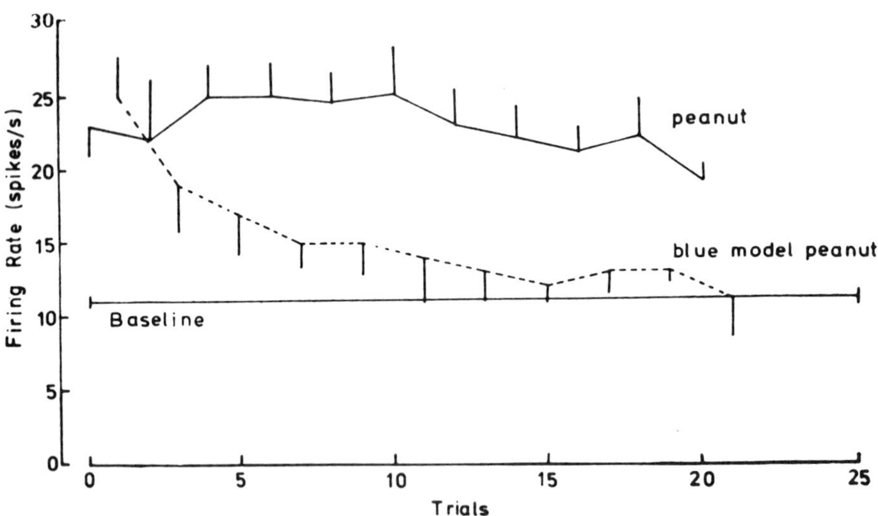

Figure 5.3 The hunger-and palatability-dependent responsiveness of hypothalamic neurons as learned in the monkey. s, sec. [From Mora *et al.* (45).]

responded to the sound associated to food delivery and not to the other one, and only when the animal was hungry (46). Also in rats, the modulation of olfactory bulb response by hunger and satiety to food-related odors may be learned. An odorant added to the food of rats since weaning enhances the mitral cell responses in the hungry adult rat. The same odorant is inefficient in controls (47).

F. Comments

These fascinating experiments did not prove that the explored neurons learn. The modulation of its activity may be due to excitatory or inhibitory afferent pathways to the neuron and/or to the local release of facilitating or inhibiting neurotransmitters.

On the other hand, the preceding results provide no answer to the fundamental question of the brain mechanisms of learning. This problem is obscured by the equivocal use of two terms: reward and/or reinforcement. The term and notion of food reward (as already mentioned) stem from specialists on instrumental learning. In operant learning, the palatability of the delivered food is the food reward. It is also called a secondary reinforcer, inasmuch as it may act as a reinforcer of a second-order conditioning. However, to call food reward the palatability of food and its role in driving the ingestive response is equivocal. The electrical stimulation of the LH probably brings about the activity of the neuronal network, which normally is activated from the same site by the combination of sensory projection with blood-borne signals. To ask whether this stimulation produces hunger or the reward, or both, makes no sense. However, rats learn to self-stimulate the LH, and it is tempting to say that it does so because it is rewarded. But, in fact, the palatability *per se,* associated in humans to the feeling of pleasure, inasmuch as this palatability effect is reproduced by the LH electrical stimulation, is "the reward." In other words, to say that a food is palatable and rewarding is a pleonasm. The fact that satiety induction abolishes both the initial palatability of a food and the LH self-stimulation proves this.

The main and unsolved question is the nature and site of brain targets responding to the postingestive reinforcer, which may be considered as the true reward. How do hunger relief and postingestive satiety act to make the associated sensory stimuli palatable, i.e., conditioned stimuli, to eat? As yet, this is totally unknown. The well-received notion of "brain rewarding systems" apparently is a somewhat rough approach to the problem. It is derived from the study of intracranial self-stimulation in various parts of the brain. When the site is known to be the key locus for the stimulation of a natural behavior, as is the case in LH, it is not reasonable to say that this site belongs to a general brain rewarding system.

Rather, it is the site of the stimulation of that behavior, including the responsiveness to specific sensory stimuli, here the palatability of foods. The pleasantness–unpleasantness dimension, otherwise called reward, or hedonic value, intrinsically depends on the specific neuronal network subserving each specific behavior. It is also the case for the learning of these responses. A unique rewarding system and one learning mechanism separated from one other and both polyvalent in subserving various behaviors are unlikely. Instead, various specific rewarding systems and learning mechanisms are integrated in and not separable from specific neurobehavioral networks. It must not be forgotten that the physiological and homeostatic learnings are always learnings of behaviors.

II. *Brain Mechanisms of Spontaneous and Learned Aversions*

Studies on the effects of lesions and on neurochemical manipulations are the only basis for our knowledge, still fragile, of brain mechanisms on unlearned and learned aversions and their learning.

A. Spontaneous Aversion

Two different effects of a lesion altering taste aversion are difficult to dissociate. The lesion can prevent the expression of aversions either unlearned or learned. In this case, the result of the lesion may be misinterpreted as an alteration of the learning process. On the contrary, some lesions sparing the manifestation of spontaneous or previously learned aversion, but preventing the acquisition of new aversions, will be clearly shown to impair the learning process.

Few works are available on one or several related structures whose lesion abolishes spontaneous or unlearned aversions such as the quinine aversion. Contradictory results on the role of septal lesions are reported. According to Rolls and Rolls (48), the septal lesion reduces neophobia, which is a spontaneous aversion to novel foods. According to Dovonick *et al.* (49), the same lesion increases the quinine aversion. On the contrary, lesion of the mesencephalic interpedoncular nuclei attenuates this aversion.

B. Learned Aversion

Convergent results demonstrate that an amygdala lesion and disconnection between the amygdala and the hypothalamus prevent the acquisition of taste aversion (50–53). However, most of these works show that the prominent effect of the lesion is a disappearance of neophobia. Hence, it may be thought that the amygdala lesion does not alter the learning process but, rather, the expression of aversions, for two reasons. Neophobia is a spontaneous aversion, and its alteration suggests that the expression other aversions, although acquired, will be prevented. The other reason is based on the common assertion that neophobia is a prerequisite of the conditioning of taste aversion. Repeated presentations of saccharin prior to the pairing with LiCl poisoning attenuate or prevent the conditioning; however, the opposite occurs after a LH lesion. In lesioned rats hardly recovered from their aphagia, aversive taste stimuli are rejected. The formation of a new conditioned taste aversion is impossible. But the same rats manifest aversions learned prior to the lesion (54, 55).

Area postrema lesions impair the learning of taste aversions (50); however, as after the LH lesion, it is an impairment of some factors required for the associative learning. In one particular experiment, rats learned a taste aversion and then half of them received the area postrema lesion. The previously learned aversion was identically manifested by lesioned and intact rats (56). The lesion *per se* seems to act as a reinforcer [unconditioned stimulus (UCS)] of aversion and may explain the hypophagia following the lesion, which would be a sensory aversion to the offered food (57). After an area postrema lesion, a conditioned taste aversion by LiCl, scopolamine, and other poisons is no longer obtained. By contrast, the reduction of food intake by amphetamine is not affected, indicating that this effect of amphetamine is not the result of a learned aversion (58).

Child *et al.* (59) reported results supporting the notion that vagal afferents in NTS and area postrema (involved in nausea and vomiting) and their links with gustatory projections play a major role in the learning of taste aversion. However, the effect of a subdiaphragmatic vagotomy varies with various UCS. The conditioned taste aversion is not prevented by LiCl but is by cholecystokinin (otherwise interpreted as a satiety effect) (60).

A cortical spreading depression performed between the conditioned stimulus (CS) presentation and the UCS prevents the conditioned taste aversion, but the cortical depression performed 5 min after the delivery of the conditioning agent does not prevent the learning. Thus, intact thalamoneocortical functions would be necessary for the bridge between the perceived sensory stimulus and the reinforcer, but not necessary in the learning effect of the latter (61). However, a different result was reported by Buresova and Bures (62), who showed that spreading depression prevents the conditioned taste aversion when performed before the CS only. Following ablation of the neocortex, the conditioned taste aversion is altered, but the spontaneous so-called preferences–aversions for NaCl remain unchanged (63). In such decorticated rats, the taste aversion is acquired more slowly. It is generalized to stimuli other than those represented by the CS.

Thus, the neocortex ablation does not prevent the conditioned taste aversion but disrupts its sensory specificity (64, 65). Following a lesion restricted to the gustatory insular cortex, a conditioned taste aversion paradigm using an odorized sweetened water as a CS induces the aversion against the odor but not against sweetness. Following olfactory bulbectomy, it is the opposite (65). A small localized lesion dissociates quinine and acid aversions. Other researchers (66) point out the importance of intact connections between the gustatory cortex and amygdala and of connections between the latter and the hypothalamus for the achievement of the conditioned taste aversion.

Like external and internal aggressions, sensory aversions, even to a bitter solution, mobilize the pituitary–adrenal axis. It was suggested that the impairment of conditioned taste aversion after amygdala lesion includes a deficit of the pituitary–adrenal function. ACTH given to rats after the lesion improves the manifestation of the learned aversion (67). A flavor made aversive by a pairing with LiCl activates the pituitary–adrenal axis. A pre-exposure to the flavor attenuates both the acquisition of aversion and the hormonal response (68).

Neurochemical approaches gave as yet inconclusive results. A 6-OH-DOPA pretreatment producing a 65% depletion of brain norepinephrine does not prevent the conditioned taste aversion (69). Norepinephrine injected into the basolateral amygdala increases neophobia, which is suppressed by the lesion. A local depletion of norepinephrine does not change the response to a novel food but blocks the conditioned taste aversion. Serotonin and a depletion of serotonin are without any effect (70).

The discovery of brain opiates opened a new avenue for research. Responses to pain, illness, aggression, and aversive stimuli are a continuum. All are exaggerated by opiate antagonists. This fact only could lead to answer the basic question: How do pain and illness induce sensory aversion?

III. *Brain Opiates and Palatability*

A. Basic Findings

Five years after the discovery by Hughes and Kosterlitz of the brain opiate system, it was demonstrated that an opiate antagonist, naloxone, strongly modified or eliminated palatability responses. Thus, it was suggested that a brain opiate release and opioergic system were the basis for palatability, also called food reward (71).

Rats were habituated to a 30-min morning presentation of water or a saccharin solution on alternate days. They manifested their preference for the

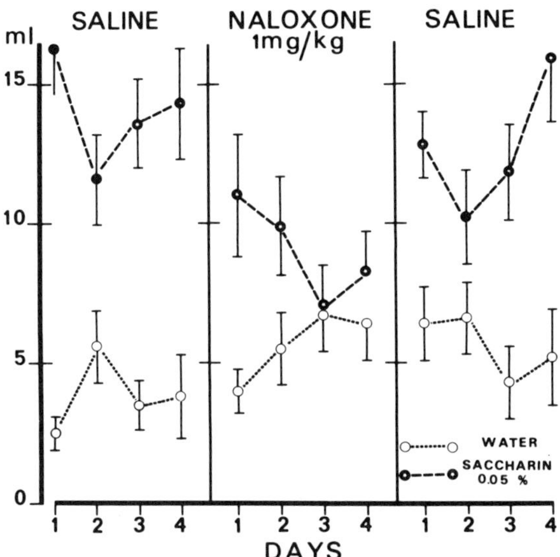

Figure 5.4 The sweet saccharin preference disappears
under the effect of a naloxone injection.

sweet solution during the last 6 days, taken as base line. Then, during 6 consecutive days, the alternate presentations continued, and rats were injected by 1
mg/kg naloxone 30 min prior to their intake of saccharin or water. Their intake of
the saccharin solution dropped sharply, that of water slightly. The difference of
intake of the two fluids was no longer significant; in other words, rats drank the
sweet solution as they drank water (Fig. 5.4). An identical abolition of the
palatability of a glucose instead of a saccharin solution was recorded. Unexpect-
edly, naloxone, tested in the same condition, exaggerated aversions. The intake
of a slightly aversive quinine solution was reduced to zero under naloxone.
Glucose made aversive by the classical conditioned taste aversion paradigm was
more aversive under naloxone injected during the extinction phase.

Based on this result, it was suggested that an opiate release was the neurochemical basis for the palatability of foods and was also involved in some
manner in food aversion. In addition, taking into account the known fact that
stress and nociception induce opiate releases and analgesia, it was proposed to
consider sensory aversion as one of these nociceptive responses. As such, it was
suggested that they were attenuated by opiate release and, therefore, exaggerated
by its blockade.

These basic data (71) were rapidly confirmed by various experimenters (72–
77). The preference curve for saccharin solutions as a function of concentration

was shown shifted to the right under naloxone (78). Contrary to the interpretation of the authors, it was not an inhibition of the aversion to high concentration of saccharin solutions by naloxone; rather, it was evidence for the required higher sweetness to obtain the optimal preference.

In a choice of saccharin versus water, the rapid appearance of the saccharin preference was blocked by naloxone, and the dose required to block the acquisition of preference increased with the saccharin concentration (79). The 1-hr intake of a sucrose solution was reduced by half following 1.25 mg/kg naloxone. This attenuation of the preference for sucrose in sham-drinking also varied with the sucrose concentration (80, 81). This naloxone antagonism of sweet preferences is not a conditioned taste aversion (82). The exaggeration of aversions was confirmed by Cooper (72).

Although nonopioid-dependent analgesia was demonstrated, analgesia tested by hot-plate or tail-flick tests indicated opioid release. It was ascertained when the effect was found naloxone-reversible. Direct and indirect evidence was then provided that the effect of the opiate antagonist on taste responsiveness was due, as suggested earlier, to a β-endorphin release elicited by both the sensory stimulation of feeding and the food nociception. In 10-day-old rats, the oral infusion of a sucrose solution increased by two times the latency of pain responses (hot-plate test). This effect was naloxone-reversible. The oral sucrose solution also reduced the vocalization of young isolated from the dam. This effect was also naloxone-reversible (83, 84).

A fascinating experiment was done by Dum et al. (85) (Fig. 5.5). Rats were habituated to a morning presentation of a highly palatable solution (chocolate milk), which they drank voraciously. Twenty minutes after the consumption, labeled ethorphin (β-endorphin analogue) was injected. Compared to controls, the binding of ethorphin was considerably reduced in the hypothalamus. The authors interpreted this finding as an effect of the occupation of binding sites by an in situ-released β-endorphin stimulated by the palatable food. In the same habituated rats, the expectancy of the palatable solution induced analgesia. It was naloxone-reversible. This effect was reduced in morphine-tolerant rats.

In later chapters devoted to experimental models of obesity, hyperphagia and obesity induced in rats by a cafeteria regimen will be described. Rats offered a multiple and permanent choice of highly palatable foods became hyperphagic and, after some weeks, obese. This occurred only if the palatable foods were changed everyday. These cafeteria-induced hyperphagia and obesity were prevented by a long-acting opiate antagonist (85, 86, 87).

Another model of obesity described later will be the so-called "tail-pinch" obesity of rats. Rats submitted to sessions of slight stress by tail-pinching ate

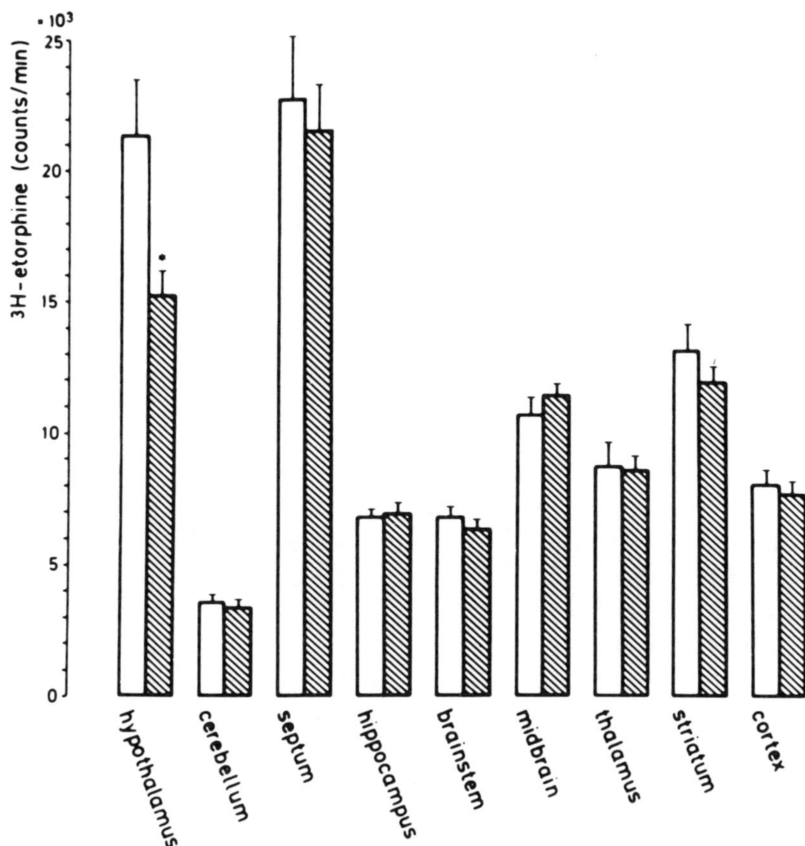

Figure 5.5 Radioactivity in different brain parts of rats 20 min after injection with ³H-etorphine (2μCi/rat). All rats were nondeprived and were given either chocolate milk (striped columns) or water (open columns) to drink freely for 20 min before sacrifice (eight rats/group). Mean ± SEM. *Statistically significant difference from controls ($p < 0.05$). [From Dum *et al.* (85)]

voraciously during and after the session. They gained weight by a repetition of the tail-pinching. They overate palatable food or sucrose solutions only (88, 89). The tail-pinch, like other stresses, induced analgesia (90). Thus, it was thought that tail-pinching induced hyperphagia because an opiate release produced overeating and its selectivity toward palatable foods. Morley and Levine (91) and Levine *et al.* (92) confirmed this by showing that the tail-pinch-induced eating was blocked by naloxone.

Like repeated injection of morphine, repeated stimulation of endogenous opiate releases induces morphine tolerance and dependence. This was demonstrated for the first time by Christie and Chesher (93), who showed that repeated stresses of the mouse induced a morphinelike withdrawal syndrome precipitated by naloxone. Is this tolerance-dependence to morphine exhibited by cafeteria and chronically tail-pinch-treated rats? The answer is yes. After 10 days of tail-pinching, a naloxone injection precipitated the morphine-withdrawal syndrome in rats (91). After 2 wk of a cafeteria regimen, and contrary to controls maintained on the stock diet, naloxone injection provoked tremor and head-shaking (94). Following 15 days of drinking a saccharin solution as the only source of fluids, rats exhibited a morphine tolerance (reduced analgesia) (95, 96).

B. The Effects of Exogenous Morphine

Together, it was ascertained that the manifestation of food palatability was associated with and presumably a cause of an endogenous β-endorphin release. This could be confirmed by the effect of morphine administrations on intakes or by endogenous opioid released by stressors other than tail-pinching.

It was generally reported that small to moderate doses of peripherally injected morphine increased food intake. Some works suggest a relation to food palatability. In a choice among the three macronutrients, the highest augmentation of intake after some hours was observed on the most palatable fat diet (97). Consistent with this result is the finding that a stress known to induce analgesia augments the intake of a high-fat diet far more than that of carbohydrate- or protein-rich diets. Naloxone reduces this differential response (98). Morphine and other selective agonists of brain opiate receptors were injected in various parts of the brain. Agonists of μ-, δ-, and κ-receptors were compared. Only the former augmented intake of a saccharin solution, and this augmentation was palatability-dependent (99). Morphine, locally injected in periventricular nuclei, the fornix, or the VMH, augments food intake (100).

The ventral tegmentum is widely acknowledged as a site for the rewarding effect of morphine. It is the only region in which rats self-inject morphine (101). The acquisition of a behavioral response rewarded by self-stimulation of ventral tegmentum is blocked by naloxone. However, the acquisition of the same instrumental response by food delivery is not blocked by the opiate antagonist. This could suggest that opioids intrinsically linked to the manifestation of acquired palatability would not be involved in the primary reinforcement of this palatability by food ingestion.

C. Circulating β-Endorphin

The blood β-endorphin could be reasonably thought to be related to food intake and specifically to the palatability of foods. This was supported by various experimenters claiming a biphasic increase of circulating β-endorphin during and after a meal, in humans and in rats (102–104). However, no correlation at all was found by others between variations of blood endorphin and food intake (105, 106). The opposite result was presumably due to an elevation of circulating β-endorphin induced by the stress of handling and of the blood-taking, mistaken as an effect of food intake. The negative result strongly suggests that opiates involved in palatability are neither of pituitary nor of intestinal origin and, rather, are synthesized and act locally in the brain. In what loci? It is probably the LH critically involved in the sensory stimulation of feeding. Morphine locally injected in the LH indeed increases the food reward. Naloxone inhibits feeding induced by LH stimulation and, furthermore, blocks the enhancement of self-stimulation by food deprivation (107–110).

D. Opiates and Palatability in Humans

Studies in humans were of importance. It is commonly assumed that hyper-stimulation of feeding by sensory stimuli is present in the obese. It is viewed either as a cause of the initial hyperphagia establishing obesity or as an effect of the achieved obesity. Later (Chapter 10), we will see that this palatability-induced hyperphagia in the obese might be considered as a food addiction, just as rats in a cafeteria regimen are actually food addicts. The palatability-reducing effect of the opiate antagonist was clearly demonstrated by Fantino *et al.* (111); 60 mg naloxone reduced the declared pleasantness of only taste and olfactory food stimuli. The effect on the scored palatability was stronger than the same effect induced by a 100-g glucose load in the stomach. Naloxone combined with the stomach load potentialized its effect. In another study, normal-weight subjects scored their preferences for creams containing various proportions of fat and sucrose. Naloxone reduced the palatability rating and lowered the intake of the fattest cream by 50%.

E. Putative Role of Other Neurotransmitter Systems in Palatability

The DOPA antagonist pimozide was shown to abolish the operant response of rats rewarded by food delivery. While this reward was postulated as an effect of food palatability, it was suggested that the dopaminergic system was involved in a unique brain rewarding system implicated in food reward among others. A dopaminergic network active in an inhibition or a facilitation of the ventral tegmentum was hypothesized as a substrate of this system. Its blockade by pimozide would induce a state called, by the author, anhedonia (112, 113).

This hypothesis was sufficient to stimulate a flow of literature on the matter (114–116). As usual in this field of neurotransmitter studies, contradictory and challenging results were presented. Arguments, apparently as convincing as those presented in favor of the DOPA theory, were proposed to assume that an adrenergic and not a dopaminergic system was concerned (117). In other studies, it was claimed that the effects of pimozide disturbed the performance of bar-pressing and not the food reward. An elaborated study by Wise *et al.* (118) attempted to prove the opposite. Of course, partisans of a role for serotoninergic involvement finally entered the debate (119, 120). While waiting for a peace between partisans and opponents, it is reasonable to take no part in the conflict.

References

1. Price, J. (1985). Beyond the primary olfactory cortex: Olfactory related area in the neocortex, thalamus and hypothalamus. *Chem. Senses* **10**, 239–248.
2. Scott, J. W., and Chafin, B. R. (1975). Origin of olfactory projections to lateral hypothalamus and nuclei gemini in the rat. *Brain Res.* **88**, 64–68.
3. Scott, J. W., and Leonard, C. M. (1971). The olfactory connections of the lateral hypothalamus in the rat, mouse and hamster. *J. Comp. Neurol.* **141**, 331–344.
4. Van der Kooy, D., Koda, L. Y., McGinty, J. F., Gerfen, C. R., and Bloom, F. E. (1984). The organization of projections from the cortex amygdala and hypothalamus to the NTS in rat. *J. Comp. Neurol.* **224**, 1–24.
5. Garcia-Diaz, D. E., Jimenez-Montufar, L. L., and Guevara-Aguilar, R. (1988). Olfactory and visceral projections to the NTS. *Physiol. Behav.* **44**, 619–624.
6. Norgren, R. (1981). The central organization of the gustatory and visceral afferent system in the NTS. In Y. Katsuki, R. Norgren and M. Sato (Eds.) Brain Mechanisms of Sensation, New York: Wiley & Sons, pp. 143–160.
7. Norgren, R. (1974). Gustatory afferents to ventral forebrain. *Brain Res.* **81**, 285–295.
8. Hamilton, R., and Norgren, T. (1984). Central projection of gustatory nerve in the rat. *J. Comp. Neurol.* **222**, 560–577.
9. Aguilar-Baguroni, H. (1976). Hypothalamic influences of the electrical activity of the olfactory pathways. *Brain Res. Bull.* **1**, 263–273.
10. Price, J. L., and Powell, T. P. (1970). An experimental study of the origin and course of the centrifugal fibers to the olfactory bulb. *J. Anat.* **107**, 215–234.

11. Pager, J. (1977). Ascending olfactory informations: Centrifugal influxes contributing to a nutritional modulation of mitral cell responses. *Br. Res.* **130**, 351–359.

12. Rogers, T., and Pelton, E. (1984). Neurons of the vagal division of the solitary nuclei activated by the PVN of the hypothalamus. *J. Autonom. Nerv. Syst.* **10**, 193–197.

13. Scott, J. W., and Pfaffmann, C. (1972). Characteristics of responses of lateral hypothalamic neurons to stimulation of the olfactory system. *Brain Res.* **48**, 251–264.

14. Oomura, Y., Ooyama, H., Yamamoto, T., and Naka, F. (1967). Reciprocal relationship of the lateral and ventromedial hypothalamus in the regulation of food intake. *Physiol. Behav.* **2**, 97–115.

15. Karadi, Z., Oomura, Y., Nishino, H., and Aou, S. (1989). Olfactory coding in the monkey lateral hypothalamus: Behavioral and neurochemical properties of odor-responding neurons. *Physiol. Behav.* **45**, 1249–1257.

16. Rolls, E. T., Sanghera, M. K., and Roper-Hall, A. (1979). The latency of the activation of neurons in the lateral hypothalamus and substantia innominata by feeding in the monkey. *Brain Res.* **164**, 121–135.

17. Norgren, R. (1970). Gustatory response in the hypothalamus. *Brain Res.* **21**, 63–77.

18. Schwartzbaum, J. S. (1988). Electrophysiology of feeding reward in the lateral hypothalamus of the rabbit. *Physiol. Behav.* **47**, 502–526.

19. Drefus, J. H., Murphy, J., and Gloor, P. (1974). Contrasting effects of two identified amygdaloid efferent pathways on single hypothalamic neurons. *J. Neurophysiol.* **33**, 237–247.

20. Rolls, E. (1972). Activation of amygdaloid neurons in rewarded eating and drinking elicited by electrical stimulation of the brain. *Brain Res.* **45**, 365–382.

21. Rolls, E. T., and Rolls, B. J. (1973). Altered food preferences after lesions in the basolateral region of the amygdala in the rat. *J. Comp. Physiol. Psychol.* **83**, 248–259.

22. Fitzgerald, R., and Burton, M. (1981). Effect of small basolateral amygdala lesion on ingestion in the rat. *Physiol. Behav.* **27**, 433–437.

23. Larue, C., and Le Magnen, J. (1973). Effets de l'interruption des voies olfacto-hypothalamiques sur la séquence alimentaire du rat. *J. Physiol. (Paris)* **66**, 699–713.

24. Larue-Achagiotis, C., and Le Magnen, J. (1974). Effets de l'association de la bulbectomie et de la lésion des noyaux ventro-médians de l'hypothalamus sur les résponses alimentaires du rat blanc. *J. Physiol. (Paris)* **68**, 81–95.

25. Black, R. M., and Weingarten, H. P. (1988). Comparison of taste receptivity changes induced by VMH lesion and stria terminalis transections. *Physiol. Behav.* **44**, 699–708.

26. Lenard, L., Hahn, Z., and Karadi, Z. (1982). Body weight changes after

chemical manipulations of rat amygdala: Noradrenergic and dopaminergic mechanisms. *Brain Res.* **249**, 95–102.

27. Pfaffmann, C., and Bare, J. K. (1950). Gustatory nerve discharges in normal and adrenalectomized rats. *J. Comp. Physiol. Psychol.* **43**, 320–324.

28. Pfaffmann, C. P., and Hagstrom, E. C. (1955). Factors influencing taste sensitivity to sugars. *Am. J. Physiol.* **183**, 651.

29. Bereiter, D., Berthoud, H., and Jeanrenaud, B. (1981). Hypothalamic input to brainstem neuron responses to oro-pharyngeal stimulations. *Exp. Br. Res.* **39**, 32–35.

30. Matsuo, Y., Shimazu, Y., and Busano, K. (1984). Lateral hypothalamus modulation of oral sensory afferent activity in the NTS. *J. Neurosci.* **4**, 1202.

31. Murzi, E. (1986). Lateral hypothalamic sites eliciting eating affect medullary taste neurons in rats. *Physiol. Behav.* **36**, 829–834.

32. Chang, F. C., and Scott, T. R. (1984). Conditioned taste aversions modify neural responses in the rat nucleus tractus solitarius. *J. Neurosci.* **4**, 1850–1862.

33. Giza, B., and Scott, T. (1987). Intraveinous glucose loads decrease gustatory evoked responses to sugars. *Am. J. Physiol.* **252**, R294–306.

34. Giza, B., and Scott, T. R. (1983). Blood glucose selectively affects taste evoked activity in the rat nucleus tractus solitarius. *Physiol. Behav.* **31**, 643–650.

35. Giachetti, I., MacLéod, P., and Le Magnen, J. (1972). Contrôle centrifuge des afférences olfactives en fonction des états de faim et de satiété chez le rat. *C. R. Soc. Biol. (Paris)* **164**, 841–846.

36. Pager, J., Giachetti, I., Holley, A., and Le Magnen, J. (1972). A selective control of olfactory bulb electrical activity in relation to food deprivation and satiety in rats. *Physiol. Behav.* **9**, 573–580.

37. Shiraishi, T. (1988). Feeding related LH neuron responses to odors depend on food deprivation in rats. *Physiol. Behav.* **44**, 591–597.

38. Burton, M. J., Rolls, E. T., and Mora, F. (1976). Effects of hunger on the responses of neurones in the hypothalamus to the sight and taste of food. *Exp. Neurol.* **53**, 508–519.

39. Rolls, E. T., Roper-Hall, A., and Sanghera, M. K. (1977). Activity of neurones in the substantia innominata and LH during the initiation of feeding in the monkey. *J. Physiol. (London)* **282**, 24P.

40. Rolls, E. T., Burton, M. J., and Mora, F. (1976). Hypothalamic neuronal responses associated with the sight of food. *Brain Res.* **111**, 53–66.

41. Rolls, E. T. (1987). Taste processing in the NTS and in the three taste cortical areas in primates. *Chem. Senses* **12**, 176.

42. Inoue, S., and Oomura, Y. (1985). Reward related neuronal activity in monkey dorsolateral prefrontal cortex during feeding behavior. *Brain Res.* **346**, 307–312.

43. Nishino, K., and Ono, T. (1986). Neural activity in the adjacent to the dorsal amygdala monkey during operant feeding behavior. *Brain Res. Bull.* **17**, 947–954.

44. Ono, T., Tamura, R., Nishijo, H., Nakamura, K., and Tabuchi, E. (1989). Contri-

bution of amygdala and LH neurons to the visual information processing of food and no food in the monkey. *Physiol Behav.* **45,** 411–421.

45. Mora, F., Rolls, E. T., and Burton, M. J. (1976). Modulation during learning of the responses of neurons in the lateral hypothalamus to the sight of food. *Exp. Neurol.* **53,** 508–519.

46. Phillips, P., and Olds, J. (1969). Unit activity motivation dependent response from midbrain neurons. *Science* **195,** 1279.

47. Pager, J. (1974). A selective modulation of the olfactory bulb electrical activity in relation to the learning of palatability in hungry and satiated rats. *Physiol. Behav.* **12,** 189–196.

48. Rolls, B. J., and Rolls, E. T. (1973). Effects of lesion in the basolateral amygdala on fluid intake in the rat. *J. Comp. Physiol. Psychol.* **83,** 240–247.

49. Donovick, P. J., Burright, R. G., Kaplan, J., and Rosentreich, L. (1969). Habenular lesion: Water consumption and palatability of fluids in the rat. *Physiol. Behav.* **4,** 45–47.

50. Ashe, F., and Nachman, M. (1980). Neural mechanisms in taste aversion learning. In J. Sprague and A. Epstein, (eds.), *Progress in psychobiology and physiological psychology,* ed. Vol. 9 (pp. 233–262). New York: Academic Press.

51. Nachman, M., and Ashe, F. (1974). Effects of basolateral amygdala lesion on neophobia and learned taste aversion and sodium appetite in rats. *J. Comp. Physiol. Psychol.* **86,** 622–643.

52. Kemble, E., and Pagel, G. (1973). Failure to form a learned taste aversion in rats with amygdala lesion. *Bull. Psychol. Soc.* **20,** 155–159.

53. Kolakowska, L., Larue-Achagiotis, C., and Le Magnen, J. (1984). Effets comparés de la lésion du noyau baso-latéral et du noyau latéral de l'amygdale sur la néophobie et l'aversion gustative conditionnée chez le rat. *Physiol. Behav.* **32,** 647–651.

54. Roth, S. R., Schwartz, M., and Teitelbaum, P. (1973). Failure of recovered LH rats to learn specific food aversions. *J.Comp. Physiol. Psychol.* **83,** 184–197.

55. Schwartz, M., and Teitelbaum, P. (1974). Dissociation between learning and remembering in rats after lesion in the LH. *J. Comp. Physiol. Psychol.* **86,** 97–108.

56. Robin, B. (1984). Recall of previously acquired taste aversion in the rat following lesion of area postrema. *Physiol. Behav.* **34,** 503–506.

57. Tomoyotsu, N., and Kenney, N. (1989). Response to palatability after area postrema lesions: Results of learned aversions. *Am. J. Physiol.* **257,** R1075–1082.

58. Berger, B. D., Wise, C. D., and Stein, L. (1973). Area postrema damage and bait shyness. *J. Comp. Physiol. Psychol.* **82,** 475–479.

59. Child, G., Novin, D. Rodgers, J., and Garcia, J. (1978). Conditioned taste aversion: Vagal and circulatory mediation of the toxic UC stimulus. *Behav. Biol.* **24,** 509–519.

60. Martin, J. R., Cheng, F. Y., and Novin, D. (1978). Acquisition of learned taste aversion following bilateral subdiaphragmatic vagotomy in rats. *Physiol. Behav.* **21,** 13–17.

61. Davis, J. L., and Bures, J. (1972). Disruption of saccharin aversion learning in rats by cortical spreading depression in the CS-US interval. *J. Comp. Physiol. Psychol.* **80,** 398–402.

62. Buresova, O., and Bures, J. (1973). Cortical and subcortical components of the conditioned saccharin aversion. *Physiol. Behav.* **11,** 435–439.

63. Kiefer, S., Carah, O., and Garcia, J. (1984). Neonatal ablation of gustatory neocortex in rat on taste aversion learning and taste reactivity. *Behav. Neurosci.* **98,** 812.

64. Braun, J. J., Slick, T. B., and Lorden, J. F. (1972). Involvement of gustatory neocortex in the learning of taste aversions. *Physiol. Behav.* **9,** 637–641.

65. Kiefer, S., Leach, L., and Bray, G. (1984). Taste amnesia following neocortex ablation: Dissociation from odor and generalisation across taste qualities. *Behav. Neurosci.* **98,** 599–608.

66. Yamamoto, T., Asuma, S., and Kawamamura, Y. (1984). Functional relations between the cortical gustatory area and the amygdala: Electrophysiological and behavioral study in rats. *Exp. Brain Res.* **56,** 23–31.

67. Burt, G. S., and Smotherman, W. P. (1980). Amygdalectomy induced deficit in conditioned taste aversion: Possible pituitary adrenal involvement. *Physiol. Behav.* **24,** 651–655.

68. Smotherman, W. P., Mariolis, A., and Levine, S. (1980). Flavor preexposure in CTA situation: Dissociation of behavioral and endocrine effects in the rat. *J. Comp. Physiol. Psychol.* **94,** 25–35.

69. Mason, S. T., Roberts, D. C. S., and Fibiger, H. C. (1978). Noradrenaline and neophobia. *Physiol. Behav.* **21,** 353–361.

70. Brosini, F., and Rolls, E. T. (1984). Role of adrenaline and serotonin in the basolateral region of amygdala in food preference and learned taste aversion in the rat. *Physiol. Behav.* **33,** 36–43.

71. Le Magnen, J., Marfaing-Jallat, P., Miceli, D., and Devos, M. (1980). Pain modulating and reward systems: A single brain mechanism. *Pharmacol. Biochem. Behav.* **12,** 707–709.

72. Cooper, S. J. (1980), Naloxone: Effects on food and water consumption in the non-deprived and deprived rats. *Psychopharmacology* **71,** 1–6.

73. Cooper, S. J., and Turkish, S. (1979). Effect of naloxone on food preference and concurrent behavior responses in food deprived rat. *Pharm. Biochem. Behav.* **33,** 17–20.

74. Cooper, S. J. (1985). Evidence for opiate receptor involvement in the consumption of a high palatability diet in non-deprived rat. *Neuropeptides* **5,** 345–348.

75. Cooper, S. J. (1985). Selective attenuation of sweetened milk consumption by opiate antagonist in male and female of Roman strain of rats. *Neuropeptides* **5,** 379–385.

76. Ostrowski, N. L., Rowland, N., Foley, T. L., Nelson, J. L., and Reid, L. D.

(1981). Morphine antagonists and consummatory behaviors. *Pharm. Biochem. Behav.* **14,** 549–559.

77. Siviy, S., and Reid, L. (1983). Endorphinergic modulation of acceptability of putative reinforcers. *Appetite* **4,** 249–257.

78. Siviy, S. M., Calcagnetti, D. J., and Reid, L. D. (1982). Opioids and palatability. In *Neural basis of feeding reward* B. G. Hovel and D. Novin (pp. 517–524.). Los Angeles: Haer Institute for Electrophysiological research.

79. Lynch, W. C. (1986). Opiate blockade inhibits saccharin uptake and blocks normal preference acquisition. *Pharm. Biochem. Behav.* **24,** 831–836.

80. Kirkham, T. C., and Cooper, S. J. (1988). Naloxone attenuation of sham-feeding is modified by manipulations of sucrose solution concentrations. *Physiol. Behav.* **44,** 491–494.

81. Kirkham, T., and Cooper, S. (1988). Attenuation of sham-feeding by naloxone is stereo-specific: Evidence for opioid mediation of oro-sensory reward. *Physiol. Behav.* **43,** 845–847.

82. Leshem, L. (1984). Suppression of feeding by naloxone in rats: A dose–response comparison of anorexia and CTA suggesting an anorectic-specific effect. *Psychopharmacology* **80,** 127–130.

83. Blass, E. M., and Fitzgerald, E. (1988). Milk-induced analgesia and conforting in 10 day-old rats: Opioid mediation. *Pharm. Biochem. Behav.* **29,** 6–13.

84. Blass, E., Fitzgerald, E., and Kehoe, P. (1987). Interactions between sucrose, pain and isolation distress. *Pharm. Biochem. Behav.* **26,** 483–489.

85. Dum, J., Gramsch, C. H., and Herz, A. (1983). Activation of hypothalamic beta endorphin pools by reward induced by highly palatable food. *Pharm. Biochem. Behav.* **18,** 443–447.

86. Apfelbaum, M., and Mandenoff, A. (1981). Naltrexone suppresses the hyperphagia induced in the rat by highly palatable diets. *Pharm. Biochem. Behav.* **15,** 89–91.

87. Mandenoff, A., Bertière, M. C., Betoulle, D., and Apfelbaum, M. (1984). Difference in the sensitivity of mu and kappa systems in cafeteria rats. *Neuropeptides* **5,** 265–268.

88. Marques, D. M., Fisher, A. E., Okrutny, M. S., and Rowland, N. E. (1979). Tailpinch induced fluid ingestion: Interaction of taste and deprivation. *Physiol. Behav.* **22,** 37–41.

89. Lowy, M. T., Maickel, R. P., and Yim, G. K. W. (1980). Naloxone reduction of stress-related feeding. *Life Sci.* **26,** 2113–2118.

90. Simone, D., and Bodnar, P. (1981). Modulation of anti-nociception by tail-pinch stress. *Life Sci.* **30,** 718–720.

91. Morley, J. E., and Levine, A. S. (1980). Stress-induced eating is mediated through endogenous opiates. *Science* **296,** 1259.

92. Levine, A. S., Wilcox, G. I., Grace, M., and Morley, J. E. (1982). Tail pinch-induced consummatory behaviors associated with analgesia. *Physiol. Behav.* **28,** 959–962.

93. Christie, M. J., and Chesher, G. B. (1982). Physical dependence on physiologically released endogenous opiates. *Life Sci.* **30,** 1173–1177.

94. Le Magnen, J. (1987). Palatability: Concept, terminology and mechanisms. In R. A. Boakes, D. A. Popplewell, and M. J. Burton (eds.), *Eating habits* (pp. 131–154) John Wiley & Sons, London.

95. Lieblich, I., Cohen, E., Ganchrow, J. R., Blass, E. M., and Bergmann, F. (1983). Morphine tolerance in genetically-selected rats induced by chronically elevated saccharin. *Science* **221,** 871–873.

96. Cohen, E., Lieblich, I., and Bergmann, F. (1985). Effects of chronically elevated intake of different concentrations of saccharin on morphine tolerance in genetically-selected rats. *Physiol. Behav.* **32,** 1041–1043.

97. Marks-Kaufman, V., and Kanarek, R. (1980). Morphine selectively affects macronutrient intake in the rat. *Pharm. Biochem.* **12,** 329–330.

98. Vaswani, K., Tejwani, G. A., and Mousa, S. (1983). Stress-induced differential intake of various diets and water by rats: The role of the opiate system. *Life Sci.* **32,** 1983–1996.

99. Gosnell, B. A., and Majchrzak, M. J. (1989). Centrally-administered opioid peptides stimulate saccharin intake in non-deprived rats. *Pharm. Biochem. Behav.* **33,** 805–811.

100. Shor-Posner, G., Azar, A. P., Filart, R., Tempel, D., and Leibowitz, S. F. (1986). Morphine-stimulated feeding: Analysis of macronutrient selection and PVN lesion. *Pharm. Biochem. Behav.* **24,** 931–940.

101. Britt, M., and Wise, R. (1983). Ventral tegmental sites of opiate reward. *Brain Res.* **258,** 275–280.

102. Getto, C. J., Fullerton, D. T., and Carlson, I. H. (1984). Plasma immunoreactive beta-endorphin response to glucose ingestion in human obesity. *Appetite* **5,** 329–335.

103. Matsumura, M., Fukuda, N., Saito, S., and Mori, H. (1982). Effects of a test meal duodenal acidification and tetragastrin on the plasma concentration of endorphin like immunoreactivity in man. *Regul. Peptides* **4,** 173–181.

104. Davis, J. M., Lowy, M. T., Yim, G. R. W., Lamb, D. R., and Malven, P. V. (1983). Relationships between plasma concentration of immunoreactive beta endorphin and food intake in rats. *Peptides* **4,** 77–83.

105. Wallace, M., Fraser, C. D., Clements, J. A., and Funder, J. W. (1980). Naloxone adrenalectomy and steroid placement: Evidence against the role of circulating endorphin on food intake. *Endocrinology* **108,** 189–192.

106. Melchior, J. C., Fantino, M., Colas-Linhart, N., Rigaud, D., Petiet, A., Laforest, M. D., and Apfelbaum, M. (1991). Lack of plasmatic β-endorphin response to gastronomic meal in healthy humans. *Physiol. Behav.,* **49,** in press.

107. Ono, T., and Oomura, Y. (1980). Morphine and enkephalin effects on hypothalamic glucoresponsive neurons. *Brain Res.* **175,** 207–212.

108. Sikdar, S. K., and Oomura, Y. (1985). Selective inhibition of glucosensitive neurons in rat lateral hypothalamic by noxious stimuli and morphine. *J. Neurophysiol.* **53,** 17–31.

109. Woods, G., and Leibowitz, S. (1985). Hypothalamic sites sensitive to morphine and naloxone: Effects on feeding behavior. *Pharm. Biochem. Behav.* **21,** 433–435.

110. Carr, P., and Simon, E. (1983). The role of opioid in feeding and reward elicited by LH electrical stimulation. *Life Sci.* **33,** 49–66.

111. Fantino, M., Horsotte, J., and Apfelbaum, M. (1986). An opioid antagonist, naltrexone reduces preference for sucrose in humans. *Am. J. Physiol.* **251,** R91–96.

112. Wise, R. A. (1982). Neuroleptics and operant behavior: The anhedonia hypothesis. *Brain Behav. Sci.* **5,** 39–52.

113. Wise, R. A. (1988). Neuroleptic attenuation of self-stimulation: Reward not performance deficit. *Life Sci.* **22,** 535–542.

114. Bailey, K., Hsiao, S., and King, J. (1980). Hedonic reactivity to sucrose in rats: Modification by pimozide. *Physiol. Behav.* **35,** 347–356.

115. Geary, N., and Smith, G. P. (1985). Pimozide decreases the positive reinforcement: Effect of sham feeding sucrose in the rat. *Pharm. Biochem. Behav.* **22,** 787–790.

116. Clifton, J., and Cooper, S. (1989). Stimulation and inhibition of food intake by the selective DOPA D2 agonist: A meal pattern analysis. *Pharm. Biochem. Behav.* **33,** 21–26.

117. Matthews, G., Gibson, H., and Booth, D. A. (1985). Norepinephrine facilitated eating: Reduction of saccharin preference in conditioned behavior preferences with increase of quinine aversion. *Pharm. Biochem. Behav.* **22,** 1045–1052.

118. Wise, R. A., Spindler, J., De With, H., and Gerber, G. J. Neuroleptic-induced anhedonia in rats: Pimozide blocks reward quality of foods. *Science* **201,** 262–264.

119. Harsing, L. G., Jr., Yang, H.-Y. T., and Costa, E. (1982). Accumulation of hypothalamic endorphin after repeated injections of anorectics which release serotonin. *J. Path. Exp. Ther.* **223,** 689–694.

120. Neill, J., and Cooper, S. (1988). Evidence for serotoninergic modulations of sucrose sham-feeding in the gastric fistulated rats. *Physiol. Behav.* **44,** 453–459.

Chapter Six

Determinants of Meal Size

By studying the normal feeding of animals and humans, we have seen the respective roles of the amount eaten in each meal and of their time patterning in determining the cumulative food intake. The intake adjustments to changes of metabolic requirements and to nutritive properties of the food was achieved in the free-feeding condition by the adjustment of postmeal satiety durations to the meal sizes. In addition, meal sizes, from their initiation to their spontaneous end, were also adjusted in various conditions in a regulatory way. This was particularly the case after food deprivation and with a palatable food, which augmented the meal without inertia. In the feeding schedule condition, which prevails in humans, the meal size becomes the only determinant of the cumulative intake and the only potential regulator of the body energy and nutritional balance. The problem of meal intake (the eating *per se*) was, and still is, the subject of numerous works. In these works, a purely behavioral study often overlooks the regulatory aspect, i.e., the mechanism by which an adjusted meal size contributes to the energy balance in combination with the mechanism of its initiation.

I. *Oral Determinants of Meal Size in Rats*

When a rat is stimulated to eat by the combined action of the systemic and sensory stimuli of the behavior, how is this initial stimulation progressively abolished by eating the food? In other words, how is eating a food satiating? The self-evident answer is that to achieve satiation, it is precisely needed to eat, i.e., to take the food in the mouth, to masticate, and to swallow bits of food successively and to fill the gastrointestinal tract by this oral intake. First of all, based on this self-evidence, it is suggested that the activity of the food at successive levels of its trajectory (the mouth, stomach, and intestine) and eventually the postabsorptive compartment could be, separately or in combination, concerned with the satiation process.

A. Time Course of Normal Oral Intake

The recording of rats eating showed that the animal eats at a constant rate from the beginning to the end of its meal in the *ad libitum* condition. On the contrary, the initial rate of eating increases according to the duration of previous food deprivation and the food palatability. Then eating rate declines until the end of the meal. However, in the absence of a possible recording of the detailed chewing–swallowing pattern in rats, such studies did not permit (as was done in human subjects) a detailed analysis of the microstructure and time course of the intake of solid foods. With a liquid food, recording the licking rate was possible (Fig. 6.1). This technique, developed by Davis and his co-workers (1, 2, 3) was used in an impressive series of works to assess the respective role of the initial palatability and of the satiating process in determining the amount eaten.

Rats lick at a constant rate of 6–7 laps/sec in short bursts of licking separated by pauses. Long bursts and short pauses at the start of the meal are the best measure of the initial stimulation. Bursts are prolonged and pauses shortened by food deprivation. In a given state of deprivation, this initial licking rate increases as a function of the palatability of the food and is a measure of this palatability level. From this initial level, bursts of licking shorten and their interval lengthens. This declining rate of licking is the measure of the satiation process and its time course. The slope of this decline and its value for a particular food is a measure of the satiating capacity of that food (1).

Two important notions emerged from these initial works and confirmed earlier suggestions. The meal size results from a balance between the level of the initial stimulation and the negative feedback provided by the food passing the mouth and filling the stomach. The higher the palatability of the food and the previous deprivation at the start of the meal, the stronger the satiating capacity of the food needed to stop the intake.

Davis *et al.* (2) and Davis and Smith (3), on the basis of other data (described below) in which changes in the licking rate were observed after gastric loads or intestinal perfusions, considered that the negative feedback (i.e., the satiating capacity of the food) is an effect of the gastrointestinal filling only; therefore, they surprisingly overlooked a lot of old and modern results that demonstrated an oral satiating action of the food passing the mouth before entering the stomach.

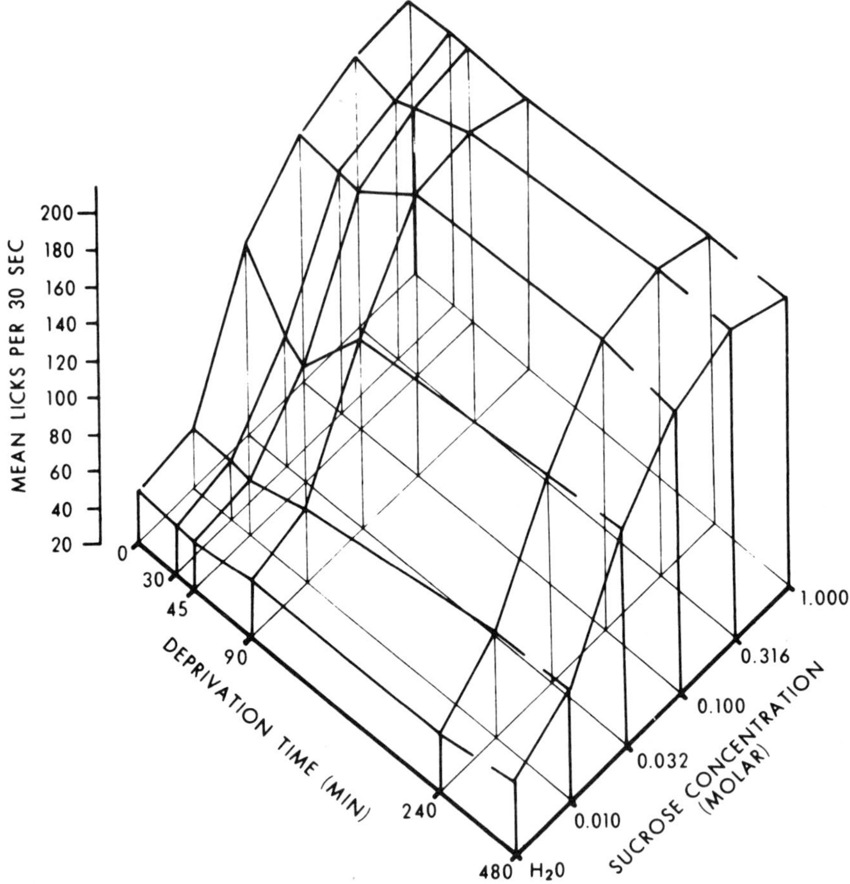

Figure 6.1 Mean licking rate: combined effect of deprivation and palatability. [From Davis and Campbell (1).]

B. The Oral Satiating Capacity of the Food

Miller and Kessen (4) were the first to demonstrate that oral intake was more satiating than food directly put in the stomach. Maze-learning was more rapid when rewarded with the oral intake of a food rather than by the same amount of the food delivered in the stomach. An instrumental response and the spontaneous oral intake of milk during 10 min were reduced by the following manipulations prior to the test (in order of increasing efficiency): 14 ml of saline tubed in the stomach, 14 ml milk in the stomach, 14 ml milk orally drunk and entering the stomach (5, 6). Another experiment demonstrated at this time that this oral-satiating capacity of a food was sensory in nature. Rats were habituated to take a 30-min morning meal of a synthetic diet, at first unaltered and then later flavored with traces of an arbitrarily chosen odorant. During the first 2 or 3 days, this flavoring reduced the intake, which was rapidly returning to base line. Following 42 days of a constant daily intake of the flavored diet, the flavor was suddenly removed. During 2 or 3 days, the 30-min intake of the unflavored diet was augmented up to 50% in some rats (7). Thus, the removal of the flavor had reduced a sensory satiating action of the food. Accordingly, it was proposed for the first time to dissociate two orosensory actions in the control of food intake: the palatability involved in meal initiation and an orosensory satiating property.

In the next step, the same researcher (8) demonstrated that this oral satiating effect of the food was sensory-specific. Satiation provided by the sensory activity of a food was specific to the sensory properties (e.g., taste, smell) of that food. The two decisive experiments revealing in 1956 this sensory-specific satiation were replicated and confirmed 20 years later by a number of other experimenters (9–11).

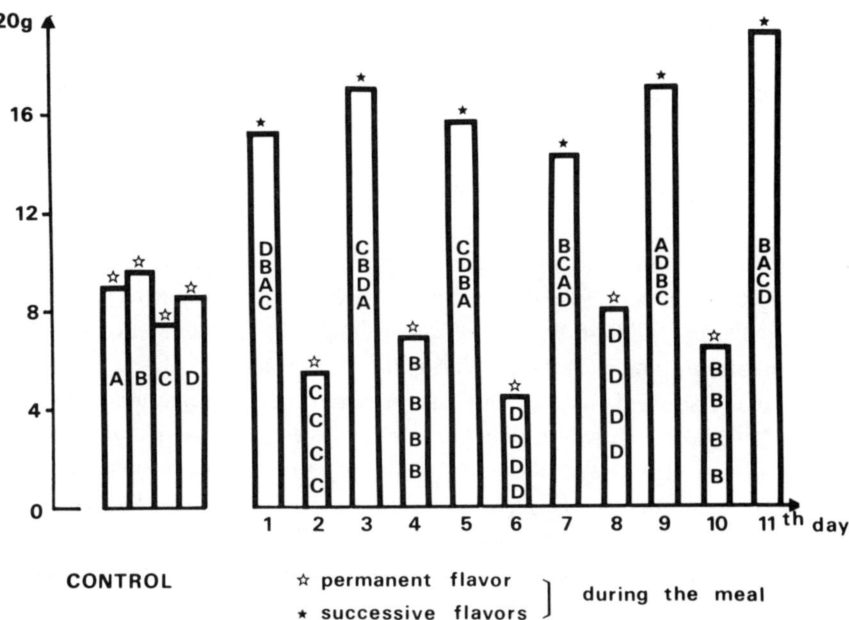

Figure 6.2 Enhanced food intake in the meal by a successive presentation of palatable foods.

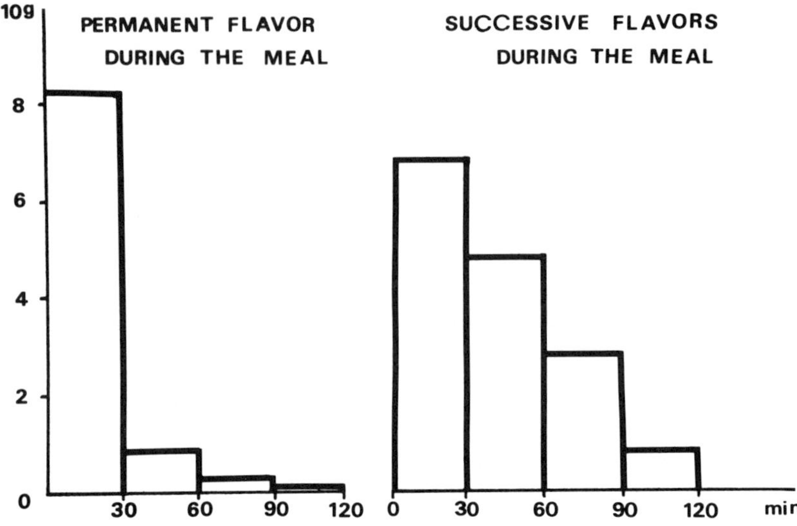

Figure 6.3 Sensory-specific satiation in rats.

Rats were habituated to take a 2-hr morning meal of a synthetic diet, at first unaltered and later flavored with either odorant A, B, C, or D. The flavored diets were offered at random during successive days. After an initial neophobic decrease of intake, the base line intake was rapidly reestablished and was identical irrespective of the flavor. Following 32 days of this habituation period, rats were presented successively the four flavored versions for 30 min during the 2-hr meal. The order of succession of the four flavors in the 30-min fraction of the meal was at random among rats and in successive tests. Under the effect of successive presentations within the meal, considerable meal overeating was observed, reaching two- or three-fold the previous control with a permanent flavor during the 2 hr. During the first 30 min, whatever the flavor, rats ate approximately the same amount of food as that eaten during the 2 hr by controls. Then, rats satiated on the first presented flavored diet renewed intake on presentation of the other flavored versions. However, the most instructive fact was that this resumed intake decreased whatever the order of flavor presentations during the successive 30 min and became almost null during the last one. Thus, the rat is orally satiated only for the food it has just eaten, but the effects of this oral sensory-specific satiation on the cumulative intake of a series of differently flavored foods are limited by the progressive filling of the stomach. The oral satiation abolishes only the contribution of palatability to the overall stimulation to eat, thus introducing only a partial satiety. It may be said that "appetite" is satiated in the mouth, and hunger in the stomach and further down (Figs. 6.2 and 6.3).

An elegant study by Schultz and Lawrence (12) corroborated these results. Satiated rats learned a maze when rewarded with a sucrose solution or a food, but only if the sucrose solution or the food were not used to satiate the rat.

Another experiment by Le Magnen (13) showed that the simultaneous

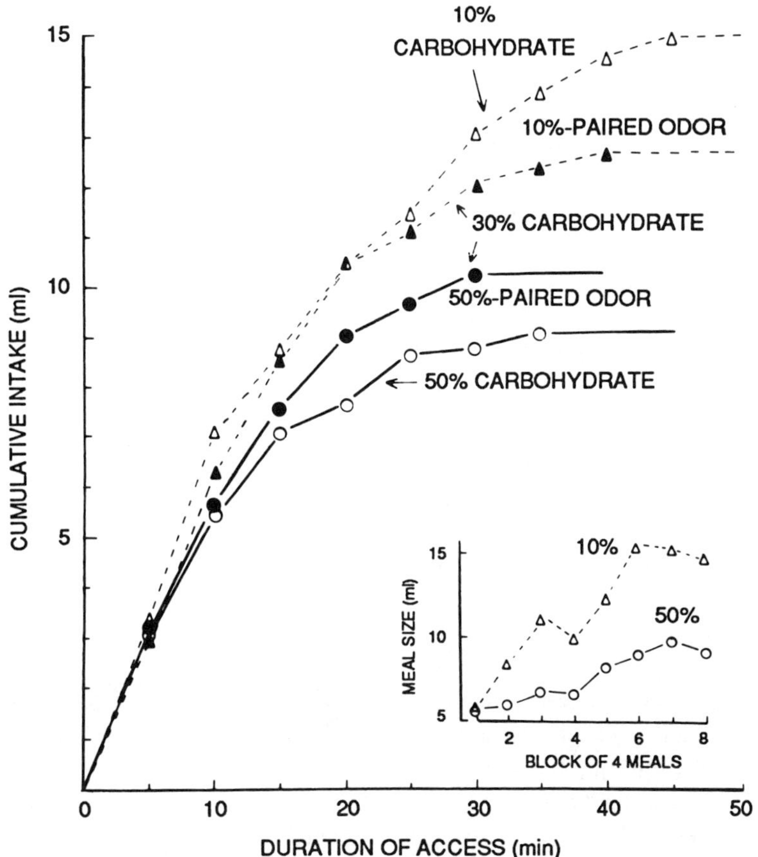

Figure 6.4 Feeding bout size and structure after repeated access to 10% and 50% odorized dextrin suspensions (Experiment 1). Inset: Differentiation of larger intake bout when 10% dextrin was presented over 16 presentations alternated with 16 presentations of 50% dextrin of another odor. Main panel: Mean cumulative intakes on the last pair of 10% and 50% dextrin presentations (open symbols) and on the subsequent pair of trials in which 30% dextrin was presented odorized with 10% paired odor or 50% odor in turn (solid symbols). [From Booth (22).]

presentations of differently flavored foods, like successive ones, induced meal overeating. Rats were offered a synthetic complete diet flavored with three different odorants—A, B, or C—during a 2-hr morning meal. After 27 days of a single-flavored diet, the 2-hr intake, regardless of the flavor and the position in the cage, was identical (12.6 g). On six successive tests, the simultaneous presentation of the three flavored versions during the 2-hr meal produced an increase of the intake by an average of 25% during the second choice. In all rats, this augmented intake was transient, and an intake identical to that of separated presentations was reestablished more or less rapidly in the 10 rats.

This result was the origin of the finding that hyperphagia is induced by a choice of various high-palatability foods, later called the cafeteria-regimen. This hyperphagia is achieved by huge meals during which rats eat the various available foods successively (Le Magnen and Devos, unpublished; 14). Thus, hyperphagia induced by the cafeteria regimen is an effect of the sensory-specific satiation.

On the basis of new observations of this phenomenon, Mook *et al.* (11) proposed a speculative theory of multiple mechanisms of satiation or satieties. It seems easier to explain the phenomenon by the respective roles of the oral and gastric components of satiation. The first one is food sensory-specific, not additive; the second one is that volume and other factors are specific and not additive when various foods are successively eaten. The participation of the two components in stopping intake makes the rat orally satiated and potentially gastrically unsatiated on the first presented food, the opposite on the last presented or last spontaneously eaten food.

C. Sham-Feeding and Real Feeding

The preceding results suggested that the oral satiating effect of the food was combined with a postoral negative feedback in a real meal. Sham-feeding experiments confirmed this notion. In such experiments, the time course and level of the oral intake of a liquid food are compared in conditions in which the ingested food either leaves the stomach through an open gastric fistula (sham-feeding) or a closed fistula (real feeding). Rats in the sham-feeding condition swallow a large amount of the liquid food (one to three times the real intake) before stopping. Nevertheless, they stop and, thus, are satiated by a larger meal. After this sham-feeding, they resume their oral intake following a pause shorter than that in the real feeding and whose duration is unrelated to the preceding amount that was sham-fed (15). A stomachal or intestinal infusion of a liquid food reduced the sham-feeding and reestablished meal sizes and meal-to-meal intervals similar to those of the real feeding (16). A duodenal infusion prior to, during, and after the sham-feeding revealed a required relationship between the pre- and postgastric inhibition. A maximal reduction of the sham-feeding occurred when the intestinal perfusion was initiated 12 min after the start of the sham-feeding (17). The sham-feeding of a glucose solution varied with the concentration according to an inverse U-shaped curve. Repeated sham-feeding of low and high concentrations flattened the curve. A gastric preload of water made the sham-feeding linearly increasing with concentrations (18).

The saccharin intake may be compared with the sham-feeding of a glucose solution. Following the real feeding of a saccharin solution as well as the sham-feeding of a glucose solution, a major behavioral symptom of satiety (the resting and sleep of the rats) is absent. A gastric preload of glucose prior to the saccharin intake as prior to the sham-feeding of the glucose solution, adds this symptom to other indices of satiety (19). It is interesting to note that in rats rendered obese by some weeks of the cafeteria regimen sham-feeding increases the intake only by 8% compared to their real feeding, whereas the same sham-feeding doubles the intake during 4 hr in nonobese controls. Thus, in such rats that are anorectic at the discontinuation of the cafeteria regimen, the sham-feeding is also inhibited (20).

D. The Conditioning of the Sensory-Specific Satiation

The preceding result suggests that sensory satiation may be conditioned. At the same time, this was also suggested by detailed experiments by Smith and Duffy (21). The inhibition by a preload of a 2-hr intake was measured during the first 10 and 30 min and during the 2 hr. An intragastric preload of an inert material reduced the first 10-min intake, as does the preload of a glucose solution. The bulk in the stomach was only responsible for this initial inhibition, and arguments are developed in favor of a conditioning of this effect. Later during the 2-hr intake, the reduction of intake by the glucose load no longer depended on the bulk.

However, a clear-cut demonstration of a conditioning of sensory-specific satiation and of its separation from the conditioned palatability was magistrally provided in an experiment by Booth (22) (Fig. 6.4). Rats were offered alternatively two starch solutions that differed in their caloric density. Each version was labeled with a different flavor. The rats rapidly exhibited an increased volume intake of the less dense solution. Thus, caloric intake of the two solutions became identical. This different volume intake satiation was sensory-specific. Rats offered separately the same solution of an intermediary caloric density flavored with one of the two flavors maintained the higher intake of the liquid food flavored with the odorant previously associated with the less dense solution. This was not due to a conditioning of the initial palatability. In a simultaneous and brief choice, no preference was manifested for either of the two flavored versions. This was due to a different and conditioned decline of the intake rate. The volume intake of the less dense solution increased because the intake rate declined more slowly until satiety (22).

What are the reinforcer and the site of its action? In a complementary experiment, Booth and Davis (23) showed that the different densities of the two flavored solutions were active in the conditioning process for the first 5 min of the intake and, thus, in the early phase of intestinal absorption. A subdiaphragmatic vagotomy did not prevent and, on the contrary, accelerated the conditioning. Therefore, a vagal afferent or efferent would not be involved. Different results on this point are reported by Deutsch and Wang (24). The higher volume intake of a flavored nutritive solution versus an otherwise flavored saline solution was rapidly acquired in rats with a linkage of the pylorus; therefore, the different caloric value of the two solutions could act as a reinforcer in the stomach alone.

In repeated sham-feeding tests, a progressive increase of the drinking rate was observed by several investigators (25–27). This effect could reflect an extinction of the conditioned oral satiation. In the study by Weingarten and Molikovsky (25), the increase of sham-feeding of two flavored sucrose solutions in successive tests was compared. The progressive increase of the sham-feeding occurred only with the flavored sucrose solution that had been previously experienced by rats in a real feeding. Thus, the progressive increase of the sham-feeding rate in repeated tests is an extinction of a previously acquired satiating capacity of the liquid food.

Finally, it seems reasonable to conclude from all these results that the persistency of the postabsorptive satiety is the primary reinforcer of the orosensory satiating capacity of the food, as it is in the conditioning of the initial palatability. This would explain in particular the fact that high-palatable and calorie-rich foods are also more rapidly satiating than low-palatability and less caloric foods. However, this does not exclude a role in the conditioning of preabsorptive and just postabsorptive events that are associated to the termination of meals, as described in the following chapters.

II. *Gastrointestinal and Systemic Factors as Determinants of Meal Size*

Studies on the orosensory contribution to the satiation process provided evidence indicating that food reaching the gastrointestinal tract played a role in determining the normal meal size. Other results suggested a role for postabsorptive events.

A. Satiating Activity of Food in the Stomach and/or Intestine

Many studies support the notion that stomach filling plays an exclusive role in satiety onset and, therefore, in determining meal size.

In rats, the largest free meals at night last an average of 10 min. It is unlikely that by the end of the meal a substantial amount of food accumulated in the stomach during the meal might have reached the small intestine and contributed to determine the end of this meal. This fact provides an *a priori* argument for authors who favor the exclusive role of the stomach versus the intestine. Snowdon (28) provided experimental arguments. The removal of the food from the stomach at the end of a meal provokes a quasi-immediate initiation of a new meal. The total gastric emptying of 3.9 ml NaCl solution (isotonic to the experimented liquid food) is achieved in 100 min. This time corresponds to the normal meal-to-meal interval after such a meal. As mentioned earlier (Chapter 2), this argues in favor of a role for the time course of gastric emptying in the intestine in the initiation, and not in the termination, of the meal. In vagotomized rats, the resumption of the meal after the gastric removal of the food is delayed. The gastric emptying of such rats is more rapid than in intact controls. The fact that this early emptying is associated with a pattern of small and frequent meals argues in favor of a role of the small intestine in meal termination. On the contrary, the effect of knotting the pylorus on meal size supports the notion of the major role of the stomach filling. The same liquid food taken orally was added in the stomach through a cannula during the meal. A precise compensation of this gastric addition was achieved by a reduction of the oral intake, a linkage of the pylorus being present or removed. Considering the effects of various volumes and amount of calories added in the stomach and their effect on the oral intake, the author concluded that nutritive properties, more than distension, are involved in the termination of small meals without any role for the duodenum (29–31). Also, according to Kraly and Smith (32), the oral intake of a liquid food was the same, with the pylorus linked or opened (32).

Many other works, however, suggest an effective role for food action in the duodenum. In one particular experiment, rats orally took either milk or a 3.2% glucose solution. At the start of intake, they were infused through a duodenal cannula by 2 ml of either 5, 10, or 15% glucose solutions. The intake of milk was reduced only by 15% glucose, and that of the low-palatability glucose solution was reduced proportionally to the amount of infused glucose. This result again points out the role of the balance between palatability and postoral events in determining meal size and also supports the role of the small intestine (33, 34).

To decide between the stomach and the small intestine, it was necessary to know the exact time course of the gastric emptying from the beginning to the end of a meal. This rate of gastric emptying of a meal was measured in rats deprived for 16 hr (35). The offered food was of various caloric density (0.25–0 kcal/ml) and of various macronutrient composition. Two conditions were observed: intake up to satiety or the meal limited to 15 ml. The rate of gastric emptying was shown to be constant in terms of kilocalories per minute and, thus, inversely proportional to the caloric density of the ingested liquid food. In another experiment, the same investigators (36) observed huge meals of rats in the sham-feeding condition. During sham-feeding, they infused the same liquid food sham-fed at the rate of 0.06–0.44 kcal/min in the duodenum. This infusion reduced the sham-feeding in a dose-dependent manner. The sham-feeding was stopped after 2–3 kcal infused in the duodenum. During a real meal, the gastric emptying rate was found to be 0.3 kcal/min. By the end of the meal, 3.8 kcal had left the stomach. The authors concluded that the caloric post-gastric load was a sufficient factor of satiation in the absence of the action of gastric distension.

Experiments on parabiotic crossing of intestinal segments did not support a role played by the intestine. In one experiment (37), 30 cm of the jejunum of a rat was bypassed to a rat partner and 30 cm of the jejunum of the latter bypassed to the former. Thus, the food eaten by the first one passed into the jejunum of its partner before returning to its intestine and vice versa. The first rat was fed 10 min before the second one and ate a normal meal compared to controls, despite the lack of signals emanating from its jejunum. The partner started to eat 10 min after receiving signals in its jejunum, due to the preceding intake of the other rat and also took a normal-sized meal.

The same discrepancies regarding the respective role of the stomach and the intestine in determining meal size are reported in the monkey. Some investigators provide strong arguments in favor of the exclusive role of stomach distension (38); others in favor of the intestine (39).

Numerous works were also carried out on the rabbit. It is difficult to take them into consideration without caution. The researcher neglected the specificity of rabbit feeding behavior, which is created by caecotrophy. 10% of the oral intake of the rabbit is not digested and, instead, is transformed into the caecotrophs, which are taken anally and reingested. The effects of duodenal infusions on oral intake, widely experimented, may interfere in an unknown manner with both the intestinal absorption and the mechanism of the caeco-trophic behavior (40).

Cholecystokinin (CCK) and its octopeptide analogue, CCK_8, injected into rats reduce meal size. It was hypothesized that the intestinal CCK release was a satiety factor. It was, of course, *a priori* excluded that the release of an intestinal hormone might explain the complex mechanism involved in the orogastrointestinal process of satiation. Furthermore, such a release could not account for the adjustment of this mechanism to the initial stimulation to eat and to the caloric value of the foods. However, this hypothesis received surprising success. In fact, CCK injection introduces conditioned taste aversion (41). At the dose used, it is highly toxic as manifested by epileptoid signs in electroencephalogram recording (42). Vagotomy eliminates its actions (43). Neglecting this fact, researchers claimed that CCK induced satiety by acting on brain CCK receptors. This was totally denied. A CCK analogue, 200 times more active than CCK_8 on these brain receptors, had no effect on feeding (Nicolaïdis, personal communication).

The classic works on the dog by Janowitz and Grossman (44) are still today of great interest, particularly in regard to the condition of a feeding schedule, rarely studied in rats. An oral prefeeding of a part of the normal meal 20 min prior to the meal proportionally reduced its size. A stomach-tubing of an inert material of 20% of the bulk of the normal meal was without any effect. Forty-four to 66% reduced the subsequent intake by 33–51% only when the tubing was made immediately prior to the habitual meal. Tubing the food in the stomach had the same effect as that of the inert material. Esophagotomized dogs were sham-fed three times per day. The sham-feeding lasted 14.5 min, versus 1.5 min in a real feeding. Satiety induced by sham-feeding lasted only 30 min. Duration of sham-feeding was not reduced by an intragastric feeding despite the volume tube-fed immediately or 3 hr prior to the sham-feeding; however, this reduction was obtained when the intragastric load was given during the sham-feeding. Inflating a balloon in the stomach reduced the sham-feeding only following a maximal distension.

B. Postabsorptive Systemic Events as Components of the Satiation Process

It is reasonable to speculate that satiety, in order to persist after its induction by eating foods, must be associated to the elimination of the systemic events that initiated the meal. In Section IV of this chapter, a possible role for the preabsorptive reflexly induced insulin release, known as the cephalic phase of insulin secretion, will be examined. Here, only the postabsorptive and just postprandial events will be considered as putative systemic factors of satiation.

The first signs of postmeal absorption, tested by the appearance of a blood radioactivity following ingestion of a labeled food, are manifested 8–9 min after the beginning of the ingestion in rats. This is the mean time of the end of a meal or at least of the approach of satiety. This is also the time of the appearance of the well-known prandial and postprandial hyperglycemia, hyperglucagonemia, and hyperinsulinemia. These events lead to the sudden rise of tissue glucose utilization, which represents the inversion of the opposite state present at meal initiation. Therefore, they may be involved in a final phase of satiation. Surprisingly, the sequence of these just postprandial blood changes is poorly understood. However, it is possible to hypothesize that a glucagon release occurs first, producing increases in hepatic glucose production and hyperglycemia, which, in turn, stimulate insulin release. Some results suggest that indeed the postprandial insulin release is involved in satiation. In adult rats, a blockade of insulin release by mannoheptulose considerably reduced the satiating effect of a premeal glucose load manifested by the subsequent low meal size. Thus, the insulin release, due to the absorption of the load like to that of a meal, would be a requisite of satiation (45).

Lindberg *et al.* (46) provide more convincing results. In rats, various doses of streptozotocine produced various degrees of glucose intolerance. Food intake was then tested during some hours of the day in animals deprived at night. The more glucose-intolerant the rats were, the more they ate. According to the authors, the absence of prandial insulin release and the lack of its effect on the metabolic utilization of ingested foods eliminated an action of insulin as a factor of satiation.

The reported results regarding the glucagon action are fully convincing. A glucagon antibody was injected at the start of the first meal in 12-hr deprived rats. The meal size increased by 61% and its duration by 75%. A reduced hyperglycemia in the hepatic vein, not in the portal one, confirms the hypothesis that the glucagon prandial release causes prandial hyperglycemia. The authors conclude that glucagon is a satiety-inducing factor in relation to prandial hyperglycemia (47). The opposite experiment confirms the result and its interpretation. Injecting glucagon at the start of a meal in deprived rats reduced the size of the meal by half (48). The hyperglycemia measured at the beginning and during the meal was augmented. The ratio of the blood glucose in the hepatic vein to the portal one was also augmented, indicating the rise of hepatic glucose production (49). Finally, Geary and his co-workers (50, 51) elegantly showed that injecting glucagon in the portal vein during the first 2 min of a spontaneous nocturnal meal reduced the meal size.

III. *Determinants of Meal Size in Humans*

Mechanisms governing meal sizes are particularly important in humans. Earlier we emphasized that the main feature of the human feeding pattern is a schedule of diurnal meals. Except in acute conditions of enhanced energy expenditures (e.g., cold, exercise), in which extra meals can be taken, this socially imposed eating habit is generally followed. As already mentioned, such a feeding schedule eliminates the main and more rapid regulation of body energy balance by food intake through the modulation of meal frequency. As shown in animal models, responses to changes of the caloric density of the food and to a chronic intragastric feeding are considerably slower and limited when the compensation can only be performed by an adjustment of intake in a single or some fixed daily meals. Consequently, the extent to which meal size will be adjusted both to caloric and nutrient contents of ingested foods and to changes of the metabolic requirements will be, in humans, the only basis for a regulation of body energy balance.

A. Sensory Factors in Human Meals

In human meals, considerable significance is given to food palatability. Sensory-specific satiation, experimented in rats, is common experience in humans. In ordinary meals made of successive courses, humans are satiated on the first course; they resume eating with a good appetite on the second or third ones. Fully satiated on the main dishes, they eat a dessert (a food sweetened by either sucrose or a noncaloric sweetener) made (it is not fortuitous) of sweetened items, the most potent sensory stimulus of eating. When a dessert is offered at the beginning of a meal, the subsequent meal in a free choice is reduced, as is the intake of sweets in the choice (52). In varied meals, humans can actually eat 1000 kcal or more in one meal.

The most instructive experimental study was done by Bellisle and Le Magnen (53), who used the recording of a chewing—swallowing pattern. With this technique (briefly described earlier), the chewing and swallowing of human subjects were recorded from the beginning to the end of a test meal. In these meals, subjects were offered pieces of bread of a constant volume made differently palatable by a layer of various flavored spreads. Meals of single food flavors were compared among them and compared to varied meals in which the subjects freely ate various offered items. Normal-weight subjects ate the highly palatable food four times more and three times longer than the least palatable one and 25% more and 33% longer than the former in varied meals. The high- and low-palatability foods of each subject, so indicated by the amount eaten, were correlated to an initial visual analogue rating of pleasantness. The chewing time was shorter and the number of masticatory movements before swallowing was fewer on a high- than on a low-palatability food (54). As in the recording of licking rates of rats, the evolution of chewing parameters clearly indicated the decline of palatability by progressive satiation. During the last quarter of a meal, compared to the first one, with an intermediary palatable food, the chewing time increased by 11.1% and the number of masticatory movements by food unit by 10.3%. Intrameal pauses also increased at the end of the meal (+17%). Thus, at the end of the meal, subjects masticated a high-palatability food as they masticated a low-palatability food at the beginning of the meal. In varied meals, decreased chewing time in the shift from a flavored food to the following one clearly indicated sensory-specific satiation. Spiegel and Shrager (54), using the same technique, confirmed that the eating rate at the start of the meal was proportional to, and therefore an objective measure of, the palatability.

As mentioned earlier, Kissileff and Thornton (55) established the cumulative curve of intake within a meal in humans. They began from the general and reasonable assumption that the cumulative intake curve results from the algebraic summation of an initial facilitatory action of hunger and palatability and of an inhibitory decelerating action provided by the orogastric satiating efficiency of the food. This approach is not very different from that of Davis in rats.

This sensory-specific satiation in humans was extensively studied by Rolls and co-workers (56–59) (Fig. 6.5). In the initial study, the pleasantness of eight foods was scored by subjects prior to and 2–30 min after eating one of them. The fall of the score was more accentuated after the food was eaten. The higher intake of mixed- versus single-flavored meals was also confirmed. When foods of different colors were offered, the decline of pleasantness was also specific to the colored foods eaten, but the successive or simultaneous presentation of these colored foods did not induce an increase in meal size. This increase (up to 60%) was only induced by variety due to different forms and flavors of the foods.

--->

Figure 6.5 The effect of eating one food to satiety on the subjective pleasantness of that food and other foods that had not been eaten. Subjects (N = 32) rated eight foods on a scale where + 2 was very pleasant and − 2 very unpleasant. After this initial rating, they ate one of the foods to satiety (four subjects per individual food). The figure shows the mean changes in subjective pleasantness when the foods were tasted again 2 min (upper figure) and 20 min (lower figure) after the end of the meal. The "total" figures show that the mean (± SEM) decrease in pleasantness was significantly greater for all the foods that had been eaten than for all of the foods that had not been eaten. [From Rolls *et al.* (56).]

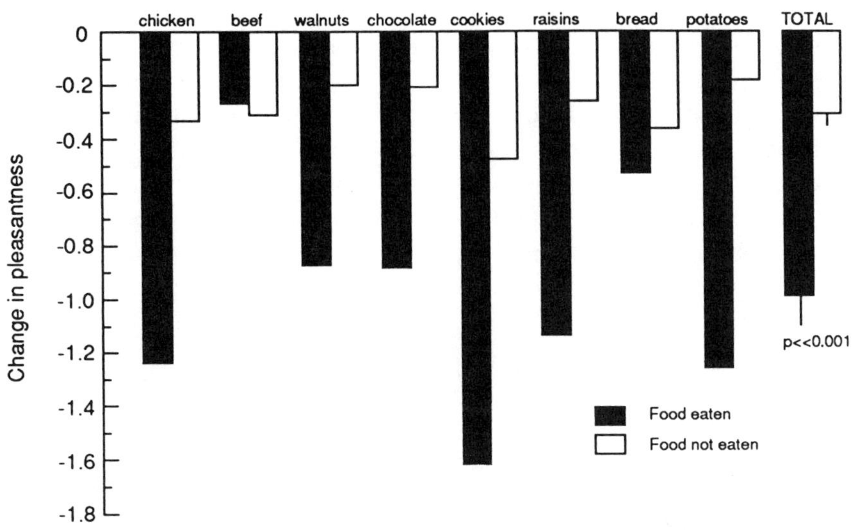

2 minutes
chicken
beef
walnuts
chocolate
cookies
raisins
bread
potatoes
TOTAL
Change in pleasantness
0
-0.2
-0.4
-0.6
-0.8
-1.0
-1.2
-1.4
-1.6
-1.8
p<<0.001
Food eaten
Food not eaten

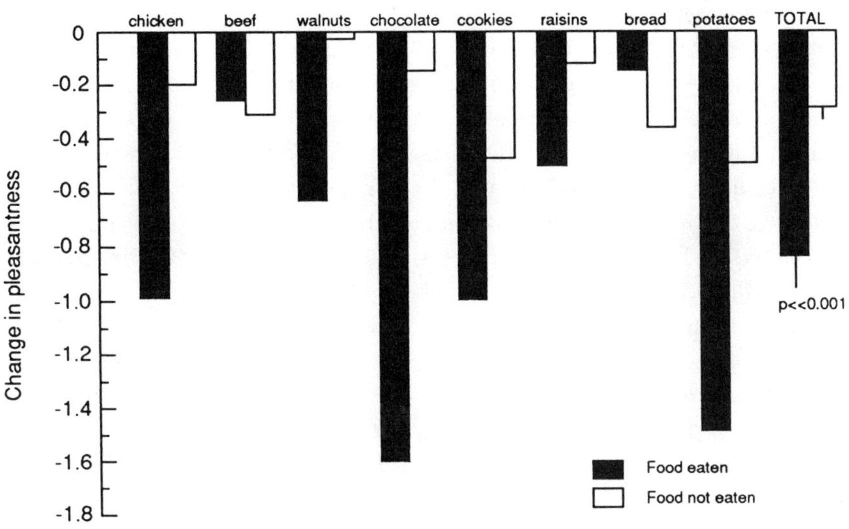

20 minutes
chicken
beef
walnuts
chocolate
cookies
raisins
bread
potatoes
TOTAL
Change in pleasantness
0
-0.2
-0.4
-0.6
-0.8
-1.0
-1.2
-1.4
-1.6
-1.8
p<<0.001
Food eaten
Food not eaten

An interesting study of the satiating efficiency of foods in humans was carried out by Kissileff *et al.* (60). Subjects received a two-course meal composed of a soup as preload followed by the test meal of various solid foods freely eaten until satiation. The total energy intake (preload + test meal) was found identical to that in a test meal not preceded by the preload. However, this total energy intake was less after a soup than a solid preload. Therefore, substituting a tomato soup for a calorically dense first course could reduce the total energy intake of that meal. The respective roles of palatability and calorie and water contents of the preload in this effect are unclear and should be determined in future studies.

B. Gastric Factors

The best studies on the role of gastric factors in satiation were by Walike *et al.* (61). Subjects received an intragastric load of a liquid food representing 20–120% of their basal meal and were offered an oral meal of the same liquid food 3–120 min after the load. The maximal depression of the oral intake was 1–5 min after the load in some subjects and 15–20 min in others. Beyond these delays, the compensation was strongly reduced. Even the maximal depression was not sufficient to compensate for the gastric preload. In all cases, the total volume and caloric intake were augmented (Fig. 6.6).

In another series of experiments, subjects ate a liquid food delivered by pump either orally in the mouth or intragastrically, or both, simultaneously. The rate of delivery was fixed by the experimenter, and the volume taken and its pattern were determined by the subject, pressing freely on a spot to command the pump. At a given speed of delivery, subjects took the same volume of the food via the mouth and via the stomach. An augmentation of the speed of delivery increased the intake before satiation, but more via the stomach than via the mouth. When 50% of the delivery was in the stomach and 50% in the mouth, subjects did not reduce their intake compared to the previous one given via the mouth alone, thus doubling their intake. They failed to augment it by the intragastric feeding neither immediately nor progressively after a dilution of the liquid food. Subjects maintained in the self-intragastric feeding during 4 days by three gastric meals per day did not lose weight but complained about the lack of oral sensations. The distinction between an oral and a gastric satiation was assessed by a hunger rating noted every 5 min, during the oral, the oral plus gastric, and the intragastric feedings. In the latter condition, the rating diminished less rapidly. It was higher at the spontaneous end of the intragastric meal than in the more rapidly and more satiating oral meal (62).

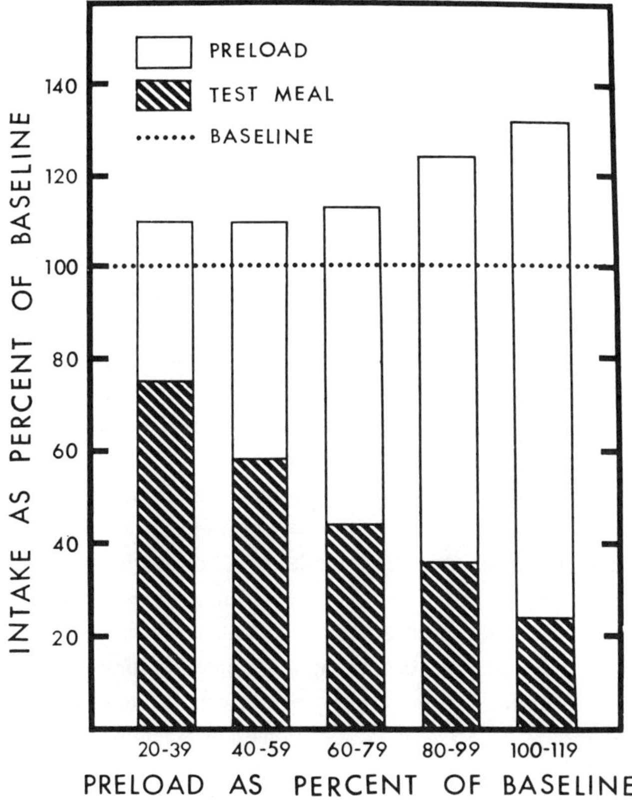

Figure 6.6 Effects of preloads on intake in the meal. [From Walike *et al.* (61).]

In a complementary study (63), the simultaneous oral and gastric feedings were shown to produce an excess of intake of 930 kcal at the termination of intake compared to the meal by the oral route alone. The subsequent free oral meals were reduced by this overeating. Finally, it was confirmed that the augmentation of the rate of eating the liquid food via the mouth of the stomach, or both, elevated the intake.

An experiment by Geliebter (64) completes our information about the role of food volume in the stomach in the satiation process. In four normal-weight subjects, a balloon was placed in the stomach and more or less filled with water. The oral intake was reduced from a volume of 400 ml of the balloon. In another experiment, the balloon was filled progressively, and the subjects scored their feeling of discomfort. The individual maximal gastric capacity was thus determined. This capacity averaged 1100 ml in the normal weight subjects and 1925 ml in obese ones. No correlation exists between the individual maximal gastric capacity and intake. Instead, a high correlation between gastric capacity and the volume of the balloon needed to reduce intake by 50% is found.

C. Conditioning of Satiation in Humans

Inasmuch as neither oral nor gastric cues provide a caloric metering of the food, studies on the adjustment of intake to changes of the caloric density of food are necessarily studies on the conditioning of the satiating capacity of foods. The delay and repetition of tests needed to obtain the caloric adjustment are signs of the conditioning process.

In the early 1960s, Campbell *et al.* (65), using the technique of a liquid food sucked from a tube (a somewhat artificial condition of eating), were the first to demonstrate the caloric adjustment of food to the dilution. They noted some delay in this adjustment. Among other experiments, the following one was particularly demonstrative (66). The intake of a conventional varied and palatable meal was measured. Sucrose in sweetened foods was suddenly replaced by aspartam. This substitution reduced the mean caloric density of the offered food by 25% and led to a compensation by an augmented volume intake after 3 days only. From day 4 to day 12, the compensation was limited to 50% and, after a long delay, stabilized to 85% of the previous caloric intake. The intake deficit was not limited to the aspartam-sweetened foods.

In another experiment on young children (67), a prefeeding, 20 min prior to the *ad libitum* test meal, was either a flavored pudding of high caloric density or the same but differently flavored pudding of low caloric density. Half of the children showed an immediate compensation. This immediate compensation was presumably not a response to the caloric density but to a preload effect. The 20-min delay was sufficient to imply a postabsorptive effect. The other half of subjects showed a compensation at the second trial only. Twelve of them maintained the difference of intake in the test meal in an extinction condition. They continued to have a greater intake after a prefeeding by the caloric-dense pudding flavored like the low caloric version was previously.

In another study (68), the subject intake was measured during 15 days in a choice of varied foods. Then, the caloric density of one-third of the offered foods was reduced. Subjects maintained their volume intake on these foods, leading to a loss of a 100-kcal intake from these foods. After some delay, this deficit was compensated for by an elevation of intake of the other foods, and the total caloric intake was reestablished.

An experiment by Wooley (69) showed the role of cognitive factors in humans and of the belief in such adjustments. Subjects received, 20 min prior to the *ad libitum* test meal, an oral preload of a liquid containing either 200 or 600 kcal. Both were presented in such a manner that they appeared either rich or poor in calories. The total caloric intake was not different as a function of the calories taken in the preloads but, rather, as a function of their appearance and of the belief of their richness by the subject. Their satiety rating was also significantly correlated to this belief.

Like in Wooley's experiment, a considerable number of works in humans are based on the use of subjective evaluation of pleasantness and unpleasantness, likes and dislikes, by category scaling or visual analogue ratings. The reliability of such evaluations and their relations to objective measures of the palatability of foods or of amounts eaten are questionable. In an exemplary study already mentioned (70), subjects were initially tested by a category scaling for their like or dislike for the same food, adulterated or not. Then, the authors looked for the relations between this rating of pleasantness and both the initial rate of eating and the amount eaten. The mean ratios of individual differences of intake to differences of rating indicated a difference of about 100 g of intake for each point of the rating. But the interindividual differences (511.4–220.0 g) indicated that such ratings are very poor predictors of the intake of the same food by various subjects.

Again, the confrontation between subjective and objective approaches obscures the result. The data are presented as *effects* of likes and dislikes and of their subjective rating on food intake, and these subjective feelings are erroneously called the palatability of the food. Instead, such studies only provide indications of the relationships between the subjectively perceived pleasantness and the palatability of the food, i.e., the sensory component of the stimulation to eat, objectively measured by the initial rate of eating in a given internal state. Based on this relationship, the pleasantness rating is unlikely to assess the contribution of the objectively measured palatability to the size of the meal. This contribution, as already said, depends on multiple factors and, particularly, on the sensory-specific satiating capacity of the food. This capacity is better assessed by a measure of the decline in the eating rate rather than by a lowering of the pleasantness rating.

D. Systemic Factors

The just postprandial insulin level suggested to play a role in the satiating process in rats is questionable in humans as a factor of satiety onset. Infusing a mixture of insulin and glucose before, during, and after a test meal did not change the meal size despite a fourfold increase of the postprandial insulin and a maintained or augmented (according to doses) blood glucose level (71). Effects of glucagon or antiglucagon injections during the meal have not been tried. Generally speaking, few attempts were made to reduce pharmacologically the meal size of humans based on the knowledge of physiological factors that determine the amount eaten in a meal. This gap is surprising because such a pharmacological action would be the remedy for meal overeating leading to obesity.

IV. *The Cephalic Phase of Insulin Release*

A. Cephalic Phase of Insulin Release in the Rat

Works in the 1930s reported that food-related sensory stimuli induced hypoglycemia in various animal species. For example, Reid (72) showed that the sniffing of meat by a hungry dog provoked an immediate fall in its blood glucose level. Later, numerous works attempted to confirm these earlier suggestions and to clarify their physiological significance in relation to the control of food intake. At first limited to a measure of blood glucose, contradictory results were reported. Either hypoglycemia or hyperglycemia were found to be associated with the oral intake of foods. Both were interpreted as preabsorptive events and suggested to be reflexly induced by the food orosensory stimulation.

When techniques of blood glucose determinations were improved and the radioimmonulogic determination of plasma insulin was available, the reality of a preabsorptive insulin release neurally induced by food-related sensory stimulation was demonstrated. The first finding was provided in the dog by Fischer *et al.* (73). They found that the oral intake of a glucose solution was rapidly followed by two successive peaks of the plasma insulin level. In a sham-feeding condition, only the first one was maintained and therefore called by the authors "neurogenic." This initial peak disappeared and the second one was preserved after a gastric glucose load.

The pioneering works in dogs were later reexamined and widely extended in rats (74) (Figs. 6.7, 6.8, and 6.9). In the first series of experiments (75), the continuous determination of blood glucose in freely moving rats was used and the immunoreactive insulin (IRI) determined every 1 min. Rats deprived for 5 hr were offered 1 ml of a 50% glucose solution during the day. They drank the solution within 1 min. At an average of 2 min after the start of drinking, plasma insulin rose from 15.5 microunit/ml as base line to 53 microunits. Following a rapid fall of this initial peak, a second rise of insulin release was observed: about

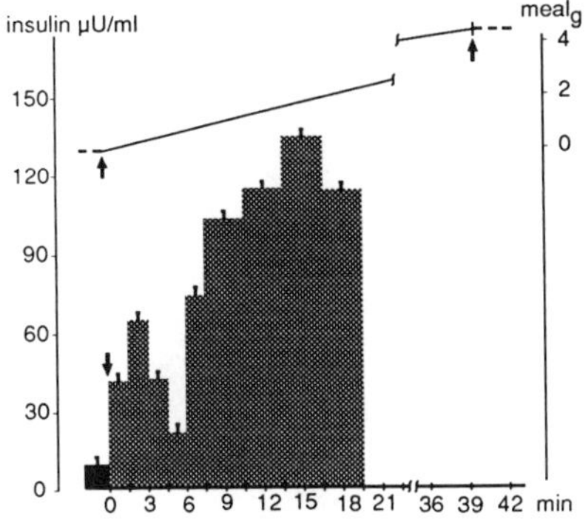
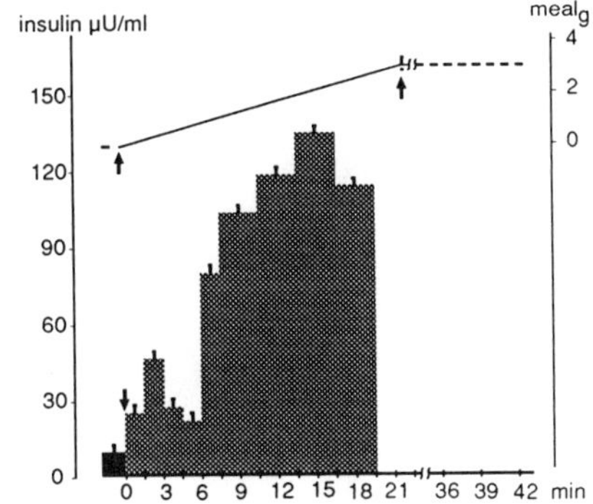
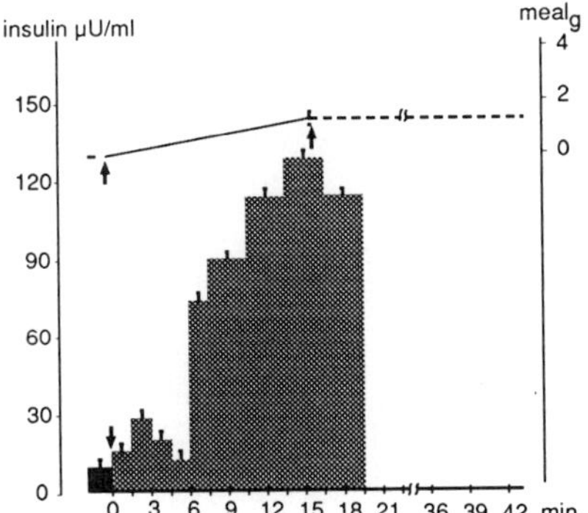

Figure 6.7 Palatability-dependent cephalic phase of insulin release, followed by postmeal hyperinsulinemia.

56 microunits 5 min after the start of the oral intake. The two peaks of plasma insulin were associated with a modest fall of blood glucose. The second one preceded the sustained postabsorptive rise of blood glucose and insulin levels. A saccharin solution substituted for the glucose solution induced the initial peak and early hypoglycemia, but the second peak disappeared. On the contrary, tubing the glucose load in the stomach eliminated the first peak and preserved the second one.

The neural command of the preabsorptive release was assessed. A subdiaphragmatic vagotomy eliminated the insulin response to oral stimulation. Isolated pancreatic islets transplanted in the kidney in totally pancreatectomized rats, achieving a functional denervated pancreas, also eliminated the cephalic phase. In addition, it was demonstrated that the preabsorptive release of insulin

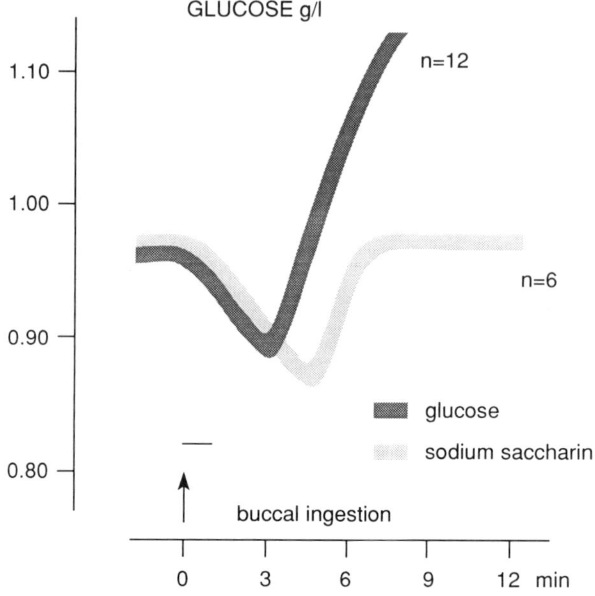

Figure 6.8 Cephalic phase of insulinosecretion following glucose or saccharin oral intakes.

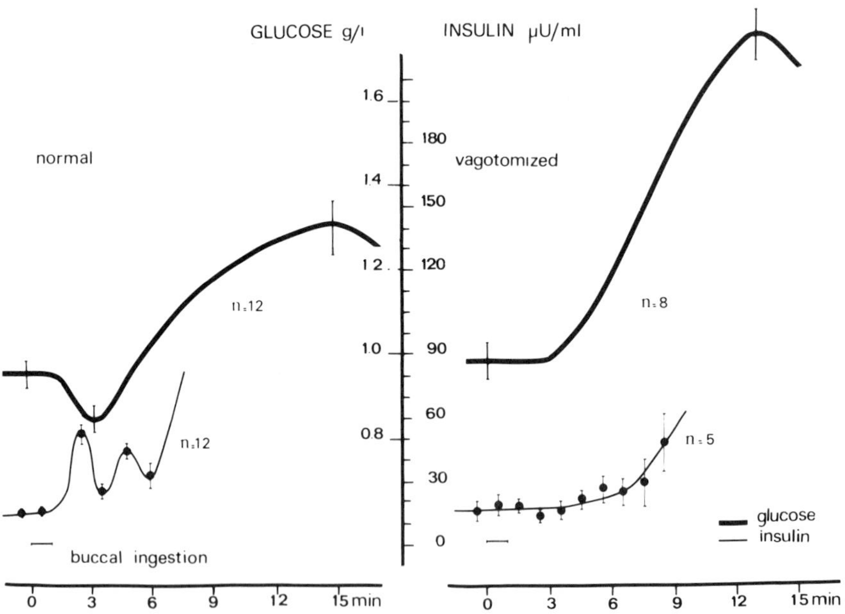

Figure 6.9 Preabsorptive insulin release compared in normal and vagotomized rats.

was palatability-dependent. The same diet was offered to rats in three versions differently flavored with noncaloric taste. Rats ate until satiety 1.5, 3.1 and 4.9 of these three flavored foods respectively. The recorded preabsorptive release was 27.7, 46.6, 65 μ V/ml, respectively (76).

Two important questions were raised by these basic findings. What is the role of these two preabsorptive insulin releases, the first one orally determined and food palatability-dependent, the second one of apparent intestinal origin? The second question concerns the relationship of the preabsorptive release in the overall mechanism that determines the meal sizes. In vagotomized rats, meals of the three flavored versions of the food were of an identical size (77). This result could not be taken as evidence that the size of differently palatable meals was entirely determined by the early insulin release. Vagotomy leads to various disturbances (particularly of gastric emptying), which presumably are a cause of the observed result.

Many other works later confirmed and extended these first findings in rats. A preabsorptive release or cephalic phase of glucagon release was shown to be associated with the insulin release (78). The latter is maintained in rats like in dogs in sham-feeding conditions. On the contrary, the early release of glucagon disappears. Thus, it was suggested that the early rise of glucagon in the blood was of intestinal origin (gut–glucagon) and possibly responsible for the second peak of pancreatic insulin release (79). The cephalic phase of insulin is blocked by atropine. It is augmented by a reduction of the sympathetic inhibitory action on the β cells. The effect of pancreas transplantation was confirmed in rats and dogs (80–82). Interestingly, rats exhibiting high or low preabsorptive insulin release were found to be highly or less hyperphagic, respectively, when submitted to the cafeteria regimen. This confirmed the enhancing effect of food palatability on pancreas release. In pancreas-transplanted rats, the injection of a small dose of insulin at the beginning of meals reduced by 50% the subsequent 50 min of prandial insulin release. In these transplanted rats, a reduction of prandial hyperglycemia was observed. It is well known that intravenous or gastric glucose is less insulinogenic than oral glucose. The origin of this difference is in the orally determined and reflexly induced insulin release (83).

The neurophysiological background of the cephalic phase has been investigated. In rats, stimulation of the tongue by glucose and sucrose solutions enhanced the neural activity in the pancreatic and hepatic branches of the vagus nerve and reduces activity in sympathetic fibers innervating the liver, the pancreas, and adrenals. Stimulation by NaCl provoked the opposite effect (84). A neurohistological study clarified the reported effects of a total subdiaphragmatic vagotomy. Sectioning the anterior and posterior hepatic and gastric branches of the nerve abolished the cephalic phase. Sectioning the celiac branch did not. The involved hepatic and pancreatic branches begin in the dorsomotor nucleus of the hindbrain. In this site, a correlation is found between the number of preganglionics of the two branches and the amplitude of the pancreatic response stimulated by each of them (85, 86).

B. Preabsorptive Insulin Release in Humans

In humans, like in rats, early works suggested a cephalic phase of insulin secretion. The visual presentation of a food to hungry subjects decreased the plasma-free fatty acid level, the most sensitive indices of an insulin release (87). Tasting a saccharin solution decreased the blood glucose level. This also suggests a reflexly triggered insulin release. IRI determinations confirmed these early suggestions. Visual or olfactory food stimulations prior to the start of the meal are indeed associated with a short rise of plasma insulin level. Its occurrence in the first minutes of the actual meal clearly indicate its preabsorptive nature (88)

Results of these various works remain questionable. The recorded peak of insulin was of a very low amplitude (some microunits/ml) above the base line. This base line was established by either 3 IRI determinations (20, 10, and 1 min prior to the test meal) (89) or 10 determinations every minute (88). Despite a statistically significant difference between the recorded rise of insulin and the base line thus established, a problem of signal : noise ratio obscured the significance of these results.

A periodic oscillation of the basal insulin was demonstrated in monkeys (90) and humans (91). In humans, spontaneous insulin cycles were exhibited with periods ranging from 11 to 14 min and amplitudes from 1.1 to 2.3 microunits/ml (91). Thus, the observed response to food stimulation could be suspected to be only a segment of the insulin cycle. To lift this suspicion, it was necessary to record these base line oscillations for a sufficient time before testing and to evaluate the responses by a statistical treatment that took into account this noise. This was achieved in the most recent experimental work by Lucas *et al.* (92) and Bellisle *et al.* (93). The minute-to-minute insulin and blood glucose was assessed during 1 hr before and 20 min after the beginning of the test meal. Subjects were tested on two foods with different palatabilities. Control tests without food presentation were carried out to observe the spontaneous fluctuations. An absolute deviation of the insulin level during the 20-min test was considered significant when it was greater than twice the standard deviation of the 60-min base line values. Preabsorptive insulin release was observed 10 times in 18 tests. The amplitude of the response represented a 17–175% increase over the preprandial mean. This amplitude was correlated to the palatability level of the offered food. However, like in preceding works, the response was shown poorly reproducible both within and among subjects.

C. Conditioning of the Cephalic Phase of Insulin Release

In animals as well as in humans, the question as to whether or not the palatability-dependent cephalic phase is conditioned like palatability is has not been clearly answered. In mice, hypoglycemia was induced at first contact with a saccharin solution, suggesting that the sweet stimulation is an unconditioned stimulus of the response. However, a repeated presentation of the solution led to the extinction of the response (94). We have seen elsewhere that an external sensory stimulus may become a conditioned stimulus of an insulin release and resulting hypoglycemia. This suggests, but only suggests, that the conditioning of the palatability of foods by postingestive events may simultaneously condition the cephalic phase.

D. Comments

The cephalic phase of insulin secretion is a particular case of a general process observed in every behavior and is classified as somatovisceral reflexes. In feeding, it is well known since the pioneering works of Pavlov (95), reexamined by numerous investigators, that food-associated stimuli reflexly induce salivary and gastric secretions. The smell or only the thinking of good foods in humans gives "l'eau à la bouche" (water in the mouth). Intestinal enzymes and pancreatic exocrine releases also were shown to be reflexly changed by sensory-food stimuli. In thirst and hydroosmotic regulation, analogous reflexes are recognized. In dehydrated and consequently oliguric rats, infusion of water in the mouth immediately increases the urinary flow (96). In dehydrated humans, drinking water reflexly induces a flush of sweating (97). In defensive behaviors, sensory stressful and generally aversive stimuli mobilize the pituitary–adrenal axis, as indicated by the brutal elevation of corticosterone and catecholamines. In sexual behaviors, the sight and smell of the female by males stimulate testosterone release.

In turn, these sensory-induced humoral and hormonal secretions elevate behavioral arousal. In humans, the smell of foods elevates the hunger feeling. The cephalic phase of insulin secretion, as mentioned earlier, elevates the stimulation to eat at the beginning of the meal and explains the famous logo "l'appétit vient en mangeant" (appetite grows as one eats). In defensive behavior, this positive feedback is obvious: External stimuli provoking fear or aggression are strengthened as such by the central action of adrenal hormones. In sexual behavior, the reflexly induced secretion of sex hormones in turn elevates sexual arousal. In the defensive and sexual behaviors involving two partners, the positive feedback plays a major role in the building up of the reciprocal behaviors and of the final performance. In defense–aggression, arousal threats and aggressive gestures reciprocally elevate arousals in the partners through the reciprocally induced hormonal releases and their central actions up to the final fight, flight, or submission. In humans, the role of such feedbacks in interindividual as well as in collective conflicts is obvious. Peace or war results from counteracting or precipitating, respectively, such positive feedbacks. In sexual behaviors, these partner interactions are conspicuous. The courtship of birds and other animals and—why not—in humans builds the sexual arousal up to the explosion of the behavior.

References

1. Davis, J. D., and Campbell, C. S. (1973). Peripheral control of meal size in the rat: Effect of sham-feeding on meal size and drinking rate. *J. Comp. Physiol. Psychol.* **83,** 379–387.
2. Davis, J., Collins, B. J., and Levine, M. W. (1975). Peripheral control of drinking: Gastrointestinal filling as a negative feedback signal, a theoretical and experimental analysis. *J. Comp. Physiol. Psychol.* **89,** 985–1002.
3. Davis, J. D., and Smith, G. P. (1988). Analysis of lick rate measures the positive and negative feedback effects of carbohydrates on eating. *Appetite* **11,** 229–238.
4. Miller, N. E., and Kessen, M. (1952). Rewarding effect of food via stomach fistula compared with those via the mouth. *J. Comp. Physiol. Psychol.* **45,** 555–569.
5. Kohn, M. (1951). Satiation of hunger from food injected directly into the stomach vs food ingested by mouth. *J. Comp. Physiol. Psychol.* **44,** 412–422.
6. Berkun, M. M., Kessen, M., and Miller, N. E. (1952). Hunger reducing effect of food by stomach fistula vs food by mouth measured by consummatory response. *J. Comp. Physiol. Psychol.* **45,** 549–557.
7. Le Magnen, J. (1956). Rôle de l'odeur ajoutée au régime dans la régulation quantitative à court terme de la prise alimentaire chez le rat blanc. *C. R. Soc. Biol. (Paris)* **150,** 136–139.
8. Le Magnen, J. (1956). Hyperphagie provoquée chez le rat blanc par altération du mecanisme de satiété périphérique. *C. R. Soc. Biol. (Paris)* **150,** 32–34.
9. Rolls, B. J., Van Duijvenvoorde, P. M., and Rowe, E. A. (1983). Variety in the diet enhances intake in a meal and contributes to the development of obesity in the rat. *Physiol. Behav.* **31,** 21–28.

10. Treit, D., Spetch, M. L., and Deutsch, A. J. (1983). Variety in the flavor of food enhances eating in the rat: A controlled demonstration. *Physiol. Behav.* **30,** 207–212.

11. Mook, D., Brane, J., and West, G. (1986). Satieties and cross specificity for three diets in the rat. *Physiol. Behav.* **36,** 887–895.

12. Schultz, Y. M., and Lawrence, D. H. (1958). Learning of a discrimination by satiated and deprived rats with sucrose incentive. *Am. J. Psychol.* **71,** 563–567.

13. Le Magnen, J. (1960). Effet d'une pluralité de stimuli alimentaires sur le déterminisme quantitatif de l'ingestion chez le rat blanc. *Arch. Sci. Physiol.* **14,** 411–419.

14. Rogers, P. J., and Blundell, J. E. (1980). Investigation of food selection and meal parameters during the development of dietary induced obesity. *Appetite* **1,** 85–88.

15. Kraly, F. S., and Smith, G. P. (1978). Combined pregastric and gastric stimulation by food is sufficient for normal meal size. *Physiol. Behav.* **21,** 405–408.

16. Liebling, D. S., Eisner, J. D., Gibbs, J., and Smith, G. P. (1975). Intestinal satiety in rats. *J. Comp. Physiol. Psychol.* **89,** 955–966.

17. Antin, J., Gibbs, J., and Smith, G. P. (1977). Intestinal satiety requires pregastric food stimulation. *Physiol. Behav.* **18,** 421–425.

18. Mook, D. G., and Brandsey, S. A. (1982). Preference–aversion drinking in the rat: The nature of postingestive satiety for sugar solutions. *Appetite* **3,** 297–307.

19. Kushner, L. R., and Mook, D. G. (1984). Behavioral correlates of oral and postingestive satiety in the rat. *Physiol. Behav.* **33,** 713–718.

20. VanderWeele, D. A., and Van Itallie, T. B. (1983). Sham-feeding is inhibited by dietary-induced obesity in rats. *Physiol. Behav.* **31,** 533–537.

21. Smith, M., and Duffy, M. (1957). Some physiological factors that regulate eating behavior. *J. Comp. Physiol. Psychol.* **50,** 601–608.

22. Booth, D. A. (1972). Conditioned satiety in rats. *J. Comp. Physiol. Psychol.* **81,** 457–471.

23. Booth, D. A., and Davis J. D. (1973). Gastrointestinal factors in the acquisition of oral sensory control of satiation. *Physiol. Behav.* **11,** 23–30.

24. Deutsch, J. A., and Wang, M. L. (1977). The stomach as a site for rapid nutrient reinforcement sensors. *Science* **195,** 89–90.

25. Weingarten, H. P., and Molikovsky, C. (1980). Taste to post-ingestive consequences. *Physiol. Behav.* **25,** 471–476.

26. Mook, D. G., Culberson, R., Gelbart, R. J., and McDonald, K. (1983). Oropharyngeal control of ingestion in rats: Acquisition of sham-drinking patterns. *Behav. Neurosci.* **97,** 574–584.

27. Van Vort, W., and Smith, G. P. (1987). Sham feeding experience produces a conditioned increase in meal size. *Appetite* **9,** 21–29.

28. Snowdon, C. T. (1969). Motivation, regulation and the control of meal parameters with oral and intragastric feeding. *J. Comp. Physiol. Psychol.* **69,** 91–100.

29. Deutsch, J. A., Young, W. G., and Kalogeris, T. J. (1978). The stomach signals satiety. *Science* **201,** 165–166.

30. Deutsch, J. A., Gonzalez, M., and Young, W. (1980). Two factors control meal size. *Brain Res. Bull.* **5,** (Suppl. 4), 55–58.

31. Deutsch, J. A., and Gonzalez, M. F. (1980) Gastric nutrient content signals satiety. *Behav. Neural Biol.* **30,** 113–116.

32. Kraly, F. S., and Smith, G. P. (1978). Combined pregastric and gastric stimulation by food is sufficient for normal meal size. *Physiol. Behav.* **21,** 405–408.

33. Campbell, S., and Davis, J. D. (1974). Peripheral control of food intake: Interaction between test diet and postingestive chemoreception. *Physiol. Behav.* **12,** 377–384.

34. Campbell, S., and Davis, J. D. (1974). Licking rate of rats is reduced by intra-duodenal and intraportal glucose infusions. *Physiol. Behav.* **12,** 357–366.

35. Kalogeris, T. J., Reidelberger, R. D., and Mendel, V. E. (1983). Effect of nutrient density and composition of liquid meals on gastric emptying in feeding rats. *Am. J. Physiol.* **244,** R865–871.

36. Reidelberger, R. D., Kalogeris, T. J., Leung, P. M. B., and Mendel, V. E. (1983). Postgastric satiety in the sham-feeding rat. *Am. J. Physiol.* **244,** R872–881.

37. Koopmans, H. S. (1975). Jejunal signals in hunger satiety. *Behav. Biol.* **14,** 109–125.

38. Wirth, J. B., and MacHugh, P. R. (1983). Gastric distension and short-term satiety in the rhesus monkey. *Am. J. Physiol.* **245,** R174–180.

39. Gibbs, J., Maddison, S. P., and Rolls, E. T. (1981). Satiety role of the small intestine examined in sham-feeding and rhesus monkey *J. Comp. Physiol. Psychol.* **95,** 1003–1015.

40. Gallouin, F. (1981). Intérêt nutritionnel et déterminisme de la caecotrophie chez le Lapin. Thèse de Doctorat d'Etat, Université Pierre et Marie Curie, Paris.

41. Deutsch, J. A., and Hardy, W. T. (1977). Cholecystokinin produces bait shyness in rats. *Nature (London)* **266,** 196.

42. De Saint Hilaire-Kafi, Z., Depoortere, H., and Nicolaïdis, S. (1989). Does cholecystokinin induce physiological satiety and sleep? *Brain Res.* **488,** 304–310.

43. Lorenz, D. N., and Goldman, S. A. (1982). Vagal mediation of cholecystokinin satiety effects in rats. *Physiol. Behav.* **29,** 599–604.

44. Janowitz, H. D., and Grossman, M. I. (1949). Some factors affecting the food intake of normal dogs and dogs with esophagotomy. *Am. J. Physiol.* **159,** 143–148.

45. Booth, D. A., and Jarman, S. P. Ontogeny and insulin dependence which follows carbohydrate absorption in the rat. *Behav. Biol.* **15,** 159–172.

46. Lindberg, N., Coburn, C., and Stricker, E. (1984). Increased feeding by rats after subdiaphragmatic subdiabetogenic streptozotocin treatment: A role of insulin in satiety. *Behav. Neurosci.* **9,** 332–345.

47. Langhans, W., Ziegler, V., Sharrer, E., and Geary, N. (1982). Stimulation of feeding in rat by intraperitoneal injection of antibodies to glucagon. *Science* **218,** 894–895.

48. Geary, N., and Smith, G. P. (1982). Pancreatic glucagon and postprandial satiety in the rat. *Physiol. Behav.* **28,** 313–322.

49. Geary, N. (1987). Food deprivation dissociates pancreatic glucagon's effects on satiety and hepatic glucose production at dark onset. *Physiol. Behav.* **39,** 507–511.

50. Geary, N., and Lesauter, J. (1989). Brief intraportal infusion of pancreatic glucagon reduces spontaneous meal size in rats. International Conference on the Physiology of Food and Fluid Intake (ICPFFI), Paris.

51. Geary, N., Langhans, W., and Scherrer, E. (1981). Metabolic concomitants of glucagon-induced suppression of feeding in the rat. *Am. J. Physiol.* **241,** R330–335.

52. Brala, P. M., and Hagen, R. L. (1983). Effects of sweetness perception and caloric value of a preload on short term intake. *Physiol. Behav.* **30,** 1–9.

53. Bellisle, F., and Le Magnen, J. (1981). The structure of meals in humans: Eating and drinking patterns in lean and obese subjects. *Physiol. Behav.* **27,** 649–658.

54. Spiegel, T. A., Shrager, E. E., and Stellar, E. (1989). Responses of lean and obese subjects to preloads, deprivation and palatability. *Appetite,* **13,** 45–85.

55. Kissileff, H. R. and Thornton, J. (1982). Facilitation and inhibition in the cumulative intake curve in man. In A. J. Morrison and P. Strick (eds.) *Changing concepts in the nervous system* 585–607. New York: Academic Press.

56. Rolls, B. J., Rolls, E. T., Rowe, E. A., and Sweeney, K. (1981). Sensory specific satiety in man. *Physiol. Behav.* **27,** 137–142.

57. Rolls, E. T., Rolls, B. J., and Rowe, E. A. (1983). Sensory-specific and motivation-specific satiety for the sight and taste of food and water in man. *Physiol. Behav.* **30,** 185–192.

58. Rolls, B. J., Van Duijvenvoorde, P. M., and Rolls, E. T. (1984). Pleasantness changes and food intake in a varied four courses meal. *Appetite* **5,** 337–348.

59. Rolls, B. J., Hetherington, M., and Burleg, V. J. (1988). Sensory stimulation and energy density in the development of satiety. *Physiol. Behav.* **44,** 727–733.

60. Kissileff, H. R., Gruss, L. P., Thornton, J., and Jordan, H. A. (1984). The satiating efficiency of foods. *Physiol. Behav.* **32,** 319–332.

61. Walike, B. C., Jordan, H. A., and Stellar, E. (1969). Preloading and the regulation of food intake in man. *J. Comp. Physiol. Psychol.* **68,** 327–333.

62. Jordan, H. A. (1969). Voluntary intragastric feeding: Oral and gastric contributions to food intake and hunger in man. *J. Comp. Physiol. Psychol.* **68,** 498–506.

63. Spiegel, T. A., and Jordan, H. A. (1978). Effects of simultaneous oral intragastric ingestion on meal patterns and satiety in humans. *J. Comp. Physiol. Psychol.* **92,** 133–141.

64. Geliebter, A. (1988). Gastric distention and gastric capacity in relation to food intake in humans. *Physiol. Behav.* **44,** 665–668.

65. Campbell, R., Hashim, S. A., and Van Itallie, T. B. (1971). Nutritive density in man. *N. Eng. J. Med.* **285,** 1402.

66. Porikos, K. P., Hesser, M. F., and Van Itallie, T. B. (1982). Caloric regulation in normal-weight men maintained on a palatable diet of conventional foods. *Physiol. Behav.* **29,** 293–300.

67. Birch, L. (1986). Conditioned and unconditioned caloric compensation: Evidence for a self-regulation of food intake in young children. *Learn. Motivation* **16**, 341–355.

68. Foltin, R. W., Fischman, M. W., Emurian, C. S., and Rachlinski, J. J. (1988). Compensation for caloric dilution in humans given unrestricted access to food in a residential laboratory. *Appetite* **10**, 13–24.

69. Wooley, S. C. (1972). Physiological versus cognitive factors in short-term food regulation in the obese and non-obese *Psychosom. Med.* **34**, 62–68.

70. Kissileff, H. P. (1984). Quantitative relationship between palatability and food intake in man. In M. Kare and J. G. Brand, (eds.) *Interaction of chemical senses with nutrition* 1986 (pp. 293–337). New York: Academic Press.

71. Woo, R., Kissileff, H. P., and Pi-Sunyer, F. X. (1984). Elevated post-prandial insulin levels do not induce satiety in normal-weight humans. *Am. J. Physiol.* **247**, R745–749.

72. Reid, C. (1943). The higher centers and the blood sugar curve. *J. Physiol. (London)* **102**, 20P.

73. Fischer, U., Hommel, H., Ziegler, M., and Jutzi, E. (1972). The mechanism of insulin secretion after oral administration: III–Investigations on the mechanism of a reflectory insulin mobilization after oral stimulation. *Diabetologia* **8**, 385–390.

74. Louis-Sylvestre, J. (1976). Preabsorptive insulin release and hypoglycemia in rats. *Am. J. Physiol.* **230**, 56–60.

75. Louis-Sylvestre, J. (1978). Relationship between two stages of prandial insulin release in rats. *Am. J. Physiol.* **235**, E103–111.

76. Louis-Sylvestre, J., and Le Magnen, J. (1980). Palatability and preabsorptive insulin release. *Neurosci. Biobehav. Rev.* **4**, (Suppl. 1), 43–46.

77. Louis-Sylvestre, J., Giachetti, I., and Le Magnen, J. (1983). Vagotomy abolishes the differential palatability of food. *Appetite* **4**, 295–299.

78. De Jong, A., Strubbe, J. H., and Steffens, A. B. (1977). Hypothalamic influence on insulin and glucagon release in the rat. *Am. J. Physiol.* **233**, E380–388.

79. Berthoud, H. R., and Jeanrenaud, B. (1982). Sham-feeding induced cephalic phase of insulin release in the rat. *Am. J. Physiol.* **242**, E280–285.

80. Berthoud, H. R., Trimble, E. R., Siegel, E. G., Bereiter, D. A., and Jeanrenaud, B. (1980). Cephalic phase insulin secretion in normal and pancreatic islet-transplanted rats. *Am. J. Physiol.* **238**, E336–340.

81. Freyse, E., Kiene, S., Brincmann, T., and Fischer, U. (1982). Plasma insulin and glucose tolerance in pancreatectomized dog after partial pancreatic transplantation. *Hormone Metabol. Res.* **14**, 521–522.

82. Strubbe, J., and Van Wachen, P. (1981). Insulin secretion by the transplanted newborn pancreas during food intake in fasted and fed rats. *Diabetologia* **20**, 228–236.

83. Berthoud, H. R. (1984). The relative contribution of the nervous system hormones and metabolites to the total insulin release during a meal in the rat. *Metabolism* **33**, 18–26.

84. Niijima, A. (1989). Effect of taste stimulation on the rate of discharges in autonomic nerve innervating visceral organs in rats. *Appetite* **12,** 227.

85. Powley, T., and Berthoud, H. R. (1989). Neuroanatomical basis of cephalic phase reflex. *Appetite* **12,** 78.

86. Berthoud, H. R., Fox, E., and Powley, T. (1980). Localization of vagal ganglionics that stimulate insulin and glucagon secretion. *Am. J. Physiol.* **258,** R340–347.

87. Penick, S. B., Prince, H., and Hinkle, L. E. (1966). Fall in plasma content of free fatty acids associated with the sight of food. *N. Eng. J. Med.* **275,** 416–419.

88. Simon, C., Schlienger, J. L., Sapin, R., and Imler, M. (1986). Cephalic phase insulin secretion in relation to food presentation in normal and overweight subjects. *Physiol. Behav.* **36,** 465–469.

89. Sjönstrom, L., Garellick, G., Krotkiewski, M., and Luyckx, A. (1980). Peripheral insulin in response to the sight and smell of foods. *Metabolism* **29,** 901–906.

90. Hansen, B. C., Lever, H., Polonsky, K., and Van Cauter, E. (1986). Pulsatility of pancreatic islet peptides. *Am. J. Physiol.* **251,** E215–226.

91. Lang, D. A., Matthews, D. R., Petto, J., and Turner, R. C. (1979). Cyclic oscillations of plasma glucose and insulin in human beings. *N. Eng. J. Med.* **301,** 1023–1027.

92. Lucas, F., Bellisle, F., and Di Maio, A. (1987). Spontaneous insulin fluctuations and the preabsorptive insulin response to food ingestion in humans. *Physiol. Behav.* **40,** 631–636.

93. Bellisle, F., Demozay, F., Blazy, B., and Le Magnen, J. (1983). Reflex insulin response associated to food intake in human subjects. *Physiol. Behav.* **31,** 515–519.

94. Deutsch, R. (1974). Conditioned hypoglycemia: A mechanism for saccharin-induced sensitivity to insulin in the rat. *J. Comp. Physiol. Psychol.* **86,** 350–358.

95. Pavlov, I. P. (1927). Conditioned Reflexes. Oxford: Oxford University Press.

96. Nicolaïdis, S. (1963). Effets sur la diurèse de la stimulation des afférences buccales et gastriques par l'eau et les solutions salines. *J. Physiol. (Paris)* **55,** 309–310.

97. Nicolaïdis, S. (1966). Nouvelles données sur le réflexe potohydrotique. *J. Physiol. (Paris)* **58,** 574–575.

Chapter Seven

Brain Mechanisms of Meal Eating

A meal being initiated, eating is sustained until the onset of satiety. Eating a given amount of the offered food is required to reach this satiety, which is an effect of the abolition of initiating stimuli. What brain mechanisms are involved in this satiating effect of eating a food in which (as shown in preceding chapters), together, the orosensory, gastrointestinal, and systemic activities of the ingested food are implicated? Another brain function in feeding to be examined is the motor side of the performance. Eating is essentially to take   food in the mouth, to masticate and salivate that food, and to swallow it. The neurophysiological basis for this motor program of eating will be overviewed in this chapter.

I. *Brain Mechanisms of Satiation*

We have seen that the size of a meal results from a balance between the initial strength of the stimulation (which initiates the meal) and the level of counteracting action of the sustained eating of the food at oral, gastrointestinal, and systemic levels. We have seen that the sensory component of the initial stimulation, i.e., the food palatability, is subserved by sensory projections in brain loci in which neuronal responses are found to be hunger-dependent. While satiation is the lifting of the initial stimulation to engage in the meal, the same network of pathways involved in systemic and sensory meal initiation will also probably account for the satiation process. This is the rule of all positive–negative feedback mechanisms.

A. Sensory-Specific Satiation in the Brain

As seen earlier, neurons in the lateral hypothalamus of monkeys are found to be activated by the sight or the taste of a food in the hungry animal only, and only when the food is known as such, i.e., palatable for the animal. The same neurons demonstrate the satiating process. Their firing, from the resting level, progressively decreased while monkeys ate and no longer responded to the sight or taste of the food when the eating came to an end. In addition, the sensory specificity of satiation, behaviorally demonstrated in rats and humans, is exhibited. The explored neuron, no longer activated by sensory stimuli from a food, was reactivated by another food, which the monkeys resumed eating (1) (Fig. 7.1). Comparable results were also obtained in rats (2, 3).

In Chapter 5, Section I, the role of centrifugal pathways as a function of hunger and of food palatability in the modulation of gustatory afferents in the nucleus of the solitary tract (NTS) and in the olfactory bulb was described. The inversion of this activation both in the NTS and in the olfactory bulb is associated to the orosensory component of satiation. Moreover, vagal afferents from the stomach to the same brain regions participate in the shift of activity from hunger to satiety. Inflating a balloon in the stomach inhibits gustatory discharges in the NTS of rats, and selectively discharges to sweet stimuli (4) (Fig. 7.2). The same distension of the stomach reverses the hunger activation in the mitral cell layer of the olfactory bulb (5). In the NTS, this effect is probably due to interactions between the rostral gustatory and vagocaudal projections. Their modulation apparently depends on centrifugal fibers. Electrical stimulation of the periventricular nuclei affects vagal responses in the caudal portion of the NTS (6).

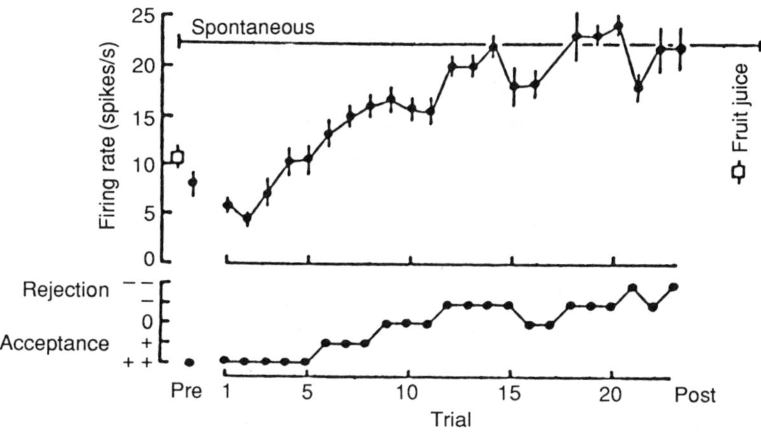

Figure 7.1 Sensory-specific satiation is manifested by the neuronal response to the sight and taste of foods in the monkey. [From Rolls (1)].

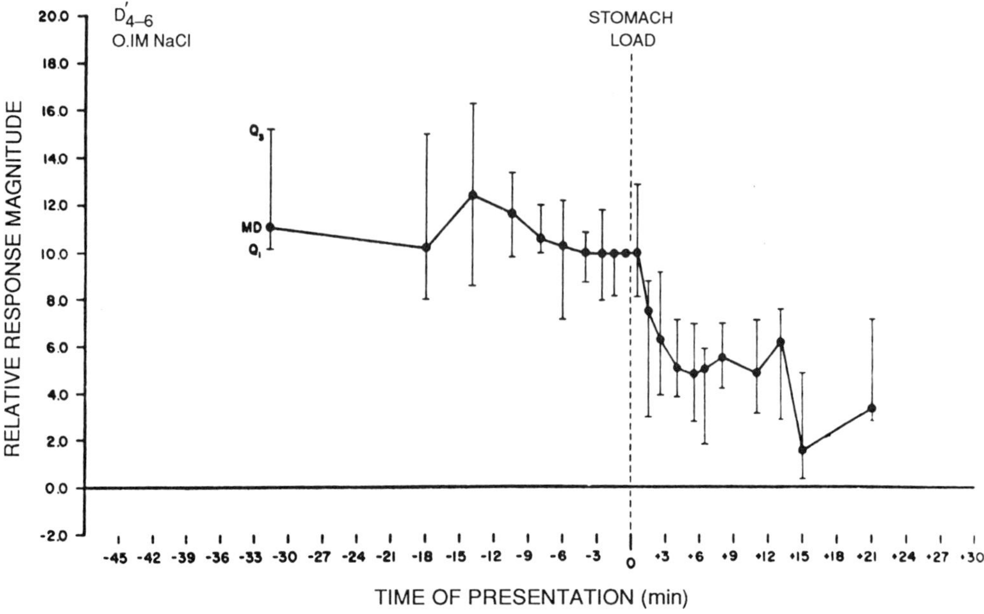

Figure 7.2 Inflating a balloon in the stomach modifies responses of NTS neurons to taste stimuli applied to the tongue. [From Glenn and Erickson *et al.* (4).]

B. The Gastrointestinal Neural Feedback

The role of gastric fullness and of neural afferent stimulated by gastric distension widely varies according to meal size, this size depending itself on various conditions in the strength of the meal stimulation. After food deprivation, the stomach is maximally filled because the initial level of positive feedback is high. In varied meals, in animals as well as humans, the sensory-specific satiation on successively or simultaneously presented foods also causes big meals, limited only by the maximal distention of the stomach. In a feeding schedule, huge meals needed to ensure the body energy balance (which probably results from a learning) also suggest the maximal distension of the gastric reservoir as a limiting factor. The role for gastric distension in satiation seems to be restricted to such big meals. The classic works of Paintal (7) and Iggo (8) described the responses of mechanoreceptors of the stomach walls transmitted to the brain via the vagal afferent fibers. Their role is not exclusive of that of other putative sensors and of other neural pathways. On the contrary, vagotomized rats, which should take big meals due to the loss of sensing the distension, take small meals.

Deutsch (9) acknowledges the distinction between large and small meals. Large meals would only be limited by gastric distension and vagal afferents. Small meals would be limited through sympathetic pathways responding to some receptors of the nutritive properties of the food. Splanchnectomy (by cutting the sympathetic afferent route) eliminates the immediate resumption of eating when the food is withdrawn from the stomach. However, this finding does not clearly support the role for sympathetic innervation in limiting small meals. The role for gastric chemoreceptors is, as yet, a hypothesis not supported by histological or electrophysiological investigations.

It is tempting to think that the involved chemoreceptors are not in the stomach and rather, are those identified in the duodenum by Mei (10) and his co-workers. The finding of these intestinal chemoreceptors was briefly mentioned early. By recording single afferent fibers of the vagus nerve in cats at the level of the plexiform ganglia, fibers were found responding to hexoses applied in the duodenum. The stimulation of the intestinal chemoreceptors by carbohydrates immediately induces an insulin release. In the jejunum, similar glucosensitive fibers were found. There they are sympathetic fibers, and their stimulation by hexoses immediately reduced plasma insulin under hyperglycemia. This effect was blocked by vagotomy and atropin treatment. In the duodenum, vagal fibers also responded to both hexoses and amino acids, or specifically to amino acids alone (11).

Thus, the macronutrient and caloric content of the intake may be detected in the small intestine through this intestinal taste. In contrasting to the success of purely hypothetical hepatic glucoreceptors, the importance of this finding of intestinal ones was, for unknown reasons, overlooked. Their functional role in signaling the levels of both emptiness and fullness of the small intestine or the rate of gastric emptying merits further investigation. Already the role they play together in meal initiation and meal termination may be suspected. The effect of their stimulation on insulin release is highly suggestive. It may be a key factor in the so-called entero–insular axis and responsible for pre- and postabsorptive pancreatic responses to the ingested food.

A number of electrophysiological studies showed that abdominal and hepatic vagal fibers project into the dorsal caudal portion of the NTS. At this level, the activity of the involved neurons interacts with area postrema, the dorsomotor vagal nucleus, and the nucleus ambiguus, the latter concerned in insulin release. In addition, it was shown that pathways from various branches of the vagus nerve run parallel to the gustatory afferent through the parabrachial relay to reach the lateral hypothalamus. There they converge to loci activated by both sensory and humoral signals (12–14).

II. *Central Mechanisms of Mastication and Swallowing*

Mastication and swallowing are centrally programmed reflexes. Both, however, are modulated by the sensory inputs eliciting the reflex motor pattern. This sensory input is generated by the food and its consistency and fluidity in the mouth. The chewing motor program is highly flexible in its temporal evolution, rhythmicity, and amplitude according to the task required to reach the product of mastication and salivation represented by a swallowable bolus. Through the latter, which represents the threshold of swallowing, chewing and swallowing are linked together. The coordination of their central mechanisms, therefore, may be suspected. Swallowing elicited by sensory stimulations in the pharynx, larynx, and esophagus is a more rigidly programmed pattern. However, it is also physiologically modulated by the consistency of the bolus. Furthermore, both central mechanisms, especially swallowing, are centrally modulated by corticofugal pathways. Investigating these central modulations of reflexes must explain phenomena in which chewing and swallowing appear to be highly dependent on hunger and satiety. We have seen that a palatable food in a hungry subject is swallowed more rapidly, i.e., after less mastication, than a low-palatability food, and the high-palatability food is also masticated more before swallowing at the approach of satiety. In forced overeating in humans, it is of common experience that swallowing becomes rapidly blocked by oversatiety and that the effort to swallow induces nausea and eventually vomiting. Thus, it is presumed that the lateral hypothalamus or closely connected structures exert a potent regulatory control on the specific systems generating the chewing and swallowing patterns.

Surprisingly enough, the motor side of feeding (mastication and swallowing, which are effectors of the behavior) is poorly understood and largely neglected by specialists of the behavior. By contrast, numerous studies on the classic neurophysiology attempted to identify the structure and organization of masticatory and swallowing centers in the brain. Unfortunately, an obvious gap exists between these neurophysiological (mainly electrophysiological) investigations and the actual functions of chewing and swallowing in feeding. Further studies, in this field like in many others, will be needed to join central neurophysiology and neurobiology of behaviors.

A. Mastication

Numerous books and extensive reviews have been published on the neurophysiological investigations of mastication (15). Only the main points will be summarized here, with some additional comments regarding the relationship of the behavioral motor output and mechanisms contributing to the behavior i.e., the hunger–satiety system.

Masticatory muscles (masseter and temporal muscles, elevators of the jaw; mylo-hyoid and anterior digastric muscle depressors) are commanded by the cranial trigeminal nerve, lingual muscles by the hypoglossal nerve. Masseter muscles are particularly rich in spindles, or proprioreceptors. The sensory input from these muscle spindles is transmitted by the trigeminal nerve. Lingual and periodontal mechanoreceptors provide the main sensory input transmitted by the hypoglossal branch. These afferents project in mesencephalic masticatory nuclei, as recognized early by Cajal. Monosynaptically, this primary projection is connected to spinal motoneurons, the bulbar motor masticatory nuclei. In addition to this brainstem system, a masticatory area is identified in the motor cortex. Recording neuronal activity in the mesencephalic nuclei, rhythmic bursts of activity were recorded, following the electrical stimulation of the cortical area or a mechanical peripheral stimulation (16).

In line with the old Sherrington reflex theories, it was thought that a programmed generator of masticatory movements was in action and could act independently of a sensory feedback from the periphery. This surprising idea was apparently supported by the fact that these rhythmic bursts persisted after a muscular paralysis. Two series of arguments may oppose this notion of a pre-programmed constant generator pattern driving the motor output. A paralysis of masseter muscles eliminated only the input from muscle spindles involved in the control of the amplitude and strength of the muscular contraction, but the main sensory input from periodontal and lingual receptors was not eliminated. On the other hand, wide variations in the amplitude and sequences of chewing with the hardness of foods, with (as shown early) hunger and food palatability, and, lastly, with individual masticatory habits could not be interpreted. Obviously, the motor output and its modulation resulted from a summation between the central generator pattern and the returned afferent input. This realistic notion is now prevailing (17).

However, considerable gaps in our knowledge about the neural control of mastication still exist. No study allows us to understand the shift from sucking a liquid or semiliquid solution to a mastication of a solid food from an unknown threshold of the solidity, requiring chewing and salivation for the food being swallowed. This appreciation of the swallowability inhibiting or initiating the mastication implies an unknown coordination between masticatory and swallowing centers. This presumed coordination between sites controlling hunger and satiety and masticatory centers is unexplored.

Finally, another aspect of mastication has been overlooked. This motor pattern in humans is widely variable among individuals. The volume of a mouthful taken in the mouth differs among individuals. With the same food and the same volume of the mouthful, the chewing pattern also will differ widely. A personal eating style is recognized. Such chewing habits highly depend on the state of dentition of the subjects, alternating left to right chewing in the same sequence or masticating on one side only or only with incisors. The complexity of food textures and consistencies is also an important variable. Some items—for example, a bit of chocolate—represent a homogeneous texture. Instead, a mouthful including both soft and hard materials and bread imposes a specific flexible and adaptative pattern (18).

Together, these facts suggest the notion that the actual chewing of foods is learned. With a background of central preprogrammed reflex patterns, a learning of chewing from infancy forms flexible and permanently adapted chewing habits in the individual. What is the reinforcer of this learning? Like in every learning it is the result, in this case, the achievement of easy and rewarding swallowing of the food as well as the optimization of the sensory-based palatability of this food.

We cannot escape the peculiar case of the apparently gratuitous mastication of chewing gum. Chewing gum holds aromas, but the long-lasting mastication not followed by swallowing seems to be *per se* the reward of this strange behavior. It is not a hunger-cutting behavior, as it is in starved people. One possibility is that this inedible mastication is the equivalent of the nonnutritive sucking so commonly observed in infancy. These totally unuseful behaviors are not at all exceptional in humans.

The last interesting question is: What is the energetic cost of mastication? Is the mastication of hard and dry low-caloric foods a source of energy expenditure that exceeds the provided calories by their ingestion? This is unknown.

B. Central Mechanisms of Swallowing

Extensive reviews of neurophysiological studies of swallowing reflexes have been published. Their overview here is inspired by the main works carried out by Jean *et al.* (19) and Jean and Car (20) in sheep and rats.

Swallowing and propelling the masticated and salivated food or a fluid from the mouth to the stomach is performed by the coordinated contractions of a series of oral, laryngeal, pharyngeal, and esophageal muscles. This motor output is generally divided into two phases: buccopharyngeal and esophageal. The sequence begins by the contraction of the myeloid muscle followed by that of the laryngeal muscle and of the middle and inferior constrictors of the pharynx. This buccopharyngeal complex is followed by the activation of muscles of the various esophageal segments. The esophageal phase excluded, the swallowing is completed in 600 msec in sheep and in 1 sec in humans. The complex can be elicited by the electrical stimulation of the superior laryngeal nerve. However, the afferent system includes the maxillary branch of the trigeminal nerve, the glossopharyngeal nerve. These nerves innervating the tongue, pharynx, larynx, and esophagus are all stimulated through mechanoreceptors. Afferent fibers reach their central target through two different pathways: the solitary tract and the spinal trigeminal pathway. They converge into the nucleus of the solitary tract and the adjacent reticular formation.

This projection has been identified as the main swallowing center in which an interneuronal network generates the sequence of motoneuron and muscle activations. These motoneurons commanding the various involved muscles are located in the rostrocaudal portion of the nucleus ambiguus and in the dorsomotor nucleus of the vagus. No direct connection exists between the afferent and efferent systems. Their relationships are organized by a neuronal network acting as a "central pattern generator" of the motor outcome of swallowing (21, 22). This complex includes two different areas: the caudal NTS and the adjacent reticular formation, the ventral reticular formation surrounding the nucleus ambiguus.

A microelectrode exploration of interneurons in the NTS by Jean identified their role in generating the sequence of swallowing. They are divided into three categories. The firing of buccopharyngeal neurons begins before and at the beginning of swallowing elicited by the stimulation of the superior laryngeal nerve. Their activity initiated in some neurons by one of two spikes is a succession of bursts. Late and very late neuron activities correspond to the esophageal phase. The sequence of activity of the various explored neurons is similar to the sequence of peripheral activation of the muscles. The caudal NTS neurons appear to be the main programming center, whereas neurons of the nucleus ambiguus areas are involved in a dispatching to the various and successively activated motoneurons. Under curarization, the entire neuronal pattern is maintained and may be stimulated independently of a muscle afferent feedback. This is true only for the neuronal pattern elicited by the electrical stimulation of the supralaryngeal nerve or by a neocortical stimulation.

Despite these fascinating neurophysiological advances, further studies are needed to account for the physiological regulation of swallowing and its link to mastication. An apparent strong facilitatory and inhibitory influence on the swallowing centers from brain sites in which the stimulation to eat is generated has not been investigated (23). Centrifugal pathways from the LH and amygdala to the NTS and to the frontal cortex are probably involved. Another gap in our knowledge is the exact mechanism of stimulations of various lingual, laryngeal, and pharyngeal nerves by the bolus ready to be swallowed. What determines the readiness to swallow, i.e., the physical nature of stimuli (semisolid or liquid)? And what are the thresholds of activation of the various peripheral nerves to be reached to turn on the reflex? This is still unknown.

References

1. Rolls, E. T. (1981). Central nervous mechanisms related to feeding and appetite. *Br. Med. Bull.*, **37**, 131–134.

2. Ono, T., Tamura, R., Nishijo, H., Nakamura, K., and Tabuchi, E. (1989). Contribution of amygdalar and LH neurons to visual information processing of food and nonfood in monkey. *Physiol Behav.* **45**, 411–421.

3. Shiraishi, T. (1988). Feeding related LH neuron responses to odors depend on food deprivation in rats. *Physiol. Behav.* **44**, 591–597.

4. Glenn, J. F., and Erickson, R. P. (1976). Gastric modulation of gustatory afferent activity. *Physiol. Behav.* **16**, 561–568.

5. Chaput, M., and Holley, A. (1976). Olfactory bulb responsiveness to food odor during stomach distension in the rat. *Chem. Senses Flavor* **2**, 189–201.

6. Rogers, T., and Pelton, E. (1984). Neurons of the vagal division of the solitary nuclei activated by the PVN of the hypothalamus. *J. Autonom. Nerv. Syst.* **10**, 193–197.

7. Paintal, A. S. (1954). A study of gastric stretch receptors: Their role in the peripheral mechanism of satiation of hunger and thirst. *J. Physiol. (London)* **126**, 255–260.

8. Iggo, A. (1957). Gastric mucosal chemoreceptors with vagal afferent fibers in the cat. *J. Exp. Physiol.* **42**, 398–409.

9. Deutsch, G., Gonzalez, C., and Young, W. (1980). Two factors control of meal size. *Br. Res. Bull.* **5**, 55–57.

10. Mei, N. (1982). Sensory structures in the viscera. In *Progress in sensory physiology*, Vol. 4 H. Autrum, D. Ottoson, E. R. Perl, R. F. Schmidt, H. Shimazu, and W. D. Willis (eds.), (pp. 1–42). Berlin: Springer-Verlag.

11. Jeanningros, R. (1981). Vagal unitary responses to intestinal aminoacids infusion in the anesthetized cat: A putative signal in protein satiety. *Physiol. Behav.* **28**, 9–21.

12. Laughton, W., Campfield, L. A., and Nelson, O. (1986). Hepatic portal and gastric afferent processing in NTS of the rat. *Neuroscience* **17**, 385.

13. Barone, F. C., Wayner, M., Aguilar-Baturoni, H. U., and Guevaras-Aguilar, R. (1979). Effects of cervical vagus nerve stimulation on hypothalamic neuronal activity. *Brain Res.* **4**, 381–392.

14. Adachi, A. (1981). Electrophysiological study of hepatic vagal projections to the NTS. *Neurosci. Lett.* **24**, 19–23.

15. Anderson, D. J., and Matthews, B. (eds.) (1976). *Mastication*. Bristol: John Wright & Sons, 270 pp.

16. Matthews, B. (1976) Reflexes elicitable from the jaw muscles in man. In D. J. Anderson and B. Matthews (eds.), *Mastication*, 139–146. Bristol: John Wiley & Sons.

17. Thexton, A. J. (1976). To what extent is mastication programmed and independent of peripheral feedback? In D. J. Anderson and B. Matthews (eds.), *Mastication* (pp. 213–219. Bristol: John Wright & Sons.

18. Lund, J. P. (1976). Evidence for a central neural pattern generator regulating the chewing cycle. In D. J. Anderson and B. Matthews (eds.), *Mastication* (pp. 204–212). Bristol: John Wright & Sons.

19. Jean, A., Car, A., and Roman, C. (1975). Comparison of activity in pontine versus medullary neurons during swallowing. *Exp. Brain Res.* **22,** 201–220.

20. Jean, A., and Car, A. (1979). Inputs to the swallowing medullary neurons from the peripheral afferent fibers and the swallowing cortical area. *Brain Res.* **178,** 567–572.

21. Jean, A. (1984). Control of the central swallowing programme by inputs from the peripheral receptors: A review. *J. Autonom. Nerv. Syst.* **10,** 225–233.

22. Car, A., Jean, A., and Roman, C. (1975). A pontine primary relay for ascending projections of the superior laryngeal nerve. *Exp. Brain Res.* **22,** 197–210.

23. Kesler, J. P., and Jean, A. (1985). Identification of the medullary swallowing in the rat. *Exp. Brain Res.* **30,** 256–273.

Chapter Eight

Regulation of Body Energy Balance and Body Weight

Many studies on eating behavior are limited to investigating some particular aspects of the behavior without any prospect of the regulatory role of the behavior. However, in most other works, it is not forgotten that we eat to be nourished. Therefore, such studies, explicitly or implicitly, are intended to find out whether or not, and how much feeding behavior participates in an overall homeostatic control of body energy and body nutritional balances. Can the neural control of food intake adjust the energy intake to energy expenditures and the steady-state conditions, and in conditions of extra expenditures such as exercise or cold exposure? Can this adjustment of the behavior satisfy the need of various macronutrients? Can it, alone or not, physiologically balance required retention of energy in, for example, growth, pregnancy, and lactation? Finally, is food intake a regulator of body weight in adulthood and, if not, what is or what are the regulators of body weight?

Answers to these questions are difficult. Many of them are still controversial. These controversies and discrepancies often originate from misleading concepts and from questionable measurements.

I. Regulation of Body Energy Content and Body Composition in the Animal Model

What is the exact meaning of body energy balance? In a steady state condition, i.e., in a state where energy retentions (such as growth) or extra expenditures (such as exercise) are excluded, a body energy balance implies that the cumulative food intake over time in terms of ingested energy matches the energy expenditure. In this case, the intake is called a normophagia. When this is not the case, and the caloric intake overexpenditures are observed in excess or in deficit, the excessive intake is reputed to be a hyperphagia, the deficient one a hypophagia. In these two cases, the large-capacity body energy store of fats plays a physiological role by storing excessive energy intake or by supplying an internal fuel to maintain the lean tissue oxidative metabolism. However, it is not so simple.

Two types of hyperphagia and hypophagia should be distinguished in their relations with the regulation of the body fat mass and, thus, of body weight. Hyperphagia will be either a cause of increase in body fats or an effect of the restoration of a previous reduction of fatness. Hypophagia will be either a cause of body fat and weight loss or an effect of a previous increase in the body fat store. In other words, a regulation of the body fat mass tending to its constancy apparently is superimposed on the control of intake and indirectly is a controller of intake. What is this lipostatic mechanism? Is this mechanism and its effects on intake sufficient to explain both the body energy balance and a constancy of body weight? This is the question.

A. Measurements: Real and Sham Concepts

Ideally, a study on the body energy balance, i.e., of balance between the energy intake and body energy expenditures, requires a simultaneous measurement of the two terms of the balance: food intake (in terms of the caloric supply by the diet) and the energy output. These simultaneous middle- or long-term measurements on the same animal were rarely achieved. In the absence of this simultaneous measure, the basal metabolic rate, instead of the total metabolic rate, of the fed and active rat is often taken into account. In separated or simultaneous measurements, various sources of errors and misunderstanding, cannot be avoided. All the energy expenditure is, of course, converted into heat output; therefore, its best measurement is a direct measure of this heat production, called a direct calorimetry. Used in human studies, this technique is never used in animal studies. The measure of oxygen consumption (VO_2) allows a calculation of energy expenditures based on the calculated thermic coefficient of the diet. This caloric value of the food is often questionable. For instance, the caloric net supply of proteins must take into account the loss by urea synthesis and excretion and, this loss being subtracted, proteins account for 4.7 kcal/g only (1). Part of the measured or calculated heat production is due to the so-called prandial thermogenesis, i.e., the short-term increase of heat production associated to the meal. We will see later the origin of this food-induced thermogenesis, partly and not exclusively due to the exothermic first metabolic handling of foods, particularly of proteins, in the liver. This extra heat production does not enter the oxidative metabolism but does enter the measured total metabolic rate of the fed animal. To interpret this measure in terms of body energy balance, it is inappropriate to subtract this part of the heat production by the food intake from the calculated caloric value of the food. If this is done, the result will be erroneously interpreted as a positive energy balance.

The concept of body energy balance, i.e., of a regulated constancy of the body energy content, solves the problem of energy retentions and of the constancy of body composition. Even in the steady-state conditions, the body energy content and the energy intake required for its maintenance would have to take into account invisible energy retentions or losses due, for instance, to the catabolic–anabolic turnover of tissues.

In the absence of a calorimetric measurement of energy expenditures, the achievement or not of a body energy balance by a hypothetic adjustment either of expenditures to intakes or of intakes to expenditures is currently assessed by a simple measure of body weight change. A weight gain in gram is taken as a measure of a positive energy balance and, thus, of hyperphagia and a weight loss as a measure of a negative energy balance and, thus, of hypophagia.

Much confusion stems from such measurements and interpretations. The most evident involves the notion and computation of feed efficiency, also called feed efficiency for growth or metabolic efficiency of the food. According to studies, it is either the ratio of the current body weight : total caloric intake in a condition compared to the previous one or the ratio of the Δ-weight : Δ-intake. When the ratio increases, the feed efficiency is said to be decreased because the amount of intake to obtain 1 g of weight gain has increased and is higher than thought or calculated to produce this weight gain. Thus, it is concluded that the excess of intake, not explained by the weight gain, measures an extra heat production, which is affirmed despite the absence of its calorimetric measure. When the ratio is decreasing and the feed efficiency for weight improved, the current interpretations are the opposite. This notion and the leading interpretations are extremely rough.

The calculation of the energy cost for growth is biased for at least three reasons. The computation is made without taking into account that all weight gain includes 71% water retention. Thus, to conclude that 1 g of weight gain is due to 1 g of deposed fats is an enormous error. The water retention being excluded, the cost of 1-g deposed fats cannot be, as currently calculated, 3–4 kcal of foods (it would be a miracle) but is of at least 10 kcal corresponding to the caloric value of 1 g fats plus the cost of its synthesis and mobilization. Another reason is in the fact that weight gain is not necessarily and exclusively a gain of body fats. In studies on energy balance and regulation of body weight, a direct measurement of the carcass lipids should ascertain a conclusion drawn from the body weight measurement. Finally, the calculation of a feed efficiency in a condition of weight loss is not usable and cannot determine whether this weight loss is due to fat oxidation and its reducing effect on food intake or to a true deficit of intake over unchanged or increased energy expenditures.

A regulation of body weight derived from the concept of body energy balance is also a difficult notion. The body weight maintenance does not depend solely on the body energy balance. This body energy balance is involved in the constancy or alteration of the body fat mass as a component of body weight but not of the body weight *per se*. No one has claimed or observed that astronauts compensate for weightlessness by considerably increasing their food intake. The main component of body weight is water, and body water content is regulated by means and behaviors other than those active in the body energy balance.

In this regulation of the body fat mass, generally called "of body weight," the main point to be considered is the functional role of body fats. This large-capacity store of energy is established during growth by an energy intake added to the actual intake needed by the oxidative metabolism. As in all stores, its functional role is to fluctuate like a ballast of energy. In other words, its functional role is to absorb an occasional short- or long-term excess of intake overexpenditures and to relay a short- or long-term deficit of the body energy balance by providing an internal fuel. Therefore, the functional role of body fats is not constant. However, a species-specific long-term constancy of body fat mass and, thus, of adult body weight over time is observed. This suggests that body fat mass is controlled by its own regulatory system limiting and correcting its overrepletion and limiting and correcting its overdepletion. Such a lipostatic mechanism is, as shown later, experimentally demonstrated. However, a limit and reparation of the overrepletion of body fats is possible only if it is associated to a reduced external intake and the limit and correction of its overdepletion if associated to an increase of intake. Thus, two types of hyperphagia and hypophagia will be distinguished. In primary hyper- and hypophagias, the excess or deficit of intake overexpenditures will cause increase and decrease, respectively, in body fats. In secondary hyper- and hypophagias, the increase or decrease in intake will be an effect of the regulatory system of body fat. I propose that these secondary hyper- and hypophagias be called over- and underphagias.

Another and very disputed point is whether or not body fats are the only buffer of the body energy balance added to the regulatory control of food intake. In an occasional or persistent condition of excess of intake overexpenditures, the regulation may possibly be operated by an extra heat regulatory production by a leak of excessive energy, like water diuresis acts to regulate body water content. In an occasional or permanent deficit of intake, a sparing of energy expenditures and a low metabolic rate also may be a regulator of body energy balance. As shown later, modern studies confirmed these suggestions, i.e., a partial and limited role of this second reensuring system in body energy balance and body weight regulations.

B. The Achievement of Body Energy Balance

The preceding chapters of this volume explained how the stimulation to eat and the intake from meal to meal in the *ad libitum* condition are permanently adjusted to both the nutritive requirements and the nutritive properties of foods. A rapid mechanism is the adjustment of the time of meal initiation and of the postprandial satiety to both the previous meal size and the metabolic food utilization from meal to meal. The second and delayed mechanism is the learning of the palatability and satiating capacity of foods allowing an anticipatory response to the nutritive properties of foods based on their sensory activities. However, as an effect of the dual (prandial and diurnal) periodicity of feeding, a permanent imbalance of body energy is present on a short- (30 or more min) and middle-term basis (12 or more hr).

In Chapter 2, Section II, the diurnal metabolic feeding cycle was described. At night in rats, a positive energy balance is recorded. The excessive intake is stored in the small-capacity store (the stomach) and body fats in the large one. A neuroendocrine background is associated with and presumably a cause of this physiological body energy imbalance. Hyperphagia is a secondary hyperphagia. It is an effector of the positive energy balance but an effect of the neuroendocrine mechanism: hyperinsulinemia, high glucose tolerance leading to fat synthesis and to the diversion of a part of ingested food toward this fat synthesis. Inhibiting this synthesis immediately eliminates hyperphagia. During the day, the opposite trend is recorded. Fat mobilization and oxidation are observed, accounting for prolonged postprandial satiety and, thus, retarded meal initiations. A neuroendocrine mechanism (glucose intolerance, low insulin responsiveness) appears early before the first intake. Hypophagia is thus a secondary hypophagia. The blockade of fat oxidation or mobilization by insulin or mercuro-acetate immediately resumes normophagia or induces hyperphagia. The main finding was the negative correlation between the nocturnal positive energy balance and the subsequent deficit. This suggested that the body energy balance was achieved on a daily basis through the diurnal balance of fat synthesis–fat lipolysis and as an effect of a lipostatic mechanism. Preventing hyperphagia or all intake at night immediately provokes hyperphagia and high fat synthesis on the subsequent day. Provoking this hyperphagia by insulin during the day immediately induces hypophagia and weight loss on the subsequent night.

In a group of rats, the achievement of the energy balance during 24 hr was tested (2). O_2 consumption and food intake were simultaneously and continuously recorded for 5 consecutive days. The residual caloric imbalance at the end of each 24-hr period (positive balance at night, minus negative during the day) was plotted against the slight weight gain or weight loss during the same 24 hr (Fig. 8.1). A significant correlation was present between the residual positive or negative balance and weight changes. This relation permitted the calculation of the caloric cost of the weight loss or gain of 1–3 g at the end of the days and the evaluation of its likelihood to be caused by a fat depot. Taking into account the water retention, this cost was of the order of 10 kcal/g synthesized fat, a value consistent with the hypothesis.

Together, these findings were the first demonstration that a physiological lipostatic mechanism does exist and is active in the nocturnal–diurnal pattern to achieve a middle-term body energy balance. They also demonstrate that this specific regulatory system of body fats indirectly influences and is permitted by the feeding mechanism. Beyond 24 hr, it is suggested that these coordinated

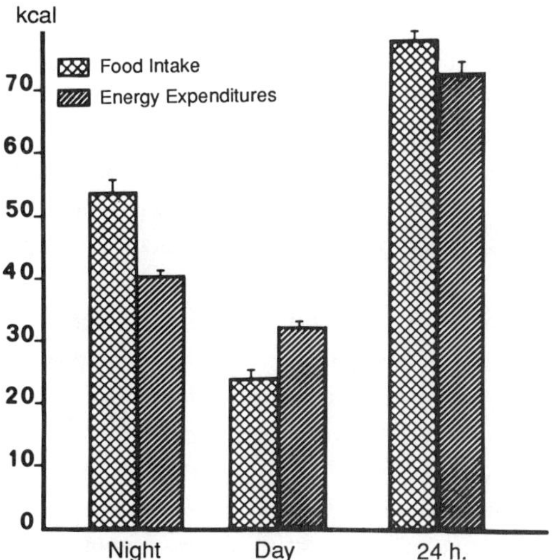

Figure 8.1 Nocturnal positive and diurnal negative energy balances.

lipostatic and feeding regulatory mechanisms are operating to achieve simultaneously the long-term body energy balance and the mean constancy of body fats.

II. *Reversibility of Induced Overweight and Underweight*

A. Reversible Obesities

Four different means are used to produce experimental overweight and adiposity. The simplest is the forced feeding of 150% or more of the previous *ad libitum* intake by gastric tubing. Another means is the already-mentioned persistent electrical stimulation of the lateral hypothalamus inducing hyperphagia and weight gain. The most common procedure is to induce hyperphagia and weight gain by long-acting insulin treatment or continuous regular insulin infusion. Finally, rats become spontaneously obese by the cafeteria regimen.

Early on, Cohn *et al.* (3), using gavage in rabbits, revealed the reversibility of induced obesity and the associated hypophagia. After 2 mo of tube-feeding, animals doubled their weight. Refed *ad libitum,* they were aphagic for 3–4 wk while they lost weight until a return to their initial weight. During this weight loss and aphagia as well as later on, they also exhibited a low level of activity.

Jen and Hansen (4) confirmed and extended this result to monkeys. Monkeys were tube-fed 100, 125, 135, or 155% of their *ad libitum* oral intake. This overfeeding, added to a residual oral intake, induced obesity. At the discontinuation of the gavage, monkeys were hypophagic for 14–50 days. The duration of the hypophagia was clearly correlated to the level of the intragastric feeding and to the induced weight gain. Young rats, tube-fed for 1 wk 150% of their previous oral intake, gained 25 g more than controls tube-fed 100% the *ad libitum* intake. Refed *ad libitum,* rats were hypophagic for 5 days. This time was exactly that of the disappearance of the elevated carcass lipids observed in a control group (5).

Hyperphagia and obesity induced by a chronic electrical stimulation of the lateral hypothalamus in rats was already described (Chapter 3). Here again, the induced overweight is lost at the cessation of the electrically induced overfeeding, and during the weight loss rats are aphagic.

As initially demonstrated by May and Beaton (6), daily protamin-zinc-insulin (PZI) administration in rats induces 24-hr hyperphagia by slightly increasing the nocturnal intake and, most of all, by eliminating functional diurnal hypophagia. In other words, the chronically induced hyperinsulinemia eliminates the lipostatic mechanism as physiologically operative in the regulation of the body energy balance within the diurnal cycle. Rats thus treated (hyperphagic and deposing fats during the day as well as during the night) become obese. This was confirmed by the simultaneous recording of O_2 consumption, CO_2 expiration, and food intake in rats on the last day of a 10-day PZI treatment (Fig. 8.2) (7). The high level of the respiratory quotient (RQ) during the day, identical to that at night, indicated the continuous 24-hr fat synthesis. The diurnal meal pattern, identical to the nocturnal one, confirmed the origin of 24-hr hyperphagia. Two days of continued recordings after the discontinuation of the insulin injections demonstrated postinsulin hypophagia and weight loss. This weight loss was substantiated by the 24-hr permanently low RQ, indicating the current lipolysis. Thus, under PZI treatment the metabolic nocturnal pattern was extended to the day. At the cessation of treatment, the normal metabolic and feeding diurnal pattern was observed at night.

A PZI treatment prolonged for 2 mo revealed an important fact (8): To maintain the rate of weight gain up to 210 g after 2 mo, the dose of insulin administered needed to increase progressively. This was the first indication that the efficiency of chronic insulin in inducing obesity (i.e., the insulin-induced fat synthesis) decreases with the induced augmentation of the body fat mass. This loss of insulin responsiveness as a function of an elevated body fat mass, otherwise confirmed, will be commented on regarding the cellular basis for the lipostatic mechanism.

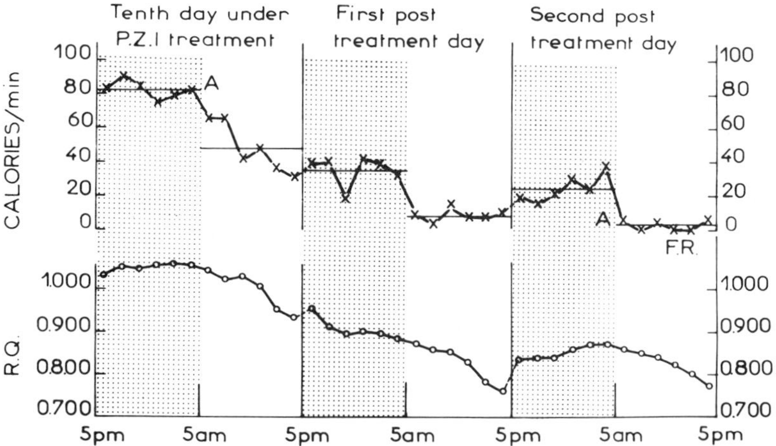

Figure. 8.2 Nocturnal and diurnal feeding and metabolic rates on day 10 of PZI treatment and on 2 days posttreatment.

While returning to the initial body weight, other evidence indicates that hypophagia is not only temporarily associated with but is also an effect of lipolysis and fat oxidation. In rats, the lipolysis during the 6 days of weight loss following PZI treatment was assessed by the observed high plasma-free fatty acids (PFFA), glycerol, and ketone body levels. Contemporary hypophagia disappears at the time of the disappearance of this blood manifestation of lipolysis (9). An extended study of this postinsulin metabolic pattern was carried out by Geary and his co-workers (10–12). In one particular study, the authors showed that after only 3 days of PZI treatment, rats were hyperglycemic after some hours of food removal at night as well ad during the day, along with the elevated PFFA and glycerol levels.

The reversibility of obesity induced by the cafeteria regimen was sometimes questioned. Returning to the previous initial body weight and the disappearance of the associated hypophagia were initially found by Sclafani and Springer (13) and confirmed by others (14); however, these results were disputed by Rolls and Rowe (15), who showed that after a long time on a cafeteria regimen, overweight was maintained despite the return to the stock diet feeding. This discrepancy was highlighted by Mandenoff *et al.* (16), who showed that the overweight induced by a relatively short period of cafeteria feeding was indeed reversible while this reversibility disappeared after months of the high-palatability food intake. This loss of the reversibility coincides with the appearance of an adipose tissue hyperplasia (newly formed adipocytes). A fat cell hypertrophy is only present as long as adiposity is reversible.

An experiment by Rogers (17) adds interesting details to this phenomenon. Two groups of rats were fed the cafeteria regimen. One group in the *ad libitum* condition became obese; the other group, restricted to its previous energy intake, did not. When both groups returned to the stock diet they were hypophagic, but the first group more and longer than the second. Thus, a part of the hypophagia is not due to obesity but to the shift from the choice of high-palatability food to the single habitual food. In the obese group, the hypophagia was due to a decrease of both meal sizes and meal number. After 3 wk, only 60% of the weight gain was lost. At this time, the meal size returned to normal, but the meal frequency remained low. The author concluded that the postcafeteria hypophagia was mainly achieved by a reduced meal frequency.

B. Regulation of Underweight

Returning to normal body weight after a weight loss and the associated hyperphagia are a mirror image of reversible overweights. This is so common that, for a long time, it was neglected as evidence for the lipostatic mechanism. During weight loss experimentally induced in rats by fasting or restrictions or by famines or undernutrition in humans, the body fat store plays its physiological role by providing an internal fuel allowing survival. Fat oxidation is substituted peripherally to carbohydrates. While PFFA is not oxidized by brain tissues, the maintenance of brain metabolism, essential for survival, is maintained by a residual food intake, neoglucogenesis, and the use of ketone bodies and amino acid breakdown products in brain oxidative metabolism (18). The total body fat depletion precedes a rapidly lethal self-combustion of proteins. This fasting metabolism has been the subject of extensive works and reviews.

The deteriorating effect of fasting is combined with a reduced metabolic rate greater than accounted for by the reduced metabolic body size (19). A striking fact is that during this phase of postfasting restoration of body fats, rats offered a choice of fat-, carbohydrate-, or protein-rich diets increased their intake of the high-fat diet (20). Following 3 days of fasting, a refeeding by carbohydrates produced within the first 3 hr an elevation of plasma insulin and, after 24 hr, a 15% rise in the fast-induced decrease of the metabolic rate (21). This is not observed in a refeeding by fats. Thyroid hormones are apparently involved in the reduction of metabolic rate by fasting and in the recovery during refeeding. Plasma tri-iodothyronine low during restriction, was found rapidly rising in the refed rat (22). This alteration of thyroid hormones would be a cause of a reduced adrenergic activation of the brown adipose tissue, contributing to the reduced metabolic rate (23).

The preceding chapters showed that a rapid and regulating response to short-term food deprivation in rats contrasted with a low response during the day and that these short-term responses also contrasted with the poor feeding response to 2 or 3 days of fasting. However, it was also shown that after a more accentuated and prolonged fasting or food restriction, refed rats, underweight by 25%, returned to their initial body weight within 2 wk and that this weight gain was indeed associated to hyperphagia. It was shown that this hyperphagia accounted for the regain of weight with calculated 8–10 kcal of excess of intake for 1 g of synthesized fat. Both hyperphagia and the rate of weight gain were lower following the same underweight induced by a total starvation compared to restriction. The difference was imputed to the impairment of insulin release and insulin responsiveness (starvation diabetes) induced by fasting and, more generally, by fat oxidation. Based on the result, it was assumed that the restoration of a normal glucose tissue utilization impaired by fasting was a requisite for both the postfast augmented intake and the rehabilitation of the normal weight (24, 25).

C. Hibernators

The weight gain–weight loss cycle of hibernators resembles the day–night cycle of rats and that experimentally induced by insulin treatments. In various hibernators (squirrels, dearmice, hamsters), the weight gain preceding hibernation is associated with hyperinsulinemia and elevation of insulin responsiveness. *In vitro* pancreatic islets are more responsive to glucose (26). Presumably, as an effect of increased insulin sensitivity of adipocytes and of the increased insulin release, the weight gain per gram of food intake is progressively augmented during this prehibernation phase and falls sharply in the subsequent period of weight loss. This elevation of the insulin-induced fat synthesis may alone account for the weight gain without an associated hyperphagia. Squirrels were either fed *ad libitum* or restricted during the period of weight gain. Squirrels in the first group were hyperphagic and accumulated fats. In the second group, squirrels restricted to the intake of the preceding period gained weight and accumulated abdominal fats like the *ad libitum* hyperphagic controls (27). Thus, the metabolic pattern is the *primum movens* of weight gain and subsequent weight loss. Hyperphagia and hypophagia, when observed, seem to be secondary to this metabolic pattern. This is reminiscent of the finding by Towby (recalled earlier) that insulin-treated rats, maintained at the same level of intake as controls by tube-feeding, nevertheless augment their fat depot (Chapter 2, Section VI).

III. *Regulation of Body Weight by the Energy Output*

A. Dietary-Induced Thermogenesis of the Brown Adipose Tissue

A series of works by Rothwell and Stock (initially observed as provocative) were rapidly confirmed by many other investigators. They gave evidence for an extra heat production by the brown adipose tissue in a dietary-induced thermogenesis, i.e., provoked by an excess of the dietary intake over the weight maintenance level. The brown adipose tissue, located in the interscapular region and along the rachis in animals, was known for a long time to be a thermogenic organ. These organs function in the nonshivering thermogenesis of cold-acclimated animals. The details of brown adipose tissue-specific cellular metabolism leading to heat production were extensively studied. Various tests of its activation were developed and are routinely used: weight and local temperature of the tissue (generally the interscapular one), cell size, volume and number (hypertrophy and/or hyperplasia), the guanosine diphosphate binding, cytochrome oxidase, etc.

In their initial princeps experiment, Rothwell and Stock (28) used rats overeating the cafeteria regimen. Using the contestable measure of the metabolic efficiency of food, they showed that the weight gain of cafeteria rats increased by 11.24 kJ/g of food eaten versus 15.29 in controls. The calculated gain of body energy per gram of intake was 30% less in cafeteria rats than in controls. A measured increase of VO_2 by 25 and 30% during the weight gain of the overeating rats was not accounted for by the cost of fat synthesis and suggested the dietary-induced thermogenesis (29). This increase of the total energy expenditures persisted at the cessation of the cafeteria regimen in rats, who were hypophagic at this time. Therefore, the increase in thermogenesis was not due to the "heat increment of foods," nor to prandial thermogenesis only. The skin temperature at the level of the intrascapular brown adipose tissue was higher than that of controls in overeating rats and, likewise, in those rats injected with noradrenaline. Postmortem, the weight of the brown adipose tissue was found to be two times higher after the cafeteria treatment, as is found in cold-acclimated rats. A correlation was found between the elevated VO_2 and the weight gain of the brown adipose tissue. By measuring the VO_2 uptake of the brown adipose tissue (30), the recorded increase in O_2 consumption in cafeteria rats was shown to be entirely due to the brown adipose tissue thermogenesis and not to other tissues. The increases in this O_2 consumption by the cafeteria regimen and by cold exposure were additive (31).

It was objected that the metabolic efficiency could not permit accurate evaluation of the fat synthesis, and short-term measurement of O_2 consumption could have been misleading; therefore, the investigators renewed their experiment by establishing the body energy balance (energy output vs. energy retention as fats) in cafeteria rats. VO_2 was measured for 24-hr periods. Lipids in the carcass were measured in controls sacrificed at different times of their growth on the cafeteria regimen (32). Thus, the investigators confirmed the augmentation of the thermogenesis by 42% by the cafeteria regimen. Only 10% of the recorded thermogenesis as due to the exothermic cost of the synthesized fats.

Thus, it seemed convincingly demonstrated that the overrepletion of body fat mass in rats overeating high-palatable foods was not prevented but, rather, slowed by an extra heat production entirely due to the activation of the brown adipose tissue. Comparable results were later provided, showing that an overnutrition by tube-feeding and the increase of body fat mass by the insulin treatment were also halted by this induced leak of energy (33).

Brown adipose tissue thermogenesis is activated by norepinephrine administration and blocked by propranolol, and later we will see that it is functionally activated from the brain by descending adrenergic fibers. Its stimulation is insulin-dependent. It is impaired in diabetic rats (34). Thyroid hormones are also implicated (35, 36).

Despite these convincing results and many replications by other authors, the findings by Rothwell and Stock were strongly questioned and tentatively denied by several investigators (37). Replicating the same measurements as those of Rothwell and Stock, they denied that the fat synthesis did not entirely account for overeating or, more precisely, of a tube-feeding by more than 100% the *ad libitum* intake. These discrepancies were the origin of a controversy that reached an extreme and unusual degree of virulence. The opponents of Rothwell and Stock were qualified as "schizophrenic" in a letter to the editor of the very serious journal *The Lancet*. The origin of these discrepancies and issuing disputes have not as yet been clarified; however, the findings of Rothwell and Stock are now universally accepted and generally confirmed.

B. Origin of Dietary-Induced Thermogenesis

What is the origin of the onset of this extra heat regulatory production? Is it the overeating *per se* or the overrepletion of body fats, which is never prevented but only stamped by the increase of energy output?

A possibility exists that the increase of body fats above its regulated constancy might bring about its own counterregulatory mechanism and, simultaneously the activation of brown adipose tissue. Later, an activation of the ventromedial hypothalamus (VMH) will be shown to govern the lipostatic mechanism and to counteract the repletion of body fats by sympathetic adrenergic pathways to the pancreas, liver, and adipose tissue itself. It will also be described that a VMH lesion, which disrupts the lipostatic mechanism, disrupts the extra heat production by the brown adipose tissue. Thus, the central command of the two systems could be brought about by the same, as-yet unsettled, stimuli from body fat repletion although this has not been experimentally demonstrated.

Despite the arguments provided by Rothwell and Stock—that the increased thermogenesis of overfed rats was entirely due to the brown adipose tissue thermogenesis—other data suggest that an increase of the prandial thermogenesis could be the source of the recorded 24-hr elevation of heat overproduction.

Glick (38) gives evidence that the meal brings about the brown adipose tissue activation. At a time during which the prandial thermogenesis is observed after a meal, the brown adipose tissue activation is assessed by the increased blood flow, local O_2 consumption, and increased guanosine diphosphate binding. The author consequently suggests that the dietary-induced thermogenesis is the cumulated thermogenesis of the successive meals, increased as a function of their size and number. The fact that cafeteria-fed rats take huge meals supports this hypothesis.

On the other hand, cafeteria-fed rats eat high-palatability foods, and new findings argue that a part of the prandial thermogenesis is palatability-dependent and increases with the palatability level. In the dog, a comparison of the prandial thermogenesis by recording O_2 consumption in sham-feeding and real-feeding condition revealed that the initial part of the thermogenesis was present in sham-feeding and real feeding but that the latter part, called digestive thermogenesis, was alone observed in tube-feeding and absent in sham-feeding. The amplitude of this cephalic phase of thermogenesis does not depend on the meal size but, rather, on the palatability of the food (39, 40). This cephalic phase of the prandial thermogenesis was latter confirmed in humans. The following experiment supports the hypothesis. For 45 days, rats received the same amount of a liquid food either orally or by tube-feeding. The latter became heavier, while the rate of fat synthesis and of glucose utilization in the brown adipose tissue was lower. The denervation of the brown adipose tissue eliminated the difference between orally fed and tube-fed rats. The author concluded that the absence of a sensory stimulation of the brown adipose tissue was responsible for the adiposity in tube-fed rats (41). Whether or not sensory stimulation is concerned in the prandial activation of the brown adipose tissue and, if so, how, is still unsettled. The intermediary of the rcflexly induced and palatability-dependent insulin release is a possibility.

IV. *Regulation of Body Energy Balance and of Body Weight in Humans*

As seen earlier, the feeding mechanism *per se* adjusts the current intake of foods to both the immediate metabolic requirement and the energetic properties of metabolizable foods. However, two reensuring mechanisms are in action to achieve a middle- or long-term body energy balance and, simultaneously, a mean constancy of the body fat mass and body weight. Fat reserves are used to buffer temporarily positive and negative balances, but their overrepletion or depletion is prevented by a lipostatic mechanism in which a lipolysis correcting the overrepletion reduces intake and lipogenesis correcting overdepletion increases intake. This mechanism is implemented on a short-term basis by an evacuation of the excess of energy as an extra heat production by the brown adipose tissue. Do the same mechanisms operate in humans? *A priori,* the impairment of these mechanisms, compared to those in animal models, may be suspected for two main reasons: Humans have little control on meal frequency. Moreover, humans eat in a permanent condition of cafeteria regimen, which, even in rats, favors a long-lasting positive energy balance and weight gain.

A. Relations of Food Intake to Constancy or Fluctuations of Body Weight

A measurement of the day-to-day food intake and of its relations with either a constancy or a change of body weight and also with measured day-to-day energy expenditures provides the first idea of the effectiveness of the three combined mechanisms without, however, discriminating the role played by each of them.

Apparently dismissing the above pessimism *a priori,* a large proportion of humans, for whom food is available at will, is commonly observed to maintain a constant body weight in adulthood over 30 or 40 years without voluntary dieting. Statistically and as a mean, a slight increase in weight with age is observed in wellnourished populations. In one particular statistic (42), 20-yr-old males weighing 65–75 kg weighed 75–90 kg when 50 yr old. The statistics include an unknown proportion of individuals who during the 30 yr strictly maintained the same body weight ± 1 or 2 kg. Based on this fact, it was easy calculating the extraordinary precision over time of the adjustment of intake to various components of expenditures. A persistent and uncorrected excess or deficit of intake of about 25 kcal/day (the equivalent of 6 g only of pure sugar) would lead to a weight gain or weight loss of 1.5 kg/yr and, thus, to 45 kg of weight gain or weight loss for 30 yr.

However, all investigators agree that adjustments of intakes to expenditures or the opposite are not achieved in humans on a daily basis. In a classic and conclusive experiment by Durnin (43, 44), the daily expenditures (between 2500 and 4000 kcal/day) and daily intake were measured for 7 consecutive days in 60 subjects. In 29 of them, a significant correlation was found between the daily energy output and the intake of one or two preceding days. Twenty-three had a mean weekly intake identical only to the mean expenditures during the week. At the end of the week, 18 subjects had a total intake higher or lower than their weekly expenditures and presumably gained or lost weight.

It is quantitatively assessed that a persistent positive energy imbalance leads to a proportional weight gain. In these cases, the body fat compartment plays its functional role in absorbing the excess. The best analysis was done by Forbes (45) (Fig. 8.3). He reexamined two old publications by Newman (N) (1902) and Gullok (G) (1922). These two gentlemen reported their measured

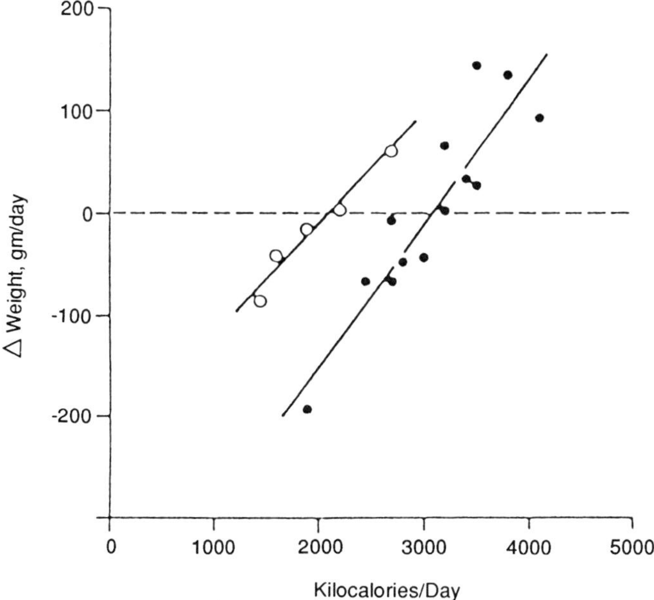

Figure 8.3 Long-term correlation between food intake and weight gain in humans. [From Forbes (45).]

daily food intake and the evolution of their body weight over a long period of time. Forbes shows that their daily intake plotted against their weight gain revealed a linear relationship between the two parameters. From the regression line, it is seen that the intake needed for a weight maintenance was 2119 kcal/day in N and 3078 kcal/day in G. The same regression allowed Forbes to determine the cost of each gram increase in the body weight: 9 kcal in N, 7.2 kcal in G. Water retention apparently was not taken into account in the last calculation. If it was, the real cost would be three times greater. This could indicate that at least two-thirds of the intake excess had been dispersed in an extra heat production limiting the body fat repletion.

The significance of the weight maintenance diet and of its difference among individuals as a function of their body weights was for a long time debated. The basal metabolic rates of humans of the same weight, age, and sex may vary from 1 to 2; in subjects of the same weight, daily intake could vary from 1 to 2 (46, 47). In 11 subjects weighing between 70 and 90 kg, a measure of the daily intake during 11 days showed no correlation between intake and body weight. However, the heavier subjects took the most calorie-dense foods. Among 71 children divided into heavy and lean, the basal metabolic rate was found lower in the first group than in the second one (48). In another study, young men of identical weight were paired on the basis of their maximal difference of intake. In each couple, no difference of basal metabolic rate was observed. A greater activity was noted in heavy eaters (49). This might suggest that above the basal metabolic rate the resting and total metabolic rate including an extra heat regulatory production occur in heavy eaters.

In this line, a study cited by James and Trayhurn (50, 51) is highly suggestive. One woman of a developing country (Gambia) ate 1300 kcal/day and, despite hard exercising and lactation, maintained a normal body weight. Nourished by 2500 kcal/day, she became obese. Women of developed countries nourished 1300 kcal/day become lean. Interestingly, the authors suggest that genetic factors are involved in selecting individuals with a low metabolic rate who can survive during infancy and can achieve their body energy balance with minimal food intake.

B. The Lipostatic Mechanism in Humans

Within 24 hr, the lipostatic mechanism operates in humans, like in rats, to compensate for the high positive balance during the day with a negative balance at night, even if this compensation remains imperfect from day to day. As already mentioned, a glucose intolerance hypoinsulinemia, observed at the beginning of the night, is the endocrine background of the nocturnal lipolysis and fat oxidation indicated by the fall of the RQ. The elevated index of lipid metabolism occurs at the end of the night, indicating the diurnal lipogenesis (52).

In another study, the daily body energy balance was established in 54 subjects in a calorimetric chamber. This 24-hr body energy balance was found to be extremely variable among subjects. The calculation from the RQ of the protein, carbohydrate, and fat balance showed that the body energy balance, either positive or negative, was correlated only to the fat balance (53). The authors concluded that a lipostatic mechanism is not operative to achieve the balance of energy intake and expenditure during the 24 hr. A separated analysis of the two half-days was, unfortunately, not made and could have shown that the lipostatic mechanism is indeed deficient on a daily basis and contributes to the body energy balance on a long-term basis only.

Several experiments of forced feeding were carried out in humans. Volunteers were asked to eat 1.5 to 2 times their *ad libitum* intake for weeks (54). All these experiments demonstrated, like in rats, the reversibility of an induced overweight. Most of the subjects gained weight and, at the cessation of the forced feeding, were hypophagic until a return to the initial body weight. However, most of these experiments also showed that the weight gain was lower than accounted for by the excess of intake.

The most original and conclusive experiment was done on nine subjects of the Massa tribe in Cameroun (55). These men were observed during a traditional cure of induced fatness by forced feeding for 65 days. The mean overeating was 264 kcal/day. The weight increased by 17 kg, including 11.5 kg of deposed fats. The mean resting metabolic rate was augmented by 43%, and the postprandial thermogenesis by 37%. The fat energy storage accounts for only 45% of the alimentary surplus.

This supports the view that, in addition to the lipostatic mechanism manifested by the reversibility of an induced obesity and the associated hypophagia, another reeinsurance of the body energy balance involves a modulation of the energy output. In another experiment (56), five subjects were fed first at a level of weight maintenance for 13 days and then by 1.7 times this level for 9 days. This overfeeding produces a 56% weight gain due to fat synthesis. The recorded elevation of the basal metabolic rate accounts for only one-third of the augmented daily expenditures. The rest is due to prandial thermogenesis, which increases proportionally to the amount eaten.

C. Dietary- and/or Prandial-Induced Thermogenesis

To know what part of the prandial thermogenesis is eventually activated in the regulation, this part, designated "facultative thermogenesis" by Jéquier, must be separated from the obligatory thermogenesis due to the first metabolic handling of ingested metabolites. According to Jéquier (57), this obligatory heat loss is 5% in the conversion of glucose to glycogen, 20% in lipids, and 25% in the conversion of proteins to metabolizable energy. In a nutritionally equilibrated meal, this heat loss is 10–14% the energy intake. A facultative thermogenesis added to this obligatory one is only observed in humans after a carbohydrate intake. The blockade by propranolol indicates this facultative thermogenesis and, in addition, demonstrates its adrenergic nature. The hyperinsulinemic–euglycemic clamp was the method of choice to control the carbohydrate-induced thermogenesis, at least via the parenteral route. In 12 normal-weight subjects, euglycemia was maintained by intravenous glucose infusions at five levels of maintained hyperinsulinemia. Thermogenesis was measured by indirect calorimetry. The thermogenesis was 5.3% of the glucose-infused energy in a physiological range of insulinemia (58). In a similar experiment, it was shown that two-thirds to three-quarters of the thermogenesis increase above the resting metabolic rate was due to the storage cost (obligatory thermogenesis) (59, 60).

Injecting propranolol (which eliminates facultative thermogenesis) permits further evaluation of its participation.

A series of new works suggests that a supplement of extra heat production or facultative thermogenesis is palatability-dependent and, thus, manifested when the food is orally ingested. The thermogenic effects of 734 kcal, either orally ingested or tubed in the stomach, were compared. The first 40-min thermogenesis was absent in the tube-feeding. In another experiment, subjects ate orally either a highly palatable meal (succession of varied foods) or an unpalatable meal (matched mixture of the same foods). Under the effect of the palatable meal, thermogenesis was elevated by 20% during the first 90 min versus 12% on the unpalatable meal. Postprandial glycemia was identical following the two meals, but plasma insulin and catecholamine levels were higher after the palatable one (61, 62). A more elaborated study confirmed this demonstration of a sensory-stimulated thermogenesis (63). O_2 consumption was compared in four conditions in nine subjects: (1) the ingestion of a palatable food, (2) visual, olfactory, and gustatory stimulations by the same noningested food, (3) gavage of the same meal, and (4) nonsensory stimulation. Following the four tests, the elevation of thermogenesis was, respectively, 12.3, 5.3, 5, and 1.1%. No difference in the thermogenic effect between an intragastric and intravenous intake was recorded (64); however, the postprandial RQ elevation was found to be identical after an oral compared to a gastric meal (65).

As in rats, the palatability-dependent cephalic phase of insulin release is suggested to play a role in this palatability-dependent thermogenesis in humans. It is tempting to speculate that its functional role is to eliminate a part of the excess of intake elicited by high-palatability food as extra heat production.

The origin of this sensory-stimulated and facultative thermogenesis and of its potential participation in the body energy balance is unknown. The brown adipose tissue cannot be involved. The thermogenic activity of this tissue, tested in cold exposure, is present during the first year of life, but the tissue disappears later. A role for muscles in the epinephrine-stimulated facultative thermogenesis was suggested. This muscular facultative thermogenesis was observed on the forearm and was associated with an increase in the local arterial epinephrine level (66).

Together, the results suggests that the body energy balance and a maintained body composition, including the fat content, are achieved in humans, with extreme precision over time, by the combined action of three mechanisms. The neurohumoral control of intake, operates mainly by the learning of palatability and satiating activities of foods, as a function of their caloric efficiencies. This slow adjusting process is supplemented within the diurnal cycle, from week to week and later by a lipostatic mechanism depleting or repleting the body fat energy reserve. This reensuring mechanism is helped by a short- or long-term modulation of energy losses.

V. *Feeding in Extra Expenditures and Energy Retention*

In addition to the basal and resting metabolic rates, the total metabolic rate includes temporary or permanent extra expenditures due to muscular activity or cold exposure. Added to the total energy output, energy retention occurs during growth and pregnancy. Lactation also is an energy extra expenditure. In these various conditions, the achievement of the body energy balance and of the physiologically commanded energy retentions requires the adjustment of food intake. In energy retentions such as growth and pregnancy, the adjustment needed is not only of the energy intake but also of specific nutrient intakes—that of fats and proteins. How is this adjustment performed?

A. Cold and Heat Exposures

Neutral ambient temperature is defined as the range of ambient temperature in which the heat loss of the body by convection and radiation is identical to the rate of metabolic heat production. Below this range, in cold ambient temperatures, the thermic gradient makes the rate of heat loss greater than the rate of heat production, thus threatening the maintenance of the constant body temperature in homeotherms. The solution is then to reduce the rate of heat loss or to increase heat production, i.e., the metabolic rate. In warm ambient temperature, above body temperature, the thermic gradient is inverted. Heat gain from the environment is added to heat production. The solution is then to avoid the heat gain and to increase the heat loss or to reduce the heat production. When the second eventuality occurs (increase or decrease of the heat production), the energy intake should be increased and decreased, respectively.

The thermoregulation by alterations of the metabolic rate is limited in various ways. In the cold, the first solution (reduction of the rate of heat losses) is achieved by vasoconstrictions of the skin and by furs in animal species living in a large thermic gradient. In humans, a behavioral thermoregulation (clothes, houses, and their heating) reduces the rate of losses so that it almost never exceeds the rate of heat production, thus requiring an elevation of the metabolic rate. Beyond the capacity of this regulation by the heat losses, a cold-induced thermogenesis does occur. Acutely, it is the result of shivering. After acclimation, a nonshivering thermogenesis occurs. Both of them must be matched by an increase of intake to avoid body weight loss. In the warm environment, thermoregulation brings about an efficient system of heat losses–sweating. The evaporation of each liter of water removes 500 kcal. If this was the only means of heat removal, 5 liters of sweat per day would be necessary in humans living in an environment higher than 37°C. In extreme heat, such as that of the desert, the loss of water to maintain body temperature may reach the catastrophic level of 1 liter/hr. The regulation of the body water content by thirst and salt intake is involved in this aspect of the thermoregulation.

Do changes of heat production and of food intake cooperate with this regulation? The increase of both the metabolic rate and food intake in the cold was demonstrated early in animal models. In rats acclimated to 3°C, the rate of glucose utilization was augmented along with that of the metabolic rate (67). This elevation of glucose utilization probably causes the concomitant increase in food intake. However, hepatic glycogen, not muscular glycogen, was reduced. Acutely, a 5°C ambient temperature immediately increases oxygen consumption and intake in some rats with a maintained body weight. In others, the elevation of intake is insufficient to prevent a weight loss during the first 6 days. O_2 consumption continues to increase parallel with thyroidal activity. Both return to basal level in 1 wk, presumably, according to the author, through the development of insulation (68). Rats at an ambient temperature of 5°C for 77 days exhibited a reduced nocturnal meal frequency and a slow increase of meal sizes, reaching a ceiling after only 2 wk. The normal meal pattern was unchanged during the day and was immediately reestablished at the return to 24°C (69). The results do not agree with those of another report (70). Simultaneous recordings on consumption and the meal pattern showed that the typical nocturnal meal pattern of large and frequent meals extended to the diurnal pattern.

At an arctic station, the daily intakes of 10 dogs were measured for 1 yr, and their variations were correlated to the ambient temperature varying between 0 and −30°C. The consumption was doubled during the winter. A negative correlation between the mean monthly temperature and the mean monthly consumption was statistically significant; this correlation was absent on a daily basis. However, the dogs lost weight during the polar winter (71) (Fig. 8.4).

Offered a choice of the three macronutrients, rats in a cold ambient temperature selected carbohydrates and augmented only the carbohydrate intake (72). This is also the case in rats injected with thyroid hormones (73). Other findings suggest that skin temperature may contribute to the increase of intake. Rats were chronically maintained at either 29° or 19°C. The latter ate more, but both groups ate more when fed at 19°C. The food temperature between 12° and 48°C had no effect (74, 75).

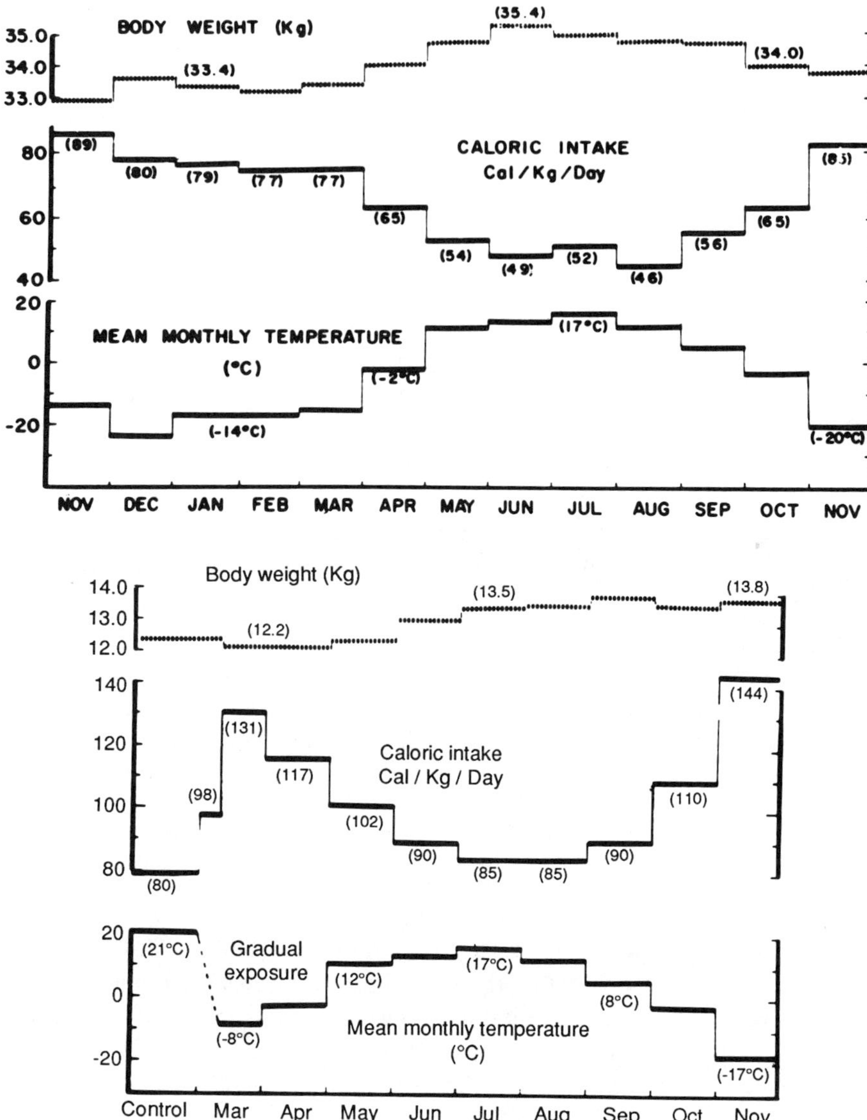

Figure 8.4 Food intake and body weight evolution in dogs placed in cold environment. [From Durrer (71).]

Few experiments have been done in humans in which behavioral thermoregulation generally avoids the call to a metabolic regulation. However, the mean consumption of a group of soldiers was compared in very warm and arctic settings. They consumed 3199 kcal/day in the desert and 4000 kcal/day in the arctic region. No increase in the selection of fats or proteins was observed (76).

The neurovegetative manifestation of thermoregulation is governed by thermosensitive neurons located in the posterior hypothalamus. Their cooling causes responses to cold, their heating to heat (77). Some works suggest that these thermosensitive brain areas also govern thyroid hormone release, responsible for the increase in metabolic rate and, in turn, food intake.

A considerable number of results demonstrates that the calorigenic activity of the brown adipose tissue is involved in nonshivering thermogenesis. This increase in energy expenditures by the augmentation of the rate of glucose utilization presumably accelerates the feeding mechanism.

Metabolic and feeding responses to heat are poorly documented. A reduction of intake resulting in a suppression of skin temperature through a lowered prandial thermogenesis may be a regulatory and perhaps learned process. Obviously, a repression of activity and of its induced extra heat is a factor of heat resistance and of the preservation of both body temperature and water content.

Rats placed at 92°F reduced their intake. They lost weight and were hyperthermic (78). According to Edholm and Goldsmith (79), in an extreme heat environment humans suppress their intake by 25%, and their energy expenditures are unchanged. They lose weight as an effect of both the negative energy balance and, presumably, unrepaired water loss.

B. Exercise

An acute and intense muscular activity can augment transitorily the metabolic rate five-fold. Chronically, the general activity in the total metabolic rate varies extremely in animal species such as humans. In humans, three types of exercise can be distinguished based on their levels, rates, and durations. The basal level of activity of awake individuals is approximately constant and identical. Added to this basal activity is the exercise of work, which, of course, varies greatly among individuals in level but, individually, is chronically performed and at a constant rate. The third category is the acute and extreme exercise of competitive sports, with an extreme variation in level, rate, and duration (e.g., the 100 m foot-race compared to the cyclist Tour de France). This latter type of exercise is essentially acute; it is generally not a permanent component of the middle-term body energy balance.

How are these low or high, chronic or acute extra expenditures balanced by food intake? Three classic works studied the feeding responses of rats in a forced exercise (80–82). Their results were convergent. Acute exercise reduced transitorily food intake and induced a weight loss. This food reduction and weight loss was later compensated for. The level of intake was restored, but rest—the cessation of the spontaneous activity—contributed to the rehabilitation of weight. Moderate work chronically imposed to the rat and separated by intervals of rest led to an overcompensation and to a middle-term weight gain. After an acute intense exercise, the duration of the reduced intake was about 18 hr and the delay of the compensation 35–40 hr.

These basic facts have been generally confirmed by a large number of works in rats, mice, and dogs (83, 84). The reduction of intake following a short, strong exercise was confirmed by Mayer *et al.* (82) to be the inverse phenomenon: Rats, rabbits, and also humans, when immobilized, do not reduce their intake and become obese (85). In a given range, the subject would be unresponsive to either an increase or decrease in energy expenditures by exercise or sedentariness. The postexercise metabolism may explain the transitory hypophagia. Muscles in untrained animals, following the use of glycogen, oxidize fats. The postexercise metabolism is typical of a postfat feeding generally associated with a transitory reduction of the intake of a carbohydrate-rich diet. Suppressed glucose oxidation, augmented fat oxidation, and lipolysis and insulin responsiveness are apparent during this postexercise phase. Insulin administration reduces this pattern in the tired muscle (86). Injection of lactate, proprionate, or acetate 30 min prior to the daily meal in the monkey diminishes the meal like forced exercise does (87).

In rats, along with the fat oxidation pattern, the urinary catecholamine excretion is elevated after exercise, and this elevation was found to be correlated to the suppression of intake. Treatment by a β-adrenergic blocker eliminated both urinary excretion of catecholamines and underfeeding (88).

In humans, an identical reduction of appetite and intake following a sudden high exercise was well documented. For instance, in a group of soldiers a hard training of 1 day led to a reduction in the same day of 1000 kcal of the voluntary food intake. The subsequent day, the intake was correlated to the amount of exercise the day before (89).

Sport medical literature provides a lot of information. It is reported that football players who trained 4 hr/day had a daily energy intake of 10,000 kcal (90). Cyclists running 15–40 km day may increase their daily intake by 25–50%. The caloric cost of a 100-km race at 40 km/hr was evaluated to be 2850 kcal. The daily intake of cyclists in the Tour de France, for example, is about 5000 kcal. In these runners of the Tour de France, the body energy balance was estimated in eight subjects for 22 days (91). Expenditures of 8900 kcal/day were found matched by an energy intake of 6000 kcal. Individually, no correlation was found between daily expenditures and daily intakes. Other reports confirmed this lack of short-term adjustment. The transitory suppression of intake after an acute effort was also reported, but this seems to vary greatly among individuals, trained and untrained subjects, and types of exercise. In athletes, the choice of the three macronutrients is not free because it is generally imposed by medical advisers. However, experimentally, it was shown that the increment of intake is made on carbohydrates and not on fats or proteins.

C. Pregnancy

Energy intake, weight gain, metabolic rate, and energy balance in pregnant rats, following the pioneering works of Brody and Kleiber, were studied in three different and elegant works by Morrison (92), Perisse and Salmon-Legagneur (93), and Champigny (94). A comparison of these works remarkably yields convergent and conclusive results.

Food intake increased as soon as the first day of pregnancy by 10–20%. This augmentation reached 50% at days 16–18 and then fell sharply before the days of parturition (95). The total metabolic rate (as measured by O_2 consumption) was augmented by 10%. Taking into account the concomitant rise of intake, the body energy balance was strongly positive until day 18: $+10.5$ kcal/day. The RQ ≥ 1 indicated the tissue synthesis and, particularly, the conversion of carbohydrates into fats. However, the elevation of metabolic rate (10%) was lower than that accounted for by the weight gain. Before parturition, the female gained 55% of her initial weight and 26.2% after parturition. In the difference, about 15% was represented by the weight of the young. Thus, at the eve of parturition, the tissue mass (mother plus fetuses) was augmented by about 42%. Referred to this weight gain and considering that the metabolic rate of fetuses per weight units is lower than it is in adults, the metabolic rate increase indicated a relative fall of the basal metabolic rate of the mother. The Morrison (92) and Champigny studies (94) agreed that one-third of the weight gain prior to parturition had been devoted to the energy retention by the formation of fetal tissues. The rest was accounted for by the weight gain of the mother (plus 27% after parturition). The body composition, tested in females sacrificed prior to parturition, did not indicate an increase in the proportion of lipids in the carcass (26%) (93). However, considering the 27% weight gain, the absolute amount of lipids and proteins synthesized in the mother's body during the gestation was also augmented by

27%. These energy retention and increase of energy stores of the pregnant female, also seen in other animal species, is thought to be a preparation to lactation.

In women, the metabolic rate increases progressively by 30% during pregnancy. Seventy-five percent of this increase would be due to the development of the uterus and to the fetus energy metabolism. According to Hytten and Leitch (96), the mean weight gain is augmented by 4 kg after 20 wk and by 12.5 kg at the end of pregnancy. Large interindividual variations were observed. At this time, the repartition of the weight gain is as follows: fetus, 3.4 kg; placenta and amniotic fluid, 1.45 kg; uterus, 0.97 kg; water retention, 2.92; energy gain stored by the mother, 3.3 kg.

This energy gain as fat deposits occurs during the first 6 mo. On the contrary, during the last quarter fat reserves are partially mobilized and oxidized. According to World Health Organization standards, the energy cost and the requirement of gestation would be 80,000 kcal, about 350 kcal/day during the two last quarters. This represents an average increase by about 15% of the normal daily expenditures. The mean increase by 4 kg of energy gain after parturition represents a storage of 30,000 kcal during gestation. Like in animal species, this elevation of body fats during pregnancy is viewed as an anticipation for postpartum lactation.

No data are available regarding the spontaneous food consumption of pregnant women. However, because the energy retention needed for both the formation of the fetus and the fat storage of the mother is effectively achieved, a free increase of approximately 15% of the daily free intake is also necessarily achieved. Although not measured, this slight hyperphagia and increased appetite of pregnant women is commonly observed. Selective cravings for some foods, particularly focused on their sensory aspects, are also known. Their mechanisms and biological significance are unknown. However, a modulation by sexual hormones of the olfactory sensitivity was presented as a possible mechanism.

D. Lactation

In cows, a daily milk production of 12.5 liters doubles the energy requirements and also doubles the spontaneous intake of food energy. Cows reaching a daily production of 100 liters are obtained by forcing the intake of high energy foods. Then the milk production is dependent on this imposed overeating (97).

In rats, milk production can only be measured by the growth rate of the litter. Inasmuch as the weight gain of the litter is proportional to the number of young, it is concluded that the daily milk ejection is also proportional and adapted to this number. The dam is hyperphagic. The respective roles of sucking, milk ejection, and hormones in stimulating intake were investigated. The ejection being prevented and the suckling by youngs being maintained, hyperphagia persisted. However, cyclic females who were submitted to suckling by foster young were not hyperphagic. Thus, suckling stimulates food intake only when mammary glands are functional. Prolactin and oxytocin injections into a mother whose young were removed or to controls did not stimulate intake. Thus, these hormones *per se* are without any effect on the stimulation of intake, independent of their action on lactation (98). In mother rats, limited to a litter of eight young, a progressive increase of the daily intake was observed. In an early study by Richter and Barelare (99), the augmentation of intake from the pregestation level was reported to be 30% before parturition and >150% at the end of lactation. The same author showed a change in the free choice of the three macronutrients by the lactating mother. The lactating rate augmented its intake of fats and proteins and relatively reduced that of carbohydrates.

However, a negative energy balance during lactation is indicated by a weight loss, mainly imputed to the loss of fats accumulated during the pregnancy. Following gestation plus lactation, mother rats still demonstrated a slight weight gain (+3.2%), accounted for by water retention.

In lactating women, daily milk production is (as is well-known) interindividually variable. Classically, it is estimated to be 300 ml at the end of the first week, 500–900 ml at the end of the first month, and 1000 ml after 6 mo. This maximal milk production would require a 30% increase of food intake. This lactation-induced hyperphagia has never been measured. However, it is certainly less than required since most women, lactating or not, lose the weight gained during pregnancy after parturition. Others women, particularly those who gained the most weight during pregnancy, do not, and from pregnancy to pregnancy become obese.

References

1. Kleiber, M. (1961). *The fire of life: An introduction to animal energetics*. New York: Wiley, 428 pp.
2. Le Magnen, J., and Devos, M. (1982). Daily energy balance in rats. *Physiol. Behav.* **29,** 807–811.
3. Cohn, M., Shrago, E., and Joseph, D. (1955). The effect of food administration on weight gains on body composition of normal and adrenalectomized rats. *Am. J. Physiol.* **180,** 503–507.
4. Jen, K., and Hansen, B. (1984). Feeding behavior during experimentally induced obesity in the monkey. *Physiol. Behav.* **33,** 863–869.
5. Drewry, M. M., Harris, R. B. S., and Martin, R. J. (1989). The effect of increased adiposity on food intake of juvenile rats. *Physiol. Behav.* **45,** 381–386.

6. May, K. K., and Beaton, J. R. (1968). Hyperphagia in the insulin-treated rats. *Proc. Soc. Exp. Biol. Med.* **127,** 1201–1204.

7. Le Magnen, J. (1976). Interactions of glucostatic and lipostatic mechanisms in the regulatory control of feeding. In D. Novin, W. Wyrwicka, G. A. Bray (eds.), *Hunger:* Basic mechanisms and clinical implications, (pp. 89–101). New York: Raven Press.

8. Hoebel, B. G., and Teitelbaum, P. (1976). Weight regulation in normal and hypothalamic hyperphagic rats. *J. Comp. Physiol. Psychol.* **61,** 189–193.

9. Carpenter, R. G., and Grossman, S. P. (1983). Reversible obesity and plasma fat metabolites. *Physiol. Behav.* **30,** 57–63.

10. Geary, N., and Scharrer, E. (1989). Meal pattern and body weight changes during insulin hyperphagia and post-insulin hypophagia. *Behav. Neural Biol.* **31,** 435–442.

11. Sharrer, E., and Geary, N. (1980). Concentration of food metabolites and liver glycogen during compensatory hypophagia. International Conference on Physiology of Food and Fluid Intake (ICPFFI), Warsaw.

12. Geary, N., Grotchell, H., and Scharrer, E. (1982). Blood metabolites and feeding during post-insulin hypophagia. *Am. J. Physiol.* **243,** R304–311.

13. Sclafani, A., and Springer, D. (1976). Dietary obesity in adult rats: Similarities to hypothalamic and human obesity syndromes. *Physiol. Behav.* **17,** 461–471.

14. Rothwell, N. J., and Stock, M. J. (1979). Regulation of energy balance in two models of reversible obesity in the rat. *J. Comp. Physiol. Psychol.* **93,** 1024–1034.

15. Rolls, B., and Rowe, E. (1977). Dietary obesity: Permanent changes in body weight. *J. Physiol.* **272,** 2P.

16. Mandenoff, A., Lenoir, T., and Apfelbaum, M. (1982). Tardive occurrence of adipocyte hyperplasia in cafeteria-fed rats. *Am. J. Physiol.* **242,** R349–351.

17. Rogers, P. J. (1985). Returning "cafeteria-fed" rats to a chow diet: Negative contrast and effects of obesity on feeding behavior. *Physiol. Behav.* **35,** 493–499.

18. Smith, A. L., Satterthwaitte, H. S., and Sokoloff, L. (1969). Induction of brain D(−)beta-hydroxybutyrate dehydrogenase activity by fasting. *Science* **163,** 79–81.

19. Boyle, T. C., Storlien, L., and Keesey, R. E. (1968). Increased efficiency of food utilization following weight loss. *Physiol. Behav.* **3,** 447–453.

20. Andik, S., Donhoffer, S., Moring, T., and Szentes, T. (1951). The effect of starvation on food intake and selection. *Acta Physiol. Acad. Sci. Hung.* **2,** 21–28.

21. Rothwell, N., Saville, M., and Stock, M. (1983). Role of insulin in thermogenic response to refeeding in 3 day-fasted rats. *Am. J. Physiol.* **245,** E150–154.

22. Harris, F., and Martin, N. (1984). Recovery of body weight from below set point in mature rats. *J. Nutr.* **114,** 1143–1150.

23. Rothwell, N., Saville, L., and Stock, M. (1982). Sympathetic and thyroid influences on metabolic rate in fed, fasted and refed rats. *Am. J. Physiol.,* **243,** R339–R346.

24. Pénicaud, L., Larue-Achagiotis, C., and Le Magnen, J. (1983). Endocrine bases for weight gain after fasting of VMH-lesioned rats. *Am. J. Physiol.* **245,** E246–252.

25. Pénicaud, L., Kandé, J., Le Magnen, J., and Girard, J. R. (1975). Insulin action during fasting and refeeding in rat determined by euglycemic clamp. *Am. J. Physiol.* **249,** B518.

26. Melnyck, L., and Boshes, M. (1980). Efficiency of food utilization during weight gain in dormice. *Physiol. Behav.* **24,** 1017–1021.

27. Dark, G., Stern, J., and Zucker, E. (1985). Adipose tissue dynamic during cyclic weight loss and weight gain of ground squirrel. *Am. J. Physiol.* **250,** R1289–1294.

28. Rothwell, N., and Stock, M. (1979). A role of brown adipose tissue in diet-induced thermogenesis. *Nature (London)* **281,** 331–334.

29. Rothwell, N., and Stock, M. (1979). Effects of continuous and discontinuous periods of cafeteria feeding on body weight, resting O_2 consumption and noradrenaline sensitivity. *J. Physiol. (London)* **261,** 59P.

30. Rothwell, N. J., and Stock, M. J. (1979). Similarities between cold and diet-induced thermogenesis in rats. *Can. J. Physiol. Pharm.* **58,** 35–45.

31. Rothwell, N., and Stock, M. A. (1981). A role for insulin in dietary-induced thermogenesis of cafeteria-fed rats. *Metabolism* **30,** 673–680.

32. Rothwell, N. J., and Stock, M. J. (1982). Energy expenditure derived from measurement of O_2 consumption and energy balance in hyperphagic cafeteria-fed rats. *J. Physiol. (London)* **324,** 59P.

33. Andrews, P., Rothwell, N., and Stock, M. (1985). Influence of subdiaphragmatic vagotomy and brown adipose tissue sympathectomy on thermogenesis in rats. *Am. J. Physiol.* **249,** E239–243.

34. Rothwell, N. J., and Stock, M. J. (1982). Neural regulation of thermogenesis. *Neuroscience* **5,** 124.

35. Tulp, O. L., Gregory, M., and Danforth, E. (1982). Characteristics of diet-induced brown adipose tissue growth and thermogenesis in rats. *Life Sci.* **30,** 825–830.

36. Tulp, O. L., Fink, N., and Danforth, E. (1982). Effects of cafeteria feeding on brown and white adipose tissue cellularity and lipogenesis and body composition. *J. Nutr.* **112,** 2250–2260.

37. Hervey, G. R., and Tobin, G. (1983). Luxous consumption, diet-induced thermogenesis, body fats: A critical review. *Clin. Sci.* **64,** 7–18.

38. Glick, A. (1984). Meal intake and brown adipose tissue in the rat. *Nutr. Behav.* **2,** 65–75.

39. Le Blanc, J., and Diamond, P. (1986). Effect of meal size and frequency on thermogenesis in dog. *Am. J. Physiol.* **280,** E144–147.

40. Diamond, P., Brondel, L., and Le Blanc, J. (1985). Palatability and post-prandial thermogenesis in dog. *Am. J. Physiol.* **248,** E75–79.

41. Saito, M. Minokoshi, Y., and Shimazu, T. (1989). Metabolic and sympathetic nerve activity of brown adipose tissue in tube-fed rats. *Am. J. Physiol.* **257,** E374–378.

42. Bray, G. A., and Campfield, L. A. (1975). Progress in endocrinology and metabolism: Metabolic factors in the control of metabolic stores. *Metabolism* **24,** 99–107.

43. Durnin, J. V. G. A. (1957). The day-to-day variations in individual food intake and energy expenditures. *J. Physiol. (London)* **136,** 34P.

44. Durnin, J. V. G. A. (1961). "Appetite" and the relationship between expenditure and intake of calories in man. *J. Physiol. (London)* **156,** 294–306.

45. Forbes, G. (1984). Energy intake and body weight: Reexamination of two classic studies. *Am. J. Clin. Nutr.* **39,** 349–350.

46. Widdowson, E. M. (1962). Nutritional individuality. *Proc. Nutr. Soc.* **21,** 121–128.

47. Miller, D. S. (1975). Thermogenesis in everyday life. In E. Jéquier (ed.), *Regulation du Bilan d'Energie* (pp. 198–208). Editions Médecine & Santé.

48. Rose, H. E., and Mayer, J. (1968). Activity, calorie intake, fat storage and the energy balance of infants. *Pediatrics* **41,** 18–29.

49. Rose, G. A., and Williams, R. T (1961). Metabolic study of large and small eaters. *Brit. J. Nutr.* **15,** 1.

50. James, W., and Trayhurn, P. (1981). Thermogenesis and obesity. *Brit. Med. Bull.* **37,** 43–48.

51. Trayhurn, P. (1986). Brown adipose tissue and diet-induced thermogenesis. In P. Trayhurn and D. G. Nicholls (eds.), *Brown adipose tissue* (pp. 299–338). London: Edward Arnold.

52. Apfelbaum, M., Botsarrton, J., and Lacatis, D. (1971). Effect of caloric restriction and excessive caloric intake on energy expenditure. *Am. J. Clin. Nutr.* **24,** 1405–1409.

53. Abbott, W., and Ravussin, E. (1988). Short-term energy balance: Relationship with carbohydrate, protein and fat balances. *Am. J. Physiol.* **255,** E332–337.

54. Sims, E. H. A., Danforth, E., Horton, E. S., Bray, G. A., Glennon, J. A., and Salans, L. B. (1973). Endocrine and metabolic effects of experimental obesity in man. *Rec. Prog. Horm. Res.* **29,** 457–496.

55. Pasquet, P., and Apfelbaum, M. (1989). Adaptation de la dépense énergétique à la suralimentation chez l'homme. Association Française pour l'Etude et la Recherche sur l'Obésité.

56. Ravussin, E., and Bogardus, C. (1982). Thermogenic response to insulin and glucose infusions in men: A model to evaluate the different components of thermic effect of carbohydrates. *Life Sci.* **31,** 2011–2018.

57. Jéquier, E. (1985). Thermogenèse induite par les aliments chez l'homme: Son rôle dans la régulation pondérale. *J. Physiol. (Paris)* **80,** 129–140.

58. Thibaud, C., and Jéquier, E. (1985). Thermogenesis induced by intraveinous infusions of glucose-insulin in healthy young men. *Int. J. Vit. Nutr.* **52,** 209–213.

59. Ravussin, E., Bungardt, B., Schutz, Y., and Jéquier, E. (1982). 24 Hr-energy expenditures and resting metabolic rate in obese, moderately obese, and control subjects. *Am. J. Clin. Nutr.* **35,** 566–573.

60. Achenson, K., Ravussin, E., Wharen, W., and Jéquier, E. (1974). Thermic effect of glucose in man: Obligatory and facultative thermogenesis. *J. Clin. Invest.* **74,** 1572–1581.

61. Le Blanc, J., Cabanac, M., and Samson, P. (1984). Reduced post-prandial heat production with gavage as compared with meal feeding in human. *Am. J. Physiol.* **246,** E65–71.

62. Le Blanc, J., and Cabanac, M. (1989). Cephalic post-prandial thermogenesis in human subjects. *Physiol. Behav.* **46,** 479–482.

63. Brondel, L. Guilland, J.-C., Mack, G., Guiguet, M., Brun, J. M., and Fantino, M. (in press). Post-prandial thermogenesis in human subjects: Role of alimentary sensory stimulation. *Physiol. Behav.*

64. Vernet, O., Nacht, C.-A., Christin, L., Schutz, Y., and Danforth, E. Jr. (1987). Beta-adrenergic blockade and intraveinous nutrient infusion induced thermogenesis in lean and obese women. *Am. J. Physiol.* **253,** E65–E71.

65. Hill, G., di Girolamo, M., and Heymsfield, S. (1985). Thermic effect of food after ingested or tube-delivered meals. *Am. J. Physiol.* **248,** E370–374.

66. Astrup, A., Bülow, J., Madsen, J., and Christensen, J. (1985). Contribution of brown adipose tissue and skeletal muscles to thermogenesis induced by epinephrine in man. *Am. J. Physiol.* **249,** E517–E515.

67. Baker, J. F., and Shaler, E. A. (1953). Carbohydrate metabolism of the rat exposed to a low environmental temperature. *Fed. Proc.* **12,** 8.

68. Cottle, W., and Carlson, L. D. (1954). Adaptive changes in rats exposed to cold. Caloric exchange. *Am. J. Physiol.* **178,** 305.

69. Leung, P. M. B., and Horwitz, B. A. N. (1976). Free feeding patterns of rats in response to changes in environmental temperature. *Am. J. Physiol.* **231,** 1220–1224.

70. Portet, R., and De Marco, F. (1979). Adaptation au froid et rythmes nychtéméraux du métabolisme énergétique et des prises alimentaires chez le rat. *Bull. Groupe Etud. Ryth. Biol.* **11,** 21–22.

71. Durrer, J. L., and Hannon, J. P. (1962). Seasonal variations in caloric intake of dogs living in arctic environment. *Am. J. Physiol.* **202,** 375–378.

72. Keshner, A., Collier, G., and Stuibh, R. (1971). Dietary self-selection at cold temperature. *Physiol. Behav.* **6,** 1–3.

73. Donhoffer, Y., and Venetzi, A. (1947). The effect of tyroxine on food intake and selection. *Am. J. Physiol.* **150 (2),** 334–336.

74. Refinetti, R. (1988). Effect of food temperature and ambient temperature during a meal on food intake of rats. *Physiol. Behav.* **43,** 245–247.

75. Kraly, F. S., and Blass, E. (1976). Mechanisms for enhanced feeding in cold in rats. *J. Comp. Physiol. Psychol.* **90,** 714–726.

76. Johnson, T. E., and Park, R. M. (1957). Environments and food intake in man. *Science* **105,** 378–379.

77. Spector, N. H., Brobeck, J. R., and Hamilton, C. L. (1968). Feeding and core temperature in the albino rat: Changes induced by preoptic heating and cooling. *Science* **163,** 286–288.

78. Brobeck, J. (1960). Food and temperature. *Rec. Prog. Horm. Res.* **16,** 459–466.

79. Edholm, O. G., and Goldsmith, R. (1966). Food intake and weight changes in climatic extremes. *Proc. Nutr. Soc.* **25,** 113–119.

80. Thomas, B. M., and Miller, A. T. (1958). Adaptation to forced exercise in the rat. *Am. J. Physiol.* **193,** 350–354.

81. Stevenson, J. A. F., Box, B. M., Feleki, V., and Beaton, J. R. (1966). Bouts of exercise and food intake in the rat. *J. Appl. Physiol.* **21,** 118–122.

82. Mayer, J., Marshall, N. B., Vitale, J. J., Christensen, J. H., Mashayekhi, M., and Stare, F. J. (1954). Exercise, food intake and body weight in normal rat and in genetically obese adult mice. *Am. J. Physiol.* **177,** 544–548.

83. Katch, V. L., Martin, R., and Martin, J. (1979). Effects of exercise intensity on food consumption in the male rat. *Am. J. Clin. Nutr.* **32,** 1401–1408.

84. Bulbulian, R. (1985). Effect of exercise duration on food intake and body composition of Swiss albino mice. *J. Appl. Physiol.* **80,** 500–505.

85. Ingle, D. J. (1949). A simple means of producing obesity in the rat. *Proc. Soc. Exp. Biol. Med.* **72,** 604–607.

86. Devlin, J. T., Barlow, J., and Horton, E. S. (1989). Whole body and regional fuel metabolism during early post-exercise recovery. *Am. J. Physiol.* **256,** E167–172.

87. Baile, C., McLaughlin, C., Zinn, R., and Mayer, J. (1971). Exercise, lactate hormones and GTG-lesion in the hypothalamus of diabetic mice. *Am. J. Physiol.* **122,** 150–158.

88. Guilland, J. C., Moreau, D., Genet, J. M., and Klepping, J. (1988). Role of catecholamines in regulation by feeding of energy balance following chronic exercise in rats. *Physiol. Behav.* **42,** 365–369.

89. Edholm, O. G., Madam, K. N., Healy, M., Wolf, H., Homdsmith, R., and Best, T. W. (1970). Food intake and energy expenditures of army recruits. *Brit. J. Nutr.* **24,** 1091–1107.

90. Short, S., and Short, W. R. (1983). Four year study of university athletes' dietary intake. *J. Am. Diet. Assoc.* **82,** 634–645.

91. Saris, W. H. M. (1990). Apport vitaminique chez les athlètes de haut niveau. In H. Mondo (ed.), (pp. 88–97). Paris: Masson.

92. Morrison, S. D. (1956). The total energy and water metabolism during pregnancy in the rat. *J. Physiol.* **134,** 650–654.

93. Perisse, J., and Salmon-Legagneur, E. (1960). Influence du niveau nutritionnel au cours de la gestation et de la lactation sur la production laitière de la ratte. *Arch. Sci. Physiol. (Paris)* **14,** 105–129.

94. Champigny, O. (1963). Echanges respiratoires et bilans azotés de la ratte gravide: coût énergétique de l'anabolisme gravidique. *C. R. Acad. Sci. (Paris)* **256,** 4755–4758.

95. Kristal, M. E., and Wampler, A. B. (1973). Food and water intake prior to parturition in rats. *J. Psychol.* **1,** 297–300.

96. Hytten, F. E., and Leitch, I. (1971). The physiology of Human Pregnancy, 2nd edition, London: Blackwell Scientific Publications.

97. Jarrige, R. (1978)./L'alimentation des ruminants. Paris: Editions de l'Institut National de Recherche Agronomique.
98. Flaming, A. S. (1976). Control of food intake in lactating rat. *Physiol. Behav.* **17,** 841–848.
99. Richter, C. P., and Barelare, B., Jr. (1938). Nutritional requirements of pregnant and lactating rats studied by the self-selection method. *Endocrinology* **22,** 15–24.

Chapter Nine

Peripheral and Central Mechanisms of the Regulation of Body Energy and Body Weight

In preceding chapters, we have seen that the notion of body energy balance is more complex than is generally thought. The achievement of the body energy balance is not separable from a simultaneous regulation of the body composition and, particularly, from that of body fats. In a condition of fat reserve depletion below the regulated level, the energy intake matches the total oxidative metabolism and energy retention for the restoration of body fats. Part of the energy intake is then diverted toward fat synthesis and is added to the normal intake needed to balance energy expenditure and lean tissues. Regarding this lean tissue, the recorded secondary hyperphagia is then a normophagia. The opposite is true in the condition of body fat repletion above the regulated level. The amount of fuel supplied by lipolysis is subtracted from the food intake so that the recorded secondary hypophagia is a normophagia regarding the lean tissue (at least the peripheral one). However, primary hyperphagia and hypophagia occur physiologically or pathologically, each limited and subsequently corrected. What are the mechanisms, peripheral and/or central, involved in this regulatory system of the fat store and of its coordination with the overall regulation of body energy balance?

I. *Cellular Self-Regulation of Adipocyte Size and Number*

It is a "vérité de la Palice" to say that a constancy of the fat mass results from an equilibrium between the overall lipogenic and lipolytic factors.

291

In addition to neural and centrally governed mechanisms, adipose tissue cell responses to hormonal lipogenic and lipolytic factors are manifestly involved in the equilibrium.

A. Adipocyte Responsiveness to Insulin, Norepinephrine, and Glucagon as a Function of Cell Size and Fat Mass

Insulin is the major agent of fat synthesis, and hyperinsulinemia a cause of fat repletion. Catecholamine released from adrenals acts both to inhibit the insulin release and to promote lipolysis together with the other lipolytic agent, glucagon. Pancreatic insulin release under hyperglycemia and catecholamine and glucagon releases under hypoglycemia are neurally and centrally modulated (see Section II of this chapter). In addition to a direct adrenergic innervation of the adipose tissue, this central modulation of pancreatic and adrenal releases will be shown later to be the main effector of the central regulation of body fats; however, strong evidence indicates that the adipocyte responsiveness to lipogenic and lipolytic hormones as a function of cell size and of the total fat mass is also involved.

Earlier it was mentioned that the dose of protamin-zinc-insulin (PZI) needed to obtain experimental obesity must be increased progressively with the progressive weight gain (1). At a constant dose of continuous infusion of regular insulin of 0.2 IU/min, the weight-promoting effect disappears after 6–7 days (2) (Fig. 9.1). Direct evidence for this reduction of the lipogenic effect of insulin as a function of the body fat repletion was experimentally provided. Rats became overweight by tube-feeding until a weight gain in several groups of 15 and 30% of their initial body weight. Then the dose of insulin needed to induce a further increase of body weight and of food intake in each group was investigated. The finding was that this dose increased when overweight (3) (Fig. 9.2).

The cellular mechanism of this resistance to insulin induced by a forced overweight or in spontaneous obesity in humans has been widely investigated. The fall of insulin-binding on membrane insulinoreceptors of adipocytes as well as of other peripheral tissues, known as a downregulation of these insulinoreceptors, is one of these mechanisms. This fall of insulin-binding will be seen as a major symptom of human obesity. However, another mechanism of this downregulation of insulin receptors, i.e., the intracellular action of insulin in the fat synthesis, is also demonstrated (4, 5).

Alterations of the lipogenic–antilipolytic action of insulin and of the lipolytic action of both norepinephrine and glucagon as a function of the adipocyte hypertrophy are widely confirmed. In rats, hamsters, and guinea pigs,

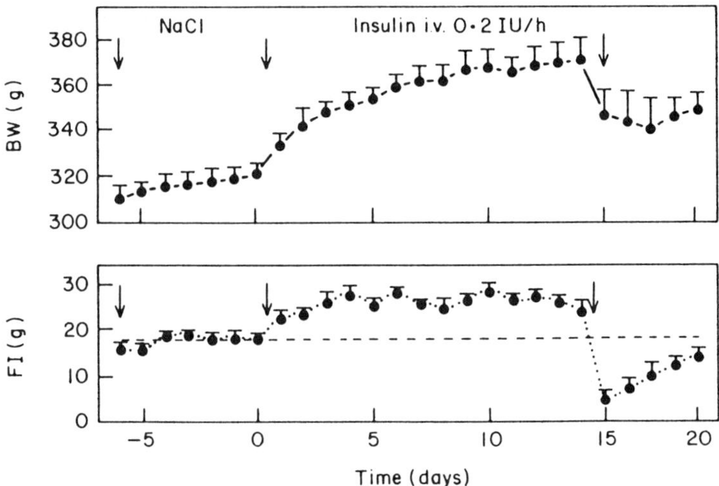

Figure 9.1 Chronic peripheral infusion of insulin: time course of the effects on food intake and body weight.

the adipocyte response to insulin decreases as the cell size increases with age. This insulin insensitivity would be due to the intracellular action of insulin. Conversely, in an *in vitro* study, the lipolytic effect of catecholamine and of adrenocorticotropin (ACTH) was shown to be augmented in large cells removed from old rats compared to small adipocytes from young rats (6–8). However, the total fat mass is involved in addition to cell sizes. In isolated adipocytes, the fat synthesis from ^{14}C-glucose in the medium increased with the cell size in the presence or absence of insulin. The same lipogenesis is recorded in cells of the same size removed from obese or nonobese rats, in the absence of insulin. In its presence, the lipogenesis was lower in cells of identical sizes from obese rats. Thus, factors other than cell size determine the reduced lipogenic action of insulin in overweight rats. Similarly, the lipolytic action of norepinephrine and ACTH augmented with both the cell size and the body weight. This was not the case with glucagon, whose lipolytic effect was found lower in heavy rats (9). In another study, the lipogenic effect of insulin and lipolytic effect of norepinephrine were found considerably greater in identically sized adipocytes from young compared to fatty old rats (10).

These actions of lipogenic and lipolytic agents on fat cells vary with locations of fat deposits. In the dog, abdominal fats are less sensitive to lipolytic agents than subcutaneous fats (11). *In vitro* and *in vivo,* the dose–response curve to norepinephrine was shown to be identical in various deposits of young rats

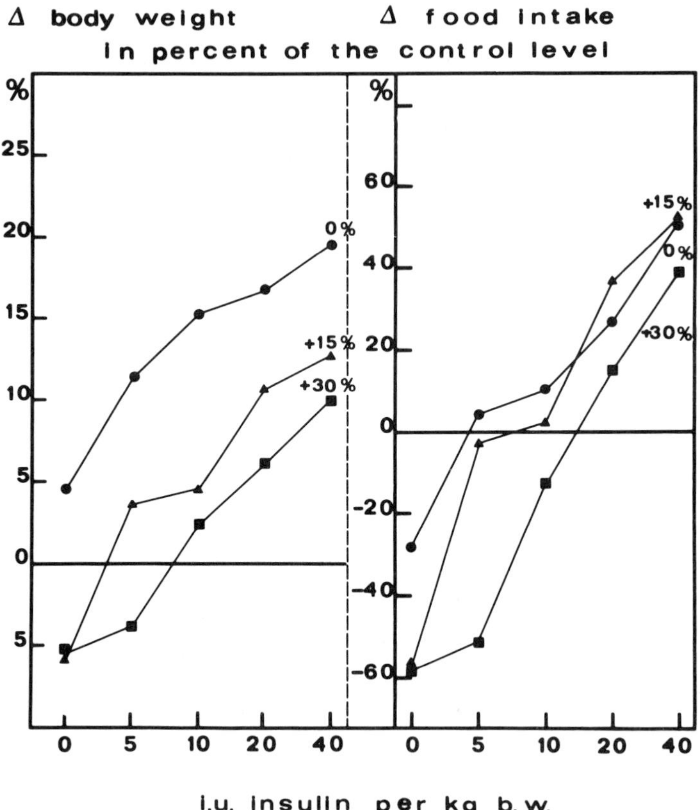

Figure 9.2 Reduced effect of long-acting insulin injection on food intake and weight gain as a function of overweight induced by forced feeding.

except in subcutaneous fats. In old fatty rats, less lipolytic activity is generalized and becomes proportional to cell size in all deposits (12).

These results are in line with others suggesting differences in the lability of fat reserves according to their localization. Some fat deposits, such as the subcutaneous fats, seem to be a labile compartment perhaps involved in rapid short-term regulatory processes such as the diurnal lipogenetic–lipolytic pattern, whereas other fat deposits are more stable and more resistant to regulatory processes.

B. Lipectomy

The surgical removal of a proportion of body fats and a subsequent regeneration or not is a test of the putative cellular self-regulatory process. The results are unclear and debated. A lipectomy in weanling rats is followed by a restoration of the body fat mass during the subsequent 7 mo. A compensatory hyperplasia, and not a cell hypertrophia, is involved in this regeneration of the species-specific fat mass (13). Osborne-Mendel rats are hyperplasic and become obese on a high-fat diet. Lipectomized rats with 25% of their adipocytes removed were less hyperphagic and became less obese on the high-fat diet than did controls. Hypertrophia of their adipocytes did not differ from that of intact controls. The author concludes that the cell size is the limiting factor of the induced obesity and the only factor in action in the control of food intake (14). In another study, a removal of 24% of body fats in nonobese rats was not followed by any changes of the size and number of remaining adipocytes nor by an increase or decrease in food intake. Thus, the cell size rather than the cell number and total fat mass would be the regulatory factor. Weight gain and the basal insulin were not changed following lipectomy (15, 16).

II. *Central Nervous System Control of Metabolism*

Abundant literature deals with the role of sympathetic and parasympathetic pathways, i.e., of the autonomic nervous system, in the control of insulin, glucagon, and somatostatin releases by the endocrine pancreas and of catecholamines by adrenal medulla and in the direct control of fat metabolism. This is a large topic of nutrition neurobiology, the regulatory control of oxidative metabolism by the central nervous system (CNS) and intermediary metabolisms. This may be considered an important aspect of neuroendocrinology, although it is rarely treated as such. These sympathetic and parasympathetic controls of visceral functions play a major role in the management of fat stores and, directly and indirectly, in feeding.

A. Some Aspects of Sympathetic and Parasympathetic Pathways in the Brain

Parasympathetic pathways, particularly vagal fibers, run parallel with gustatory afferents. The two pathways have a common relay in the nucleus of the solitary tract (NTS), the suprabrachial pontine nuclei, and then converge to the lateral hypothalamus (LH), the central nuclei of amygdala, and the bed nuclei of the stria terminalis. Moreover, direct monosynaptic connections are recognized between vagal afferents and the two latter limbic structures. Sympathetic afferents coming from the intestine also travel parallel with vagal pathways. Their terminals in the hypothalamus are not clearly identified (17). Horseradish peroxidase in the pancreas allowed Berthoud *et al.* (18) to localize cell bodies of vagal fibers afferent to the pancreas in the dorsomotor nucleus of the vagus and the nucleus ambiguus. Monosynaptic connections link the LH, dorso-motor nucleus of the vagus (DMV), nucleus ambiguus (NA), and the nucleus of the solitary tract (NTS). Similar connections also are labeled between the ventromedial hypothalamus (VMH), DMV and NA. Thus, it seems that the VMH and adjacent areas together receive from and send to the periphery both sympathetic and parasympathetic impulses.

B. Neural Control of the Endocrine Pancreas

Works by Porte *et al.* (19, 20) provide the main initial finding. In dogs and humans, insulin secretion was shown to be inhibited by norepinephrine infusion. Norepinephrine released by sympathetic fibers to β-cells acts on α-adrenergic receptors. In dogs, insulin release in the duodeno-pancreatic vein was measured under electrical stimulation of either sympathetic or parasympathetic fibers afferent to the pancreas. Sympathetic stimulation inhibited the insulin release, and parasympathetic stimulation activated it. The latter was blocked by atropine.

Girardier *et al.* (21) experimented on the perfused pancreas of dogs. The rate of increases in insulin concentration in the outflow of the perfusate augmented linearly when the concentration of glucose in the perfusate was varying from 100 to 300 mg%. Within the same range, glucagon releases diminished. In the atropinized pancreas, a low frequency stimulation of sympathetic nerves suppressed the rate of insulin secretion by 40% and augmented that of glucagon by 44%.

The antagonist action of sympathetic adrenergic fibers and of parasympathetic fibers releasing, respectively, norepinephrine and acetylcholine (ACh) on pancreatic insulin release was elegantly studied by Campfield and Smith (22, 23) and Campfield *et al.* (24). In isolated Langerhans islets of rats, they measured *in vitro* the 30-min insulin release in the incubating medium at various concentrations of glucose and various proportions of norepinephrine and/or of ACh. The thresholds and slopes of increased inhibition by norepinephrine and of facilitation by ACh as a function of their respective concentrations were determined at various levels of glucose stimulation. The antagonist action of the two neurotransmittors was striking for each of them, a level at which the action of the other was totally occluded.

In another study, it was argued that the vagus nerve activity affects insulin release and, consequently, the short-term glucose tolerance after oral- but not intravenous-induced hyperglycemia (25). Recordings of discharges from peripheral nerves were instructive. An intravenous glucose infusion in rats induced an augmentation of activity in the pancreatic branch of the vagus nerve and a reduction of it in the adrenergic adrenal nerve. Injecting insulin produced the opposite effects, with effects in the two nerves proportional to the induced hypoglycemia. Like insulin, 2-deoxy-D-glucose (2-DG) provoked a strong elevation of activity in the adrenal nerve and a reduction in the vagus nerve. A small dose of 25 mg/kg glucose intrajugularly administered did not change activity in the pancreatic vagus nerve nor in the adrenergic adrenal nerve. However, the same dose injected into the portal vein increased activity in the first of these nerve fibers and reduced it in the second one. Based on this result, the author argues that the neural modulation of the blood glucose level implies a role of pancreatic and hepatic branches of the vagus nerve and of the adrenal nerve (26). In humans, it is demonstrated that epinephrine increases the hepatic glucose production through an α-adrenergic inhibition of insulin secretion and reduces the glucose clearance through an β-adrenergic activity on tissue uptake (27). In several works, it was shown that stimulation of the vagal branch of the pancreas increased both insulin and glucagon secretions. Marliss *et al.* (28) and Kaneto *et al.* (29) concluded that sympathetic activation inhibited insulin and stimulated glucagon secretion, whereas parasympathetic stimulation increased both of them. These contrasting effects of stimulating the two nerves is confirmed in dogs, cats, and sheep (30).

C. Electrical Stimulations of Hypothalamic Sites

Effects of hypothalamic stimulation on insulin and glucagon secretions, on blood glucose levels, and lastly on lipogenesis and lipolysis provide evidence indicating an important role for the VMH in the origin of adrenergic effects on peripheral organs.

In unanesthetized rats, hypoglycemia and hypoglucagonemia were induced by electrical stimulation of both the VMH and LH. Stimulations in the two sites did not change plasma insulin (31). Stimulations of the VMH as well as of the dorsomedial nuclei and arcuate nuclei produced at a low intensity of stimulation a rapid (first 3 min) hyperglycemic response. This does not depend on adrenals nor on the pancreas and is attributed to a splanchnic action on the liver. A greater intensity of stimulation induces hyperglucagonemia and hypoinsulinemia, the latter depending on adrenals and leading to prolonged hyperglycemia (32) (Figs. 9.3 and 9.4). According to Steffens (33), stimulation of the posterior hypothalamus, contrary to that of the VMN, produces hyperinsulinemia and hypoglycemia. Morrissey *et al.* (34) confirmed this effect of a stimulation of the

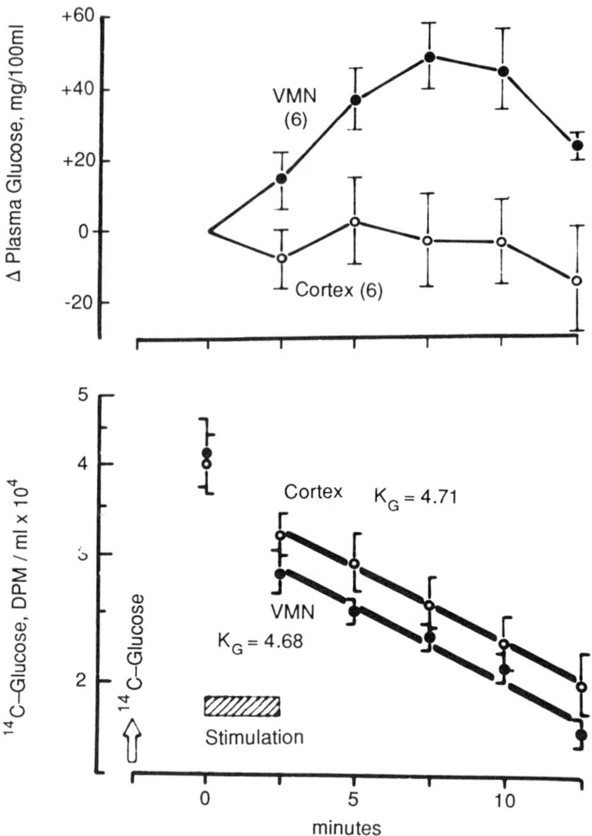

Figure 9.3 Plasma ^{14}C-glucose disappearance rates in rats during and after cerebral cortex and ventromedial hypothalamic nucleus (VMN) stimulation. Means ± SE of individual determinations are shown. Glucose decay constant (K_G) was based on a single regression analysis of all experimental points in each group. Plasma glucose levels at onset of stimulation were 113 + 6 mg/100 ml (VMN) and 107 + 3 mg/100 ml (cortex). [From Froshman and Bernardis (32).]

posterior hypothalamus and attributed it to the sympathetic activation of adrenals. Stimulation of mammillary bodies (posterior hypothalamus) elevates plasma-free fatty acids (PFFA) by 60% without changing blood glucose, whereas stimulation of the VMH produces hyperglycemia without elevation of PFFAs. The authors interestingly suggest that different sympathetic pathways coming from the two sites affect, respectively, the adipose tissue and the liver (35). In baboons, the cooling of the preoptic area by a thermode induced a chronic sympathetic activation (36). This activation initially induced a reduction of in-

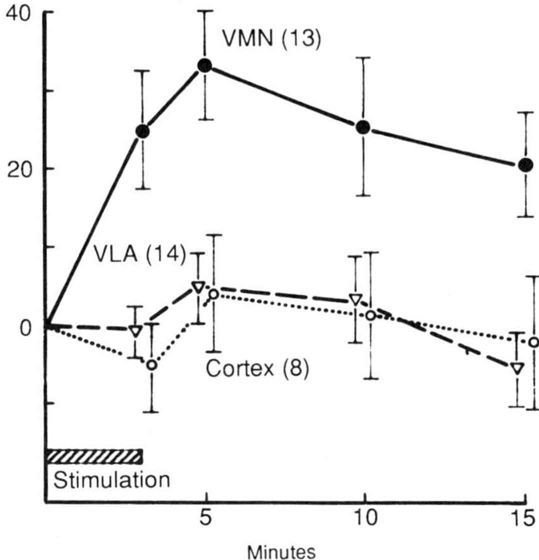

Figure 9.4 Effect of electrical stimulation of ventromedial hypothalamic nucleus (VMN), ventrolateral hypothalamus (VLA), and cerebral cortex on plasma glucose means ± SE are shown. Numbers of animals in each group are in parentheses. Initial plasma glucose levels of three groups were 142 + 5, 137 + 6, and 134 + 8 mg/100 ml, respectively. [From Froshman and Bernardis (32).]

sulin secretion, later recovered, and an increase of glucagon releases. Plasma triiodothyronine fell progressively; however, food intake and body weight were not affected.

D. Neural Control of Gastric Secretions

Electrical stimulation of the VMH inhibited gastric secretions. They were facilitated by LH stimulation. This LH facilitation was vagally mediated. Insulin and 2-DG injections strongly stimulated gastric secretion. When both were injected in rats, via the carotid, the firing rate increased in the gastric branch of the vagus nerve. This neural activity was transitorily lowered by glucose administered by the same route. In another work, stimulation of the caudal portion of the VMH was shown not to inhibit but, rather, to increase gastric secretion. When the secretion was elevated by 2-DG, the VMH stimulation reduced it. The researchers interpreted this intriguing result by the hypothesis of a heterogeneity of the VMH, which would include both gastric inhibitory and facilitatory sites (37).

E. Hypothalamic Control of Lipogenesis–Lipolysis

Experiments on the effects of VMH lesions and others, examined in the next sections, provided definitive evidence that the VMH and sympathetic descending pathways to the pancreas, liver, and adipose tissues govern the lipostatic mechanism. Effects of electrical stimulations were consistent with this conclusion. VMH stimulation and not that of the LH elevated PFFAs and glycerol, indicating the stimulation of lipolysis. It is not prevented by medullectomy and, thus, was not an effect of an adrenal catecholamine secretion. The authors concluded that this is due to the activation of sympathetic fibers innervating the adipose tissue. However, they showed that the same VMH stimulation induced hyperglucagonemia, which could be an intermediary of the induced lipolysis (38). This VMH stimulation-induced lipolysis could be the reason that old reports on a reduction of food intake by the VMH stimulation led to a longlasting and historical error in considering this site a "satiety center."

In rabbits, contrary to rats, lipolysis induced by VMH stimulation was prevented by adrenalectomy and, thus, depended on the catecholamine action on the adipose tissue (39).

F. Vagotomy

A complete subdiaphragmatic vagotomy eliminated activities of afferent and efferent fibers to and from the gastrointestinal tract, pancreas, and liver. Its parallel action on visceral and feeding responses was very instructive. An earlier and more rapid gastric emptying is associated with early and elevated postabsorptive hyperglycemia and insulinemia. This pattern was designated the "dumping syndrome" of vagotomy. In the first weeks following the vagotomy, the insulinogenic index was reduced; i.e., β-cells of the pancreas were less responsive to glucose than normals were. The diurnal cycle of plasma insulin was abolished. This hyporesponsiveness was confirmed *in vitro*. The hypoinsulinimic level during the day was extended to the night (40).

Together these symptoms are instructive. The early gastric emptying accompanied with an early and exaggerated prandial hyperglycemia is consistent with the viewpoint (otherwise supported) that the rate of gastric emptying determines the rate of intestinal absorption. The resulting rapid flow of metabolites in the portal vein suggests a cause of the augmented hyperglycemia. The disappearance of the preabsorptive insulin release may also contribute to this postprandial high glucose level. The relatively low insulinic response to glucose supports the notion of a parasympathetic tonic activity on insulinosecretion.

The parallel effect on the daily feeding pattern is still more instructive. Vagotomized rats take small meals throughout the day. Small meals normally present in during the day are also present during the night. The meal frequency is augmented during the two phases so that the meal pattern becomes identical in the two periods. In other words, vagotomy eliminates the diurnal cycle of food intake. These features of the feeding pattern compared to those of the altered neuroendocrine pattern provide fascinating confirmation of our early assumption regarding the determinants of meal size and of the prandial as well as diurnal periodicity of feeding. Small meal sizes support the notion that the timing and level of the prandial and just postprandial hyperglycemias are involved as a systemic component of the satiation process. As mentioned earlier, the elevation of the meal frequency compensating for the smaller meals indicates that the mechanism by which the postprandial satiety duration is correlated to the preceding meal size is unaltered. Most of all, the disappearance of the diurnal cycle of feeding along with that of the nocturnal hyperinsulinic and diurnal hypoinsulinic cycle provide new evidence for the causal relationships between the two neuroendocrine and behavioral cycles. Unfortunately, the lack of calorimetric studies prevents the assumption that the lipogenic–lipolytic diurnal pattern is also abolished in such rats. However, sufficient data are available to suggest the existence of a sympathetic–parasympathetic diurnal cycle as the main basis of the overall metabolic and feeding pattern. A high parasympathetic–low sympathetic tonus seems to be present at night in rats and during the day in humans; a low parasympathetic–high sympathetic tonus conversely seems to be present during the day in rats and during the night in humans.

It is of interest to note that most of these symptoms of vagotomy are transitory. Six to eight weeks after surgery, the responsiveness of isolated pancreatic islets *in vitro* to ACh is totally abolished, but a 2% elevation of insulin responsiveness to glucose is established as a functional compensation of the disappearance of the neural modulation. Then, the diurnal cycle of feeding is reestablished (41).

III. *Ventromedial Hypothalamic Lesion*

Hetherington and Ranson (42) discovered that an electrolytic lesion of the ventromedial nuclei of the VMH in rats produced hyperphagia and obesity. This outstanding finding was historically the starting point of the interest for the neurobiology of feeding and nutrition and of studies on the etiology of obesity in the subsequent decades. More than 1500 publications for 40 years were devoted to the metabolic and behavioral effects in the lesioned rat and other species (e.g., rabbit, dog, cat, monkey, bird) in which it was confirmed. Initially, most of these works focused on induced overeating. It was postulated that the primary effect of the lesion was a disruption of a major feeding mechanism and that the resulting obesity was an effect of hyperphagia. Because rats overate, it was also postulated that the rat was hyperphagic because it cannot be satiated by eating foods. Therefore, the VMH was claimed to be a "satiety center."

Gordon Kennedy (43) was the first to question this well-admitted notion and to suggest, on the basis of pioneering experimental evidence, that the destruction of the VMH eliminates a mechanism involved in the control and regulation of the body fat mass. Finally, this genius insight of Kennedy has been fully confirmed. The lesion introduces a metabolic pattern including hyperinsulinemia and a suppression of lipolytic responses leading to the incapacity of the lipostatic system to counteract a prevailing lipogenesis. Hyperphagia is an effect and not a cause of this unrestrained lipogenetic trend. Thus, the VMH is not at all involved in the central feeding mechanism. It is the central regulator of the body fat reserves through its connections with mainly sympathetic descending pathways. The elimination of this system by the lesion gives much evidence for the neurohumoral components of the system along with other studies detailed in subsequent chapters. Here, regarding the metabolic and behavioral effects of the lesion, only the most conclusive studies, among many other elegant works, have been deliberately selected and will be briefly outlined.

A. Metabolic Effects

One of the main features of the VMH-lesioned rat is hyperinsulinemia. In *ad libitum* feeding, the basal intermeal plasma insulin is elevated. Most of all, the hyperinsulinemia is prandial. The prandial–postprandial insulin is, for the same amount of food eaten, four to five times the level of intact rats, reaching 200 microunits/ml (44–46). The absolute value of the preabsorptive insulin release is also elevated (47, 48). However, this level would not be augmented relative to the basal elevated level (49). The hyperinsulinemia compared to intact controls is slightly augmented at night and strongly during the day in the initial phase of weight gain (50). As one of the effects of this diurnal hyperinsulism, substituted for the normal hypoinsulinism at this time, the low glucose tolerance is eliminated. The disappearance rate of injected glucose becomes identical to that observed at night. This is observed in 5-hr deprived rats in which a possible causal action of the overeating is eliminated (51) (Fig. 9.5).

The hyperinsulinemia is independent of the overeating, whereas the overeating is dependent (at least partially) on the hyperinsulinemia. The increased

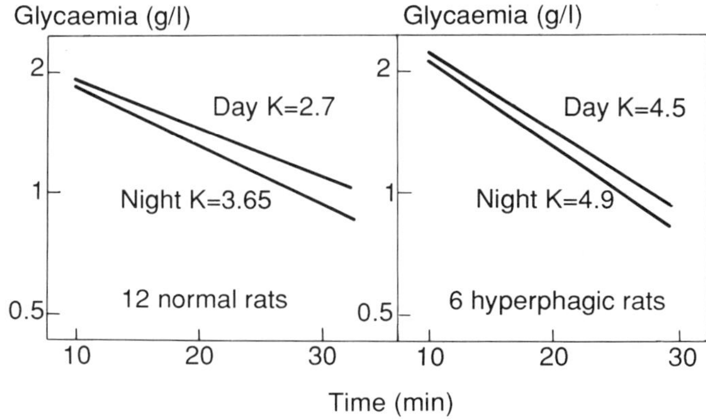

Figure 9.5 Identical nocturnal and daytime glucose tolerance in hypothalamic hyperphagic rats.

plasma insulin is present as soon as 10 min after the operation in the rat recovering from anesthesia and before eating (52). In rats operated on as weanlings, hyperinsulinemia is present several days before the appearance of hyperphagia (53–55). Fully convincing was the experiment by Hustvedt and Lovo (56) and Hustvedt (57). Lesioned rats were at first pair-fed with intact controls. Despite the same intake, they developed hyperinsulinemia and accumulated fats more than their controls did. Placed later in the *ad libitum* feeding, they manifested hyperphagia, and its levels among rats were found correlated to their hyperinsulinemia during the initial pair-feeding.

The prandial blood glucose is relatively low. Thus, the high level of prandial insulin is not due to an exaggerated contemporary hyperglycemia. In the perfused pancreas, the basal and stimulated glucagon releases are also augmented after the lesion. This hyperglucagonemia could be a cause of the weak hypoglycemia of these rats. Prandial hyperglycemia would no longer reach the level needed to induce satiety. However, it was otherwise demonstrated that neither the satiating nor the satiety effects of food were altered in VMH rats.

Confirming a pioneering finding by Han *et al.* (58), an initial insulin responsiveness and enhanced carbohydrate metabolism were demonstrated in lesioned rats. The hyperinsulinemic–euglycemic clamp revealed that glucose utilization is augmented during the first week. This enhanced insulin responsiveness is limited to the adipose tissue. Six weeks after the lesion, a reduction of muscular glucose utilization at a physiological level of hyperinsulinemia is observed. This late insulin resistance in muscles occurs, while the insulin responsiveness of adipose tissue returns to normal (59). This initial elevation of insulin responsiveness of adipose tissue and late development of a muscular resistance is a primary effect of hyperinsulinemia. In intact rats, a chronic hyperinsulinemia induced by a chronic glucose infusion also induces a greater insulin sensitivity of the adipose tissue and develops insulin resistance in muscles (60). The development of a resistance to insulin and loss of the enhanced insulin sensitivity of adipocytes are the first symptoms at the end of the dynamic phase of weight gain and of the later stabilization of the overweight, called the static phase of the lesion-induced obesity.

Lipogenesis attested by the increase in fat deposits and by adipocyte hypertrophy is, with the associated water retention, the cause of weight gain during the dynamic phase. However, the main effect of the lesion is not a stimulated lipogenesis but, rather, a suppression of the lipolytic mechanism normally compensating for fat repletion. This impairment of lipolysis is an effect of both the antilipolytic action of insulin and the result of a direct action of the lesion on adipose tissue. The initial evidence for the impaired lipolysis as a cause of the obesity was provided by showing the disappearance of the diurnal lipogenic–lipolytic pattern in VMH rats. Recording O_2 consumption and CO_2 expiration revealed that the elevated respiratory quotient at night, indicating lipogenesis, was maintained at this high level throughout the day. The diurnal lipolytic pattern and its neuroendocrine background disappeared. Rats become obese because a diurnal lipolysis no longer compensates for the normal nocturnal lipogenesis, i.e., because the lipostatic mechanism, as manifested within a diurnal cycle, is eliminated. Fed only at night or during the day, rats become less obese and nevertheless gain weight. This proves that lipogenesis indeed is stimulated more than it is in normal rats placed in the same condition but also that this increased lipogenesis is not counteracted in the subsequent 12 hr of fasting (Fig. 9.6). This impaired lipolysis is confirmed after fasting, after exercise, and in cold ambient temperature (61). Furthermore, the increase of PFFAs and glycerol by 2-DG is suppressed after the lesion. As early as 1966, a reduced response to the epinephrine-induced lipolysis of adipocytes of lesioned rats was tested *in vitro* (62) and later confirmed (63, 64). Thus, the disrupted lipolysis is not solely a consequence of the permanent hyperinsulinemia, which in PZI-treated intact rats also induces the sustained weight gain.

An important finding was the following: VMH rats and intact controls were fed intragastrically 200% of the basal intake of the latter. After 30 days, lesioned rats had deposed 25% more fats than controls did. This was not only due to the lower metabolic rate recognized in VMH rats (65). Eleven fewer calories were required for the same energy retention in lesioned rats compared to controls. This strongly suggests that the extra heat production, which in tube-overfed intact rats

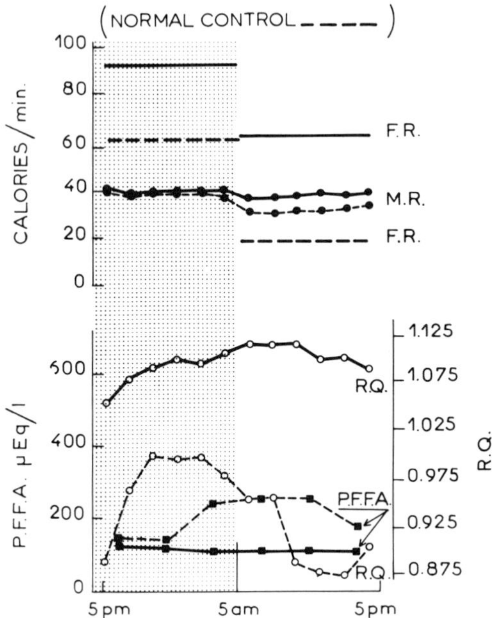

Figure 9.6 Disappearance of the diurnal lipogenic-lipolytic metabolic pattern of VMH-lesioned rats.

limits obesity, had also been impaired by the lesion (66). We will see later that the VMH lesion alters the activation of the brown adipose tissue, even in the non-shivering thermogenesis (67).

After some weeks of sustained weight gain, lesioned rats, reaching the static phase and ceiling of obesity, defended this new equilibrium between lipogenic and lipolytic factors. In rats that were overweight beyond this level by gavage or PZI, the so-induced obesity was reversible. At the cessation of the treatment, rats returned to the static level and, meanwhile, like intact rats, were hypophagic (1). Thus, at this level it seems that the stimulated lipogenesis reaches its limits, presumably due to insulin resistance and to adipocyte hypertrophy, and that the remaining lipolytic capacity may then again operate to maintain the equilibrium.

The dynamic phase of sustained weight gain results from a loss of the reversibility of obesity seen in intact rats and of its mechanism, i.e., the lipostatic mechanism. A wonderful experiment provided clear-cut evidence for this notion.

Colchicine, a blocker of synaptic transmission, was locally injected in the VMH. A dose-dependent hyperphagia and weight gain were manifested. The effect lasted some days followed by hypophagia and weight loss until a return to the initial weight. Rats pair-fed during the colchicine action and prevented from overeating and weight did not exhibit later hypophagia in *ad libitum* feeding (68) (Fig. 9.7).

All the symptoms described above strongly suggested that their common origin was a reduced activity of sympathetic fibers descending efferent pathways innervating the gastrointestinal tract, the endocrine pancreas, and the liver. Inoue and co-workers (69–72) provided and gathered a series of evidence for this notion: increased gastric activity and secretions, reduction salivary glands, and less serum DOPA-hydroxylase under sympathetic stimulation. All are signs of a reduced sympathetic tonus. Most of all, the hyperinsulinemic response to glucose seems to be the result of a suppression of the inhibitory sympathetic action on insulin secretion by β-cells. An *in vitro* study of responses of isolated islets of Langerhans to norepinephrine and ACh in the medium supports this notion. Islets taken from lesioned rat the first week after lesioning showed a reduced insulin release under norepinephrine. Taking into account the antagonistic action of norepinephrine and ACh, this *in vitro* insulin secretion indicates that the ACh-stimulating activity was augmented (73). It is of interest to note that 32 days after the lesion (about the time of appearance of the static phase), the picture is reversed (74); islets are then more sensitive to norepinephrine and less responsive to ACh.

Other reports support the notion of a reduced sympathetic tonus following the lesion: The total norepinephrine turnover is reduced in VMH rats. At the level of the adipose tissue the reduced norepinephrine activity associated with hyperinsulinemia participates in the accelerated and not corrected fat repletion (75, 76).

Afferent adrenergic fibers to adipose tissues are activated by β-adrenergic receptors in the stimulation of lipolysis and repression of lipogenesis. Lefèvre *et al.* (77) provided direct evidence for this control of lipolysis by sympathetic descending pathways. A unilateral denervation of lumbar fat deposits in rats led to an augmented weight of these fats relative to the contralateral one after 2 days of fasting. It is tempting to think that the VMH lesion is the equivalent of a bilateral sympathetic denervation of all body fats.

Inasmuch as a parasympathetic desinhibition of vagal parasympathetic activity seemed to be involved in hyperinsulinemia, it was of interest to examine the fate of the VMH syndrome after vagotomy and in diabetic rats. A striking fact is that both manipulations suppress but do not eliminate the hyperphagia and weight gain syndrome of lesioned rats. After the streptozotocin-induced di-

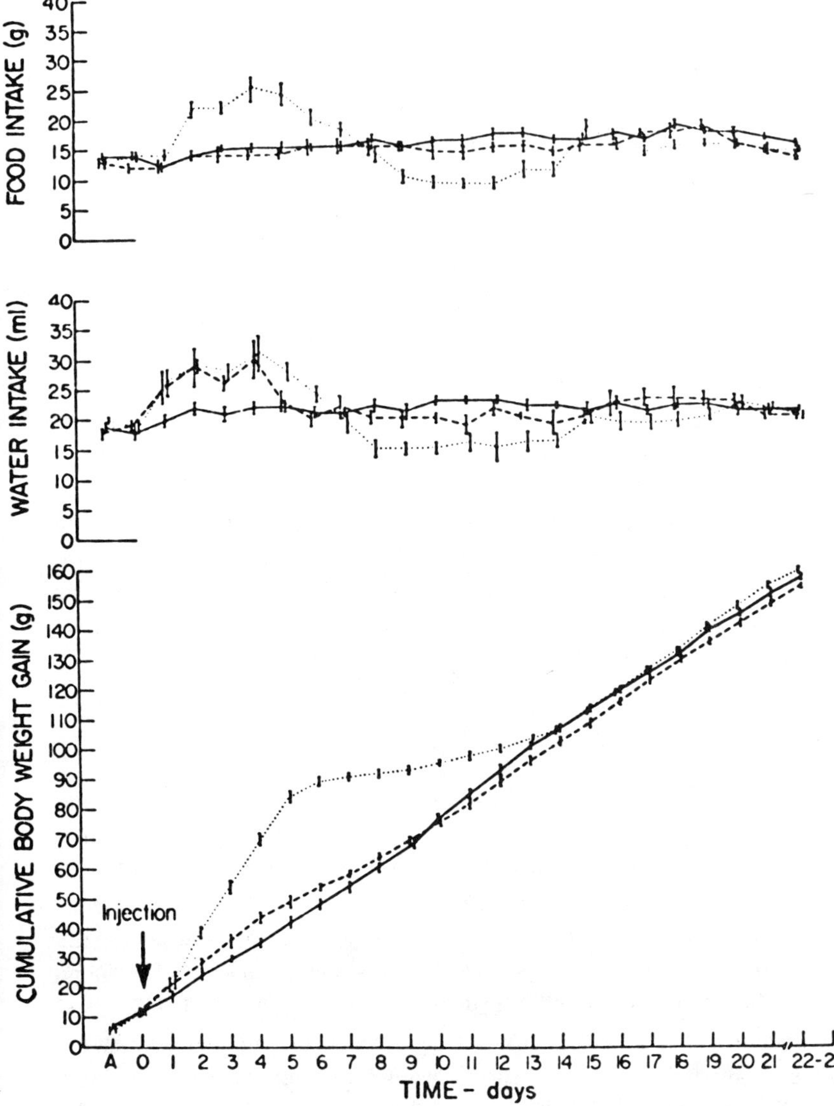

Figure 9.7 Transient hyperphagia after colchicine injection into the VMH. ________, Saline (1μl); ----------------, Colchicine (2μg/1μl)*; ··············, Colchicine (2μg/1μl).

abetes, which prevented hyperinsulinemia (78), or in diabetic rats treated by transplanted islets (69), about 20% of the weight gain of lesioned controls was maintained. Vagotomy performed after the lesion, like an atropine treatment, suppressed the hyperphagia and weight gain and the lesioned rats returned to the weight of vagotomized nonlesioned controls (79). Vagotomy 40 days prior to the VMH lesion only suppresses the hyperphagia and weight gain (80); performed after the lesion, it did not. When hyperinsulinemia is eliminated, this preserved hyperphagia and weight gain seem to confirm that, in addition to the lipogenic antilipolytic action of an increased insulin secretion, the VMH lesion produces the disruption of the sympathetic activation of lipolysis as a component of the lipostatic mechanism.

A doubt remains however. Total elimination of the effects of the lesion by a pharmacological blockade of the parasympathetic activity by atropine associated with a pharmacological stimulation of the sympathetic activity was never reported. Conversely, a mimicked VMH syndrome in intact rats by an ACh analogue associated with a β-adrenergic blocker such as propranolol was also not reported. Hyperphagia and weight gain following a guanetidine-induced sympathectomy were not observed either. Perhaps this doubt can be addressed by pharmacological treatments, which presumably do not reproduce the time course and the complex network of the action of physiological effectors.

B. The Ventromedial Hypothalamic Lesion–Hyperphagia

Hundreds of works dealing with hyperphagia induced by the VMH lesion and intending to demonstrate a role for the VMH in feeding have now lost their interest. Overeating is a trivial adjustment of the body energy balance by food intake, in which the unaltered mechanisms ruling meal sizes and meal frequency operate normally. The sequestration of a part of ingested calories in the energy retention as fats is added to the previous intake. Hypothalamic rats respond normally to prandial and periprandial loads, to chronic intragastric feeding, and to dilution of foods. Satiating as well as satiety mechanisms are therefore intact. Studying the meal pattern (Fig. 9.8) shows at night as well as during the day a highly significant postprandial correlation, better than that in intact rats. The meal : postmeal interval ratio (onset ratio) is very high, indicating that the end of satiety and meal initiation are precipitated by the dual utilization of food: oxidative metabolic rate + fat synthesis. The most significant fact is the disappearance of the diurnal periodicity of feeding. Lesioned rats eat during the day as they eat at night. The diurnal elimination of lipolysis and fat oxidation replaced by a fat synthesis indentical to that present at night is responsible for this change

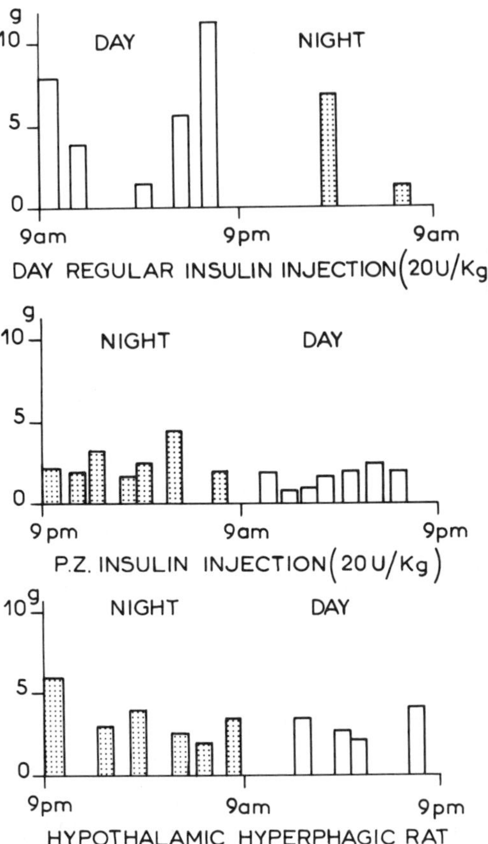

Figure 9.8 The meal pattern of hyperphagic hypothalamic rats is identical to that induced by insulin administration during daytime in normal rats.

of the meal pattern, which creates 24-hr overeating. As mentioned earlier, the same diurnal pattern is immediately produced by PZI injection or insulin infusion mimicking the permanent hyperinsulinemia of lesioned rats or by blockers of fat oxidation. Another significant difference between intact and lesioned rats is the diurnal response to short-term food deprivation. The response is poor compared to that of the night in intact rats. It is identical in lesioned rats, proving that the bad response of intact rats is due to their metabolic pattern at this time.

Numerous works described an increased responsiveness of lesioned rats to palatable as well as to aversive foods and, therefore, to sensory properties of foods. This is designated (one wonders why) the "finickiness" of VMH rats. However, it was clearly demonstrated that this effect does not depend on the

VMH. According to the extension of the electrolytic lesion, finicky not hyper-phagic, and hyperphagic not finicky rats might be found (81, 82). Disconnections between the amygdala and the LH associated with a large lesion of the VMH are possible causes of this finickiness.

IV. *Ventromedial Hypothalamic Glucoreceptors*

As already noted, the study by Jean Mayer and his group on gold-thio-glucose (GTG)-induced hyperphagia and obesity in mice and rats provided one of the most significant findings in the field. Their identification by this technique of insulin dependent glucoreceptors in the VMH area is today fully confirmed, whereas interpretations of the functional role for such glucoreceptors diverged from those initially proposed by the authors.

A. Location and Origin of Central Nervous System Cell Necrosis by Gold-Thioglucose

In 1949, Brecher and Waxler (83) reported that injecting GTG in mice (1 mg/g) induced hyperphagia and obesity resembling those induced by the VMH lesion. Later it was shown in mice and by local implants in rats and cats that (1) GTG acted as a neurotoxic inducing cell necrosis, (2) the induced necrosis was mainly limited to the VMH complex, and (3) the necrosis was related to a competition between the cell glucose and GTG uptake and indicated, better than 2-DG labeling, a specific high rate of glucose uptake of these cells.

At first contested, the VMH location of necrosis in mice was fully con-firmed by the use of radioactive GTG (84). A histologic and autoradiographic examination of lesions at various times following the injection showed the lesion and the gold deposition in glial oligodendritic cells in the VMH. Systemic injection of radioactive phlorizin maximally labeled the VMH area in which GTG induced the lesion (84). The intensity of the radioactive gold labeling and the degree of the induced hyperphagia were correlated to the blood level of GTG in the 3 min following the injection (85). A positive correlation was also shown between the extent of lesions produced by various doses of GTG and the induced hyperphagia and weight gain. This correlation was maximal for necrosis located in the premammillary bodies and preoptic areas (86). In rats, the intraperitoneal injection of the same dose used in mice is lethal. However, in rats, implants of GTG in the VMH induced hyperphagia and weight gain (87, 88, 89). Moreover, the intraperitoneal injection of the small dose of 0.5 mg/g to 24-hr fasted rats also induced hyperphagia (90).

A cell necrosis in the LH by a systemic injection of GTG was not reported. However (as already mentioned regarding LH glucosensitive sites), the local implant of the neurotoxin in rat induces aphagia, suggesting the destruction of these sites. This limitation of the GTG-induced necrosis to the diencephalon was later revised. Powley (91) demonstrated that, in addition to the hypothalamic area, GTG also induced circumscribed cell necrosis in the dorsomotor vagal nucleus, the caudal portion of the nuclei of tractus solitarius and in area postrema. Thus, it was suggested (in agreement with other data) that specific glucoreceptors were present in the hindbrain.

Convergent and fully convincing results suggest that, as 2-DG is toxic by blocking glycolysis and competing with the cell glucose uptake, GTG is highly toxic through the uptake of thioglucose driving the toxic gold into the cell like a Troy Horse. With 2-DG, this is true for all cells with high doses. That the doses of GTG (lower than the threshold of this general lethal action) induce necrosis in localized cells only clearly indicates (and more specifically than with 2-DG) a particular glucose affinity or rate of uptake of these cells.

A large number of results confirmed the notion that GTG-lesioned cells are indeed cells highly sensitive to their glucose supply. Gold-thiosorbitol, gold-thioglycerol, or gold-thiomalate did not induce hyperphagia (92). The competition with glucose and glucose analogue is obvious. An intraventricular or systemic injection of phlorizin prevents the GTG lesion (93). 2-DG or 5-thioglucose given prior to GTG reduces or prevents the GTG lesion and the induced obesity (94, 95). The metabolic effects of intraventricular 2-DG (rise of PFFA hyperglycemia) are considerably reduced by a previous GTG injection (96). The GTG lesion is prevented by a high blood glucose level or glucose administration like the lesion of pancreatic β-cells by alloxan is also preserved. This strongly indicates the competition with glucose of the GTG uptake. In addition, it is so suggested that GTG acts like a diabetogenic agent and, again, that the VMH glucoreceptors are similar in some aspects to the β-cell receptors (97).

B. Insulin Dependency of Glucose Uptake by Ventromedial Hypothalamic Glucoreceptors

Debons *et al.* (98) presented considerable evidence for the dependence of the GTG-VMH lesion on insulin; therefore, they suggested the insulin dependency of the lesioned cells (exceptional in the brain). GTG does not induce hyperphagia and obesity in diabetic mice. The autoradiographically localized gold deposition was reduced in the VMH of these diabetics. Administration of insulin in diabetics reestablished the GTG potency. In diabetic rats, the local implant of GTG in the VMH is inactive. It is also reestablished by insulin (89).

Following a systemic injection of radioactive insulin, a maximal radioactivity is localized in the same oligodendritic cells lesioned in the VMH by GTG. This suggests that brain insulinoreceptors (otherwise localized in the same region; see later) are associated to the insulin dependent glucoreceptors. Phenformin, an insulinlike antidiabetic drug, also reestablishes the vulnerability of diabetic mice to GTG (99). This highly significant co-localization of brain insulin-binding sites and of GTG-sensitive cells is also indicated by Powley (91) in the area postrema dorsomotor vago-nuclei and NTS.

Thus, the glucose uptake of GTG-lesioned cells is strongly suggested to be insulin dependent; however, this does not seem to be the case for the glucose uptake tested in the same region by ^{14}C-2-DG. Insulin was injected unilaterally in the VMH and saline in the contralateral side, and then ^{14}C-2-DG was systemically injected. No assymetry in the ^{14}C-2-DG labeling was observed. Intraventricular insulin did not augment the 2-DG-labeling in nondiabetic rats. In diabetics, the labeling following systemic insulin + ^{14}C-2-DG is equivalent to that by saline + 2-DG. The labeling by 2-DG was, on the contrary, 20 times augmented by insulin in insulin-sensitive peripheral tissues such as the diaphragm (100). Elegant experiments by Goodner and Berrie (101) and Goodner *et al.* (102) confirmed this result.

Oomura (108, Chapter 3) using the micropipette technique to perform local injections of glucose and/or of insulin and the simultaneous recording of the cell electrical activity confirmed that cells in the VMH are neurally activated by glucose and that this activity is potentiated by insulin, contrary to similar LH glucosensitive sites. However, this work gives no further indication on the functional role of these glucoreceptors, neither on their dose-dependent responses to local or peripheral glucose levels.

C. Hyperphagia and Obesity in Mice Induced by Gold-Thioglucose Lesion

Hyperphagia and resulting obesity induced by GTG resemble, but do not exactly reproduce, those induced by the VMH electrolytic lesion. Hyperphagia is not immediately apparent following the GTG injection in mice. No clear static phase is observed. The gain of weight is continuous up to senescence (103). Contrary to that of the VMH-lesioned rat, the dark–light cycle of food intake is maintained (104). However, like in the VMH lesion, hypophysectomy and adrenalectomy prevent the GTG lesion (105).

D. Metabolic Effects of the Gold-Thioglucose Lesion

Surprisingly enough, little attention was given to the metabolic consequences of the GTG lesion in the VMN and other brain sites. Nevertheless, at
least two works provide strong evidence for an impaired lipolysis, identical to
that observed after the surgical VMH lesion. Thereby, these works support the
notion of an involvement of the lesioned cells in a lipostatic mechanism. The
lipolysis induced by central 2-DG was strongly reduced after the GTG lesion
(96). Two weeks after the lesion, the *in vitro* adipocyte lipolysis under epinephrine was shown to be significantly reduced (106).

V. *Insulin Responsiveness of the Ventromedial Hypothalamus as a Putative Lipostatic Mechanism*

Strong experimental evidence, reviewed in preceding chapters, supports
the notion that descending sympathetic pathways afferent to the pancreas, the
liver, adrenals, and the adipose tissue are in action, at least to modulate the
respective effectiveness of lipogenic and lipolytic factors on adipocytes. Thereby, these efferents limit the physiological fluctuations of the body fat mass as a
buffer to the body energy balance and, thus, ensure the long-term and mean
constancy of body weight. The effect of both the activation and a paralysis or
lesion of the VMH gave evidence for a role of this area in governing this
regulation. However, major questions remain unsolved: What is the afferent
signal? How is the brain (which does not accumulate fats) informed of the
increase or decrease in the peripheral fat mass, allowing it to give orders for their
limits? Afferent sympathetic or parasympathetic pathways from body fat deposits
to the brain are unknown and apparently do not exist. Only one serious hypothesis was put forward based on two series of facts. Glucoreceptors are present in
the VMH and their glucose uptake is probably (but not certainly) insulin-dependent. Hyperinsulinemia is the universal correlate of the increase in peripheral fats
and of all forms of spontaneous and experimental obesities. Thus, the hypothesis
was that chronic hyperinsulinemia was both one cause of the body fat overrepletion and the signal acting to generate, in an undefined manner, the regulatory
response of the VMH. Arguments in favor and against such hypothesis will be
examined here.

A. Acute and Chronic Central Action of Insulin

The central action of insulin on lipolysis–lipogenesis and on food intake
appears to be the opposite of its action on peripheral insulin-sensitive tissues.

This is supported by results showing that acute and most of all chronic local or intraventricular administrations of insulin indeed induce weight loss and inhibit food intake. As mentioned earlier, insulin, locally and acutely injected into the VMH, inhibits food intake, whereas, on the contrary, injecting insulin antibodies elicits feeding (see Chapter 3, Section I). However, these effects of intra-VMH acute administration of insulin on food intake are difficult to reconcile with other data that show that the same local insulin injections induce an immediate hypoglycemia apparently due to a stimulated pancreatic insulin release (107–109).

The initial work of chronic insulin administration was carried out by Woods *et al.* (110). In the baboon, 10 days of intraventricular administration of insulin induced weight loss and hypophagia. After the cessation of injection, the return to the initial body weight was observed except with the higher dose, which was followed by an incomplete recovery.

The use of an implanted osmotic minipump substituted for injection through a chronic cannula allowed Brief and Davis (111) to fully confirm these results in rats. Three doses of insulin (5.0, 9.5, and 10.0 microunits/hr) were injected in the third ventricle for 7 days. A dose-dependent reduction of food intake was recorded with the two higher doses, and this reduction occurred at night. The observed weight loss was not entirely accounted for by the reduced intake. At the cessation of treatment, the return to a normal intake and to initial body weight was slow (7 days). No hyperphagia was associated to this recovery of body weight.

Another original technique was the implantation of isolated pancreatic islets in the third ventricle (112). This brain-implanted pancreas reduced body weight. Various other authors confirmed these results. In the work of Plata-Salaman *et al.* (113), the effects of acute and chronic infusions of insulin into the third ventricle were compared. Both reduced intake, but the acute effect was observed only at night, suggesting a role for a repressed lipogenesis and an elicited lipolysis during this period.

The relation of the effect of intraventricular insulin to the time of day is strikingly highlighted by Nagai and Nakagawa (114). A minipump infusion of 1–10 microunits/day induced a dose-dependent reduction of the night : day ratio by decreasing the nocturnal and increasing the diurnal intake. The 24-hr intake is unchanged. Insulin infusion into the suprachiasmatic nuclei has an identical and even more accentuated effect. This suprachiasmatic nuclei infusion induces a fall of plasma insulin and of blood glucose during the day, their elevation at night (115).

B. Brain Insulin Receptors

Brain tissue is not insulin-sensitive. Glucose uptake by central nervous cells and glucose transport through the blood–brain barrier are not insulin-dependent. However, the finding by autoradiography of an insulin binding in some brain sites, i.e., of brain insulin receptors, was not too surprising, primarily because binding sites of a large number of gut and other peripheral hormones were found in the brain by the same autoradiographic techniques or others. The functions of most of these receptors (CCK, gastrin, VIP and so on) are absolutely unknown. The second reason was the likelihood that the presumed insulin dependency of some localized brain areas such as the VMH could be associated with the presence of insulin receptors in the same area.

Havrankova *et al.* (116) and Van Houten and Posner (117) described for the first time insulin-binding sites. Havrankova and co-workers found them distributed in all parts of the brain with a maximal density in the olfactory bulb, cortex, preoptic region, and anterior and posterior hypothalamus and a lower density in the cerebellum and mesencephalon (Fig. 9.9). Van Houten and Posner localized more precisely high-affinity insulin receptors in circumventricular areas of the hypothalamus, arcuate nuclei, and zones of contacts with the medium eminence. In these areas, the receptors were seen by the authors to be located in both capillary walls and axon terminals. The localization of capillary walls, i.e., at the level of the blood–brain barrier, was confirmed. According to Partridge (118), at the level of privileged sites, they could affect the glucose transport inside glucosensitive cells. Corp *et al.* (119) tested the insulin-binding of brain tissue samples by autoradiography *in vitro*. In the hypothalamus, they found a stronger binding in arcuate nuclei, dorsomedial nuclei, and periventricular areas than in both the VMH and LH. The final step was the isolation and biochemical identification of the receptor. From capillaries of pigs and calves, Haskell *et al.* (120) isolated the receptor. It is a polypeptide not different from the insulin receptor of an insulin-sensitive peripheral tissue such as the liver.

C. Regulation of Brain Insulin Receptors in the Control of Peripheral Body Fats

It was hypothesized that VMH insulin receptors would be regulated like the insulin-binding of peripheral cells are. A complementary hypothesis was that a change of the insulin activity on the glucose uptake of VM glucoreceptors could be a cause of either the activation or inhibition of the VMH lipostatic system.

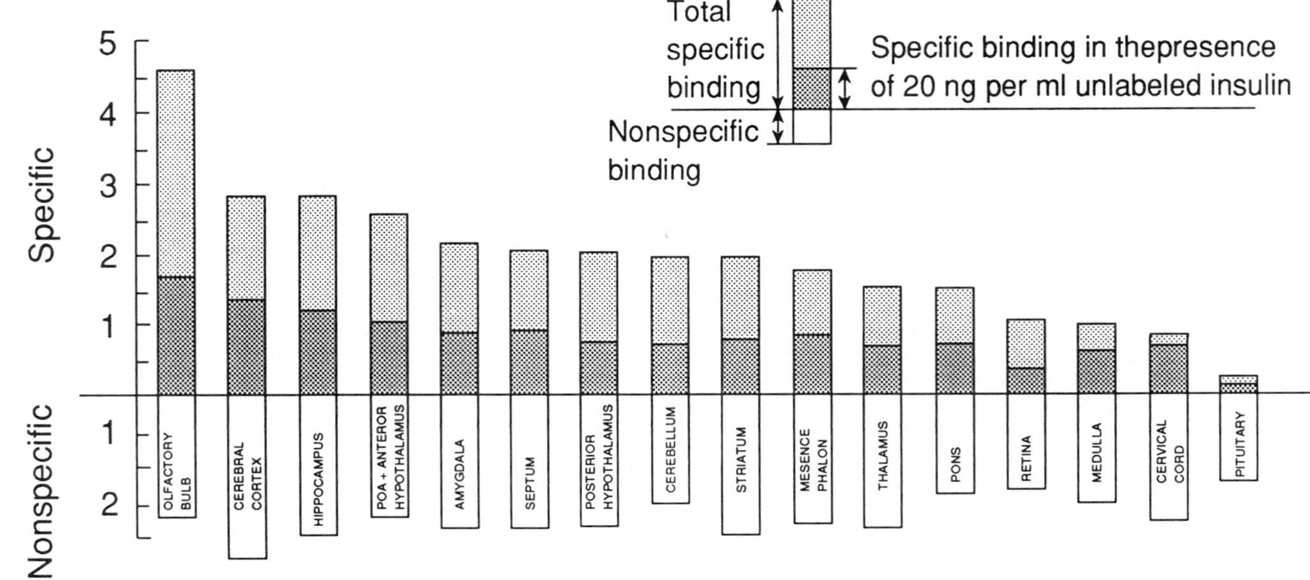

Figure 9.9 Regional distribution of brain insulinoreceptors. [From Havrankova *et al.* (116)]

This hypothesis seems to be confirmed by various reports. Indeed, hypothalamic insulin-binding is altered by hyper- and hypoinsulinemia and as a function of increased or decreased body weight, but it is apparently in an opposite way to that expected and demonstrated in peripheral insulin-sensitive cells.

According to Melnyck and Martin (121), insulin-binding is reduced in the LH compared to the VMH of the fed rat. It is reduced only in the VMN by 14 days of food restriction and weight loss. This suppressed binding in the ventromedial nucleus is associated with a considerable drop of circulating insulin in these restrained rats. Neither reduction nor augmentation of the binding are found in hypoinsulinic streptozotocin diabetic rats (122). Hyperinsulinemia induced by PZI treatment does not change ventromedial insulin-binding despite the increase in the local free insulin concentration in the hypothalamus. A slight increase in the maximal binding capacity is measured but is compensated for by an inverse change of the affinity (123). *In vitro,* insulin-binding of hypothalamic tissue is found to be stronger during the phase of weight gain in the ground squirrel. This elevated binding is highly correlated to the higher glucose-stimulated insulin secretion and to the glucose uptake of the adipose tissue during the same phase.

The finding of the highest density of insulin receptors in the olfactory bulb was intriguing. In genetically obese rats, a reduced binding is found in the external plexiform layer of the olfactory bulb of fa-fa obese compared to fa-fa and FA-FA nonhyperphagic, nonhyperinsulinic lean rats (124). The same result is reported in the olfactory bulb as a whole (125). Thus, the modulation of insulin-binding by circulating insulin would be opposite in the olfactory bulb and in the hypothalamus. Therefore, an inverse modulation of VMH insulin receptors by hyper- and hypoinsulinemia and associated weight gains and weight losses, if not entirely ascertained, is supported by these results. Further studies will be needed to assess whether or not an increased insulin-binding in the VMH by hyperinsulinemia might be responsible for a higher glucose uptake by insulin-sensitive glucoreceptors activating the VMH and its repressive activity on the ongoing weight gain. Such a possibility is consistent with the weight loss and reduced food intake induced by the chronic intraventricular administration of insulin. It is also consistent with the fact, already mentioned, that injecting insulin antibodies in the VMH produces hyperphagia.

Such a regulation of VMH insulin receptors by circulating insulin is suspected to operate in the diurnal lipogenesis–lipolysis cycle. In humans, insulin-binding on circulating monocytes is elevated at night and low during the day. Binding after lunch is half its level at dawn. The decrease during the day is parallel to the elevation of plasma insulin; the increase at night is parallel to the elevated PFFA and ketone bodies (126). A contemporary opposite trend could be in action centrally.

One experiment was designed to investigate the possibility that the progressive modulation by the nocturnal hyperinsulinism in rat could be a cause of the subsequent diurnal metabolic and feeding pattern. The hypothesis was that the immediate inversion of this diurnal pattern following a nocturnal fast could be due to the lack of this nocturnal modulation. Rats fasted at night received a continuous intravenous infusion of a small dose of regular insulin. Thus infused, their plasma insulin was compared to fasting saline-infused controls. The recording of their free-feeding pattern on the subsequent day showed a reduction of the cumulative 12-hr intake (127) (Fig. 9.10).

D. Lipid Metabolism in the Hypothalamus

Brain tissues do not synthesize and accumulate metabolizable fats and do not oxidize circulating PFFA like peripheral tissues do; however, some works suggest the opposite as well as a role for fat oxidation or fat storage in the hypothalamus in the regulation of energy balance, lipolysis, and lipogenesis.

In diabetic rats, fat oxidation is elevated in the LH during the initial period of hypophagia and high PFFA level, like in healthy rats in a negative energy balance (128). In another experiment (129), two groups were tube-fed 50 or

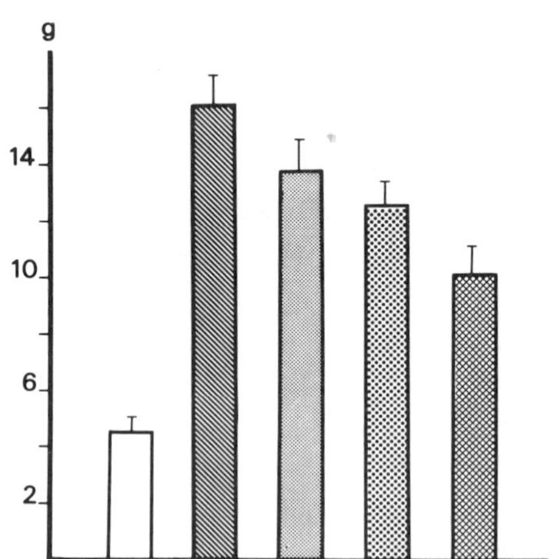

Figure 9.10 Insulin perfusion during a nocturnal fast reduces the effect of this fast on the subsequent diurnal food intake.

160% their normal intake until a 30% weight loss in the first group and a 30% gain in the second. Seventeen days of hyperphagia in the first group and 40 days of hypophagia in the second were observed during the subsequent return to the initial body weight. *In vitro* PFFA oxidation in the ventral LH was more elevated during this phase in hyperphagics than in hypophagics, the former regaining weight, the latter losing weight, and proportionally to the percentage of loss and gain of weight at the start of the refeeding. The authors concluded that fat oxidation varies as a function of the state of energy balance and of its regulation and not of the contemporary intake. It is difficult to evaluate this abstracted conclusion in terms of the role of the hypothalamus and of the LH in bringing about regulatory lipolysis and/or lipogenesis. However, the result, although of interest, is paradoxical because fat oxidation would be elevated in the hypo-thalamus at a time of high fat synthesis and reduced at a time of peripheral fat mobilization and oxidation.

The idea that some cells in the hypothalamus could be witnesses of the peripheral body fat changes by their own parallel accumulation or depletion of fats was often expressed. Only one tentative confirmation of his speculation was achieved. ^{14}C-glucose was tubed into the stomachs of rats. Some hours later, the radioactivity of various biochemically separated fractions was investigated in biopsies of the LH and VMH. The total radioactivity was higher in the VMH and its part segregated in a lipid fraction was also greater in the VMH. Further study is needed to consider the meaning of this promising result (130).

E. Parabiosis

Parabiotic rats sharing their blood through parabiosis were a good preparation to test humoral factors generated by a partner made obese on its normal partner. The basic experiment was done by Fleming (131). A normal partner coupled with a VMH-lesioned rat is strongly hypophagic and loses weight. The body fat content is reduced. Thus, factors acting to repress the weight gain generated by the lesioned rat and unable to act in this rat are manifested in its partner in which these factors can act on an intact VMH. In another study, one of the parabiotics was made progressively obese by four sessions per day of LH electrical stimulation for 2 wk. The normal partner reduced its free intake and lost weight as much as the obesity progressed in the stimulated partner. At the beginning of stimulations, glucagon and blood glucose were elevated and insulin was normal in the stimulated rat, while no change of these parameters was observed in its partner, except a fall of glycemia. The authors concluded that neither glucagon and insulin nor the blood glucose level could be the factor transmitted to the normal partner and responsible for its hypophagia and weight loss (132).

The most convincing experiment was carried out by King (133). A cross-circulation by a peristaltic pump was used instead of the surgical parabiosis. A normal rat was cross-perfused by the blood of a partner made either obese (by a high-fat diet), normal, or lean (by restriction). Paired with the obese, the rat after 30 min of cross-circulation was hypophagic; paired with the lean, it was hyperphagic compared to its response when paired with the normal-weight partner.

F. Tentative Conclusion

The existence of a lipostatic mechanism acting to limit fluctuation of body fat masses in a narrow range and to ensure a long-term constancy of body weight is fully demonstrated. Through its indirect action on the feeding control system, this mechanism is a component of the regulation of body energy balance. It is associated with another reensuring mechanism of this body energy balance represented by an extra heat production or by its inhibition by the brown adipose tissue. This lipostatic mechanism involves a reduced sensitivity of repleted adipocytes to lipogenic factors, mainly of insulin, and the increase of this responsiveness in depleted adipocytes. But the generation of these lipogenic and lipolytic agents is centrally regulated as a function of the body fat mass increase or decrease. The VMH plays a critical role in this regulation. It is activated by the increase of peripheral body fats. A role of hyperinsulinemia and of VMH glucose and insulin receptors in this activation is presumed. This activation brings about descending sympathetic and perhaps parasympathetic pathways that modulate hyperresponsiveness to lipogenic and hyporesponsiveness to lipolytic agents. Further study will be needed to know whether or not and how a change of the VMH glucose and insulin receptor functions are involved in this inhibition or inversion of the VMH activity. The experimental or pathological disruption of these two reensuring mechanisms of the body energy balance leads to obesity or to permanent leanness. The following chapters will presumably strengthen our knowledge about the physiological normal mechanism.

References

1. Hoebel, B. G., and Teitelbaum, P. (1966). Weight regulation in normal and hypothalamic hyperphagic rats. *J. Comp. Physiol. Psychol.* **61,** 189–193.
2. Larue-Achagiotis, C., and Le Magnen, J. (1985). Effect of long-term insulin on

body weight and food intake: Intravenous versus intraperitoneal routes. *Appetite* **6**, 319–329.

3. Le Magnen, J. (1976). *Hunger*. Cambridge: Cambridge University Press.
4. Olefsky, J. M., and Saekow, M. (1978). Effects of dietary carbohydrate content on insulin binding and glucose metabolism by isolated rat adipocytes. *Endocrinology* **103**, 2257–2263.
5. Beck-Nielsen, H., and Pedersen, O. (1978). Diurnal variation in insulin binding to human monocytes. *J. Clin. Endocrinol. Metab.* **47**, 385–390.
6. Di Girolamo, K., and Melinger, P. (1972). Glucose metabolism and responsiveness to bovine insulin by adipose tissue from three mammalian species. *Diabetes* **21**, 1151–1161.
7. Di Girolamo, M., and Esposito, J. (1975). Adipose tissue blood flow and cellularity in the growing rabbit. *Am. J. Physiol.* **229**, 107–112.
8. Di Girolamo, M., and Owens, J. L. (1976). Glucose metabolism in enlarged fat cells: Enhanced in large fat cells from older rats to epinephrine and A.C.T.H. *Horm. Metabol.* **87**, 645–651.
9. Hansen, F., Nielsen, T., and Gliemann, D. L. (1974). Influence of body weight and cell size on lipogenesis and lipolysis of isolated rat fat cells. *Eur. J. Clin. Invest.* **4**, 411–418.
10. Holm, G., Jacobson, B., and Smith, U. (1975). Effect of age and cell size on adipose tissue metabolism. *J. Lip. Res.* **19**, 461–464.
11. Ballard, K., and Rosell, S. (1969). The unresponsiveness of lipid metabolism to canine mesenteric adipose tissue to biogenic amines and to sympathetic nerve stimulation. *Acta Phys. Scand.* **774**, 442–448.
12. Litman, Y. (1979). Regional difference in the control of lipolysis in human adipose tissue. *Metab. Clin. Exp.* **28**, 1190–1195.
13. Faust, G. M., Johnson, N., and Hirsch, J. (1977). Surgical removal of adipose tissue alters feeding behavior of rat. *Science* **197**, 391–395.
14. Faust, G. M., Johnson, N., and Hirsch, J. (1977). Adipose tissue regeneration following lipectomy. *Science* **4301**, 491–492.
15. Kral, G. G. (1975). Surgical reduction of adipose hypercellularity in man. *Scand. J. Plast. Reconstr. Surg.* **9**, 140–143.
16. Dark, G., Forger, N., Stern, J., and Zucker, E. (1985). Recovery of lipid mass after removal of adipose tissue in ground-squirrel. *Am. J. Physiol.* **249**, R73–78.
17. Powley, T. L., and Berthoud, H. R. (1989). Neuroanatomical bases of cephalic phase reflexes. *Appetite* **12**, 78. [Abstract.]
18. Berthoud, H. R. Fox, E. A., and Powley, T. L. (1990). Localization of vagal preganglionics that stimulate insulin and glucagon secretion. *Am. J. Physiol.* **258**, 160–168.
19. Porte, D., Jr., Woods, S. C., Chen, M., Smith, P. H., and Ensinck, J. W. (1975). Central factors in the control of insulin and glucagon secretion. *Pharm. Biochem. Behav.* **3**, (Suppl. 1), 127–134.

20. Porte, D. Jr., Smith, P. H., and Ensinck, J. W. (1976). Neurohumoral regulation of pancreatic islet A and B cells. *Metabolism* **25,** (Suppl. 1), 1453–1457.

21. Girardier, L., Seydoux, J., and Campfield, L. A. (1976). Control of A and B cells in vivo by sympathetic nervous input and selective hyper- or hypoglycemia in dog pancreas. *J. Physiol. (London)* **56,** 801–814.

22. Campfield, L. A., and Smith, F. J. (1980). Modulation of insulin secretion by the autonomic nervous system. *Brain Res. Bull.* **5,** (Suppl. 4), 103–107.

23. Campfield, L. A., and Smith, F. J. (1983). Neural control of insulin secretion: Interaction of norepinephrine and acetylcholine. *Am. J. Physiol.* **244,** R629–R634.

24. Campfield, A., and Smith, F. (1980). Modulation of insulin secretion by the autonomic nervous system. *Br. Res. Bull.* **5,** 103–107.

25. Trimble, E., and Berthoud, H. (1981). Importance of cholinergic innervation of the pancreas on glucose tolerance in the rat. *Am. J. Physiol.* **246,** E337–346.

26. Niijima, A. (1975). The effect of 2-deoxy-D-glucose and D-glucose on the efferent discharge rate of sympathetic nerves. *J. Physiol. (London)* **251,** 231–243.

27. Rizza, R., Haymond, M., Cryer, P., and Gerich, J. (1979). Differential effects of epinephrine on glucose production and disposal in man. *Am. J. Physiol.* **237,** E356–362.

28. Marliss, E. B., Girardier, L., Seydoux, J., Wolheim, C. B., Kanazawa, Y., Orci, L., Renold, A. E., and Porte, D., Jr. (1973). Glucagon release induced by pancreatic nerve stimulation in the dog. *Clin. Invest.* **52,** 1246–1259.

29. Kaneto, A., Miki, Z., and Kosaka, K. (1974). Effects of vagal stimulation on glucagon and insulin secretion. *Endocrinology* **95,** 1005–1009.

30. Bloom, S. R. Vaughan, N. J. A., and Russell, R. C. G. (1974). Vagal control of glucagon release in man. *Lancet* **7880,** 546.

31. Dubuc, P. U., Leshin, L. S., and Willis, P. L. (1982). Glucose and endocrine responses to hypothalamic electrical stimulation in rats. *Am. J. Physiol.* **242,** R220–226.

32. Frohman, L. A., and Bernardis, L. J. (1971). Effect of hypothalamic stimulation on plasma glucose, insulin and glucagon levels. *Am. J. Physiol.* **221,** 1596–1603.

33. Steffens, D., and Morrissey, S. (1975). Hypothalamic stimulation induces acid secretion, hypoglycemia and hyperinsulinemia. *Am. J. Physiol.* **228,** 1296–1299.

34. Morrissey, S. M., Stephens, D. N., and Webber, D. E. (1977). The role of the posterior hypothalamus in hypoglycemia and gastric secretion. *J. Physiol. (London)* **266,** 85–89P.

35. Barkai, A., and Allweis, C. (1972). Effect of electrical stimulation of the hypothalamus on plasma free fatty acid concentration in cats. *J. Lipid Res.* **13,** 725–732.

36. Goodner, C. J., Koerker, D. J., Werrbach, J. H., Toivola, P., and Gale, C. C. (1973). Adrenergic regulation of lipolysis and insulin secretion in the fasted baboon. *Am. J. Physiol.* **224,** 534–539.

37. Ishikawa, T., Nagata, M., and Osumi, Y. (1983). Dual effects of electrical stimula-

tion of ventromedial hypothalamic neurons on gastric acid secretion in rats. *Am. J. Physiol.* **245**, G265–269.

38. Shimazu, P., and Ishikama, K. (1981). Modification by hypothalamus of glucagon and insulin secretion in the rabbit. *Endocrinol.* **108**, 605–612.

39. Kumon, A., Takahashi, A., Hara, T., and Shimazu, T. (1976). Mechanism of lipolysis induced by electrical stimulation of the hypothalamus in the rabbit. *J. Lip. Res.* **17**, 551–558.

40. Campfield, L. A., Smith, F., and Le Magnen, J. (1983). Altered endocrine pancreatic function following vagotomy: Possible behavioral and metabolic bases for assessing completeness of vagotomy. In J. G. Kral, T. L. Powley, and C. McC. Brooks (eds.), *Vagal nerve function: Behavioral and methodological considerations* (pp. 283–300). Amsterdam: Elsevier Science.

41. Campfield, L. A., Smith, F. J., and Eskinazi, R. E. (1984). Glucose responsiveness and acetylcholine sensitivity of pancreatic B-cells after vagotomy. *Am. J. Phys.* **246**, R985–993.

42. Hetherington, A. W., and Ranson, S. W. (1942). The spontaneous activity and food intake of rats with hypothalamic lesions. *Am. J. Phys.* **36**, 609–616.

43. Kennedy, G. C. (1953). The role of fat depot and hypothalamic control of food intake in the rat. *Proc. R. Soc. (London) B* **140**, 578–692.

44. Hales, C. N., and Kennedy, G. C. (1964). Plasma glucose, non-esterified fatty acid and insulin concentrations in hypothalamic hyperphagic rats. *Biochem. J.* **90**, 620–624.

45. Steffens, A. B. (1967). Blood glucose levels and food intake in normal and hypothalamic hyperphagic rats. *Acta Physiol. Pharm. Neerl.* **14**, 524–525.

46. Steffens, A. B., Mogenson, G. J., and Stevenson, J. A. F. (1972). Blood glucose, insulin and free fatty acids after stimulation and lesions of the hypothalamus. *Am. J. Physiol.* **222**, 1446–1452.

47. Louis-Sylvestre, J. (1976). Preabsorptive insulin release and hypoglycemia in rats. *Am. J. Physiol.* **230**, 56–60.

48. Christophe, J., and Mayer, J. (1960). Glucose uptake as a function of food intake in normal and hypothalamic hyperphagic rats. *Metabolism* **9**, 932–937.

49. Storlien, L. H. (1985). The role of the VMH area in periprandial glucoregulation. *Life Sci.* **36**, 505–514.

50. Pénicaud, L., Rohner-Jeanrenaud, F., and Jeanrenaud, B. (1986). In vivo metabolic changes as studied longitudinally after ventromedial hypothalamic lesions. *Am. J. Physiol.* **250**, E662–668.

51. Le Magnen, J., Devos, M., Gaudiliere, J. P., Louis-Sylvestre, J., and Tallon, S. (1973). Role of a lipostatic mechanism in regulation by feeding of energy balance in rats. *J. Comp. Physiol. Psychol.* **84**, 1–23.

52. Rohner-Jeanrenaud, F., and Jeanrenaud, B. (1980). Consequences of ventromedial hypothalamic lesions upon insulin and glucagon secretion by subsequently isolated perfused pancreas in the rat. *J. Clin. Invest.* **65**, 902–908.

53. Frohman, L. A., Goldman, J. K., and Bernardis, L. L. (1972). Metabolism of

intravenously injected ^{14}C-glucose in weanling rats with hypothalamic obesity. *Metabolism* **21**, 799–805.

54. Frohman, L. A., Bernardis, L. L., Schnatz, J. D., and Burek, L. (1969). Plasma insulin and triglyceride levels after hypothalamic lesions in weanling rats. *Am. J. Physiol.* **216**, 1496–1501.

55. Frohman, L. A., Goldman, J. K., and Bernardis, L. L. (1972). Studies on insulin sensitivity in vivo in weanling rats with hypothalamic obesity. *Metabolism* **21**, 1133–1142.

56. Hustvedt, B. E., and Lovo, A. (1972). Correlation between hyperinsulinemia and hyperphagia in rats with ventromedial hypothalamic lesions. *Acta Physiol. Scand.* **84**, 29–33.

57. Hustvedt, B. E. (1966). The effect of V.M.H. lesion on metabolism and insulin secretion of rats on a controlled feeding regimen. *Nutri. Metab.* **20**, 294–301.

58. Han, P. W., Feng, L. Y., and Kuo, P. T. (1972). Insulin sensitivity of pair-fed hyperlipemic, hyperinsulinemic obese-hypothalamic rats. *Am. J. Physiol.* **223**, 1206–1209.

59. Pénicaud, L., Kinebanyan, M. F., Ferré, P., Morin, J., Kandé, J., Smajda, C., Marfaing-Jallat, P., and Picon, L. (1989). Development of VMH obesity: In vivo insulin secretion and tissue insulin sensitivity. *Am. J. Physiol.* **257**, E255–260.

60. Laury, M. C., Pénicaud, L., Ktorza, A., Benhaim, H., Bihoreau, M. T., and Picon, L. (1989). In vivo insulin secretion and action in hyperglycemic rats. *Am. J. Physiol.* **257**, E180–184.

61. Nishizawa, Y., and Bray, G. (1978). Ventromedial hypothalamic lesions and mobilization of fatty acids. *J. Clin. Invest.* **61**, 714–721.

62. May, K. K., and Beaton, J. R. (1966). Metabolic effects of hyperphagia in the hypothalamic–hyperphagic rat. *Can. J. Physiol. Pharmacol.* **44**, 641–650.

63. Harrell, E. H., and Remley, N. R. (1973). The immediate development of behavioral and biochemical changes following ventromedial hypothalamic lesions in rats. *Behav. Biol.* **9**, 49–64.

64. Kasemsri, S., Bernardis, L. L., and Schnatz, J. D. (1972). Fat mobilisation in adipose tissue of weanling rats with hypothalamic obesity. *Hormones* **3**, 97–104.

65. Lovo, A., and Hustvedt, B. E. (1978). Early effects of feeding upon hormonal and metabolic alterations in adult rats with ventromedial (VMH) lesions. *Horm. Metab. Res.* **10**, 304–309.

66. Walgren, M. C., and Powley, T. L. (1985). Effects of intragastric hyperalimentation on pair-fed rats with ventromedial hypothalamic lesions. *Am. J. Physiol.* **248**, R172–180.

67. Seydoux, J., Rohner-Jeanrenaud, F., Assimacopoulos-Jeannet, F., Jeanrenaud, B., and Girardier, L. (1981). Functional disconnection of brown adipose tissue in hypothalamic obesity in rats. *Pflüg. Arch.* **390**, 1–4.

68. Avrith, D., and Mogenson, G. H. (1978). Reversible hyperphagia and obesity

following intracerebral microinjections of colchicine into the ventromedial hypothalamus of the rat. *Brain Res.* **153,** 99–107.

69. Inoue, S., Bray, G. A., and Muller, Y. (1978). Transplantation of pancreatic B-cells prevents development of hypothalamic obesity. *Am. J. Physiol.* **235,** E266–271.

70. Inoue, S., and Bray, G. A. (1979). Autonomic hypothesis for hypothalamic obesity. *Life Sci.* **25,** 561–566.

71. Inoue, S., and Bray, G. A. (1980). Role of the autonomic nervous system in the development of ventromedial hypothalamic obesity. *Brain Res. Bull.* **5** (Suppl. 4), 119–126.

72. Inoue, S., Muller, Y. S., and Bray, G. A. (1983). Hyperinsulinemia in rats with hypothalamic obesity: Effects of autonomic drugs and glucose. *Am. J. Physiol.* **245,** R372–378.

73. Campfield, L. A., and Smith, F. J. (1983). Alteration of islet neurotransmittor sensitivity following ventromedial hypothalamic lesion. *Am. J. Physiol.* **244,** R635–640.

74. Smith, F. J., and Campfield, L. A. (1986). Pancreatic adaptation in VMH obesity: In vivo compensatory response to altered neural input. *Am. J. Physiol.* **251,** R70–76.

75. Van der Tuig, J. G., Knehans, A. W., and Romsos, D. R. (1982). Reduced sympathetic nervous system activity in rats with ventromedial hypothalamic lesions. *Life Sci,* **30,** 913–920.

76. Van der Tuig, J. G., Kerner, J., and Romsos, D. R. (1985). Hypothalamic obesity, brown adipose tissue and sympathoadrenal activity in rats. *Am. J. Physiol.* **248,** E607–617.

77. Lefèvre, N., *et al.* (1973). Effect of denervation on the metabolism and the response to glucagon of white adipose tissue of rats. *Horm. Metab. Res.* **4,** 248–250.

78. Goldman, J. K., Schnatz, J. D., Bernardis, L. L., and Frohman, L. L. (1972). In vivo and in vitro metabolism in hypothalamic obesity. *Diabetologia* **8,** 160–164.

79. King, B. M., Carpenter, R. G., Stamoutsos, B. A., Frohman, L. A., and Grossman, S. P. (1978). Hyperphagia and obesity following ventromedial hypothalamic lesions in rats with subdiaphragmatic vagotomy. *Physiol. Behav.* **20,** 643–651.

80. King, B. M., and Frohman, L. A. (1982). The role of vagally-mediated hyperinsulinemia in hypothalamic obesity. *Neurosci. Biobehav. Rev.* **6,** 205–214.

81. Corbit, J. D., and Stellar, E. (1964). Palatability food intake and obesity in normal and hyperphagic rats. *J. Comp. Physiol. Psychol.* **58,** 63–67.

82. Weingarten, H. P., Chang, P., and Jarvie, K. R. (1983). Reactivity of normal and VMH-lesion rats to quinine adulterated foods: Evidence for negative finickiness. *Behav. Neurosci.* **97,** 221–223.

83. Brecher, G., and Waxler, S. H. (1949). Obesity in albino mice due to single injection of goldthioglucose. *Proc. Soc. Exp. Biol. Med.* **70,** 498–501.

84. Debons, A. F., Silver, L., Cronkite, F. P., Johnson, H. A., Brecher, G., Tenzer, D., and Schwartz, I. L. (1962). Localization of gold in mouse brain in relation to gold-thioglucose obesity. *Am. J. Phys.* **202,** 743–750.

85. Edelman, M., Schwartz, I. L., Cronkite, E. P., and Livingston, L. (1965). Studies of ventromedial hypothalamus with autoradiographic techniques. *Ann. N. Y. Acad. Sci.* **131,** 485–501.

86. MacBurney, L. L., Liekert, F., and Ferry, J. H. (1975). Relationship between food intake, lipid metabolism and hypothalamic damage in gold-thioglucose obesity. *Texas Press Biol.* **121,** 403–414.

87. Sandrew, B. B., and Mayer, J. (1973). Hyperphagia induced by intrahypothalamic implants of mercury thioglucose. *Physiol. Behav.* **10,** 1061–1063.

88. Smith, C. J. V., and Britt, D. L. (1971). Obesity in the rat induced by hypothalamic implants of gold thioglucose. *Physiol. Behav.* **7,** 7–10.

89. Smith, C. J. V. (1972). Hypothalamic glucoreceptors—The influence of gold thioglucose implants in the ventromedial and lateral hypothalamic areas of normal and diabetic rats. *Physiol. Behav.* **9,** 391–396.

90. Wagner, J. W., and De Groot, J. (1963). Effect of goldthioglucose injections on survival, organ damage and obesity in the rat. *Proc. Soc. Exp. Biol. Med.* **112,** 33–37.

91. Powley, T. L., and Laughton, W. (1981). Neural pathways involved in the hypothalamic integration of autonomic responses. *Diabetologia* **20,** 378–387.

92. Mayer, J., and Marshal, N. B. (1956). Specificity of gold-thioglucose for ventromedial hypothalamic lesions and obesity. *Nature (London)* **178,** 1399–1400.

93. Debons, A. F., Krimsky, I., From, A., and Pattinian, H. (1974). Phloridzin inhibition of hypothalamic necrosis induced by gold thioglucose. *Am. J. Physiol.* **226,** 574–578.

94. Likuski, H. J., Debons, A. F., and Cloutier, R. J. (1967). Inhibition of gold thioglucose induced hypothalamic obesity by glucose analogues. *Am. J. Physiol.* **212,** 669–676.

95. Brown, D. F., McGuirck, J. P., Larsen, S. P., and Minter, S. D. (1989). Beta-thioglucose inhibits gold thioglucose lesions in the ventromedial hypothalamus. *Physiol. Behav.* **46,** 369–372.

96. Muller, E., Frohman, L., and Mocchi, D. (1973). Drug control of hypothamamus and inhibition of insulin secretion due to centrally administered 2DG. *Am. J. Physiol.* **224,** 210–217.

97. Debons, A. F., Krimsky, I., and From, A. (1973). Modification of alloxan-induced diabetes: Correlated changes in hypothalamic satiety center. *Am. J. Physiol.* **224,** 862–869.

98. Debons, A. F., Krimsky, I., and From, A. (1970). A direct action of insulin on the hypothalamic satiety center. *Am. J. Physiol.* **219,** 938–943.

99. Briese, E., and Murzi, D. (1981). Reserpine prevents gold-thioglucose hypothalamic lesion in mice. *Experientia* **36,** 52.

100. Di Rocco, R. J., Yeomans, J., and Van Itallie, T. B. (1980). Insulin does not enhance uptake of ^{14}C-deoxyglucose in the ventromedial nucleus of the hypothalamus. *Brain Res. Bull.* **5,** (Suppl. 4); 43–54.

101. Goodner, C. J., and Berrie, M. A. (1977). The failure of rat hypothalamic tissues to take up labeled insulin in vivo or to respond to insulin in vitro. *Endocrinology* **101,** 605–617.

102. Goodner, C. J., Hom, F. G., and Berrie, M. A. (1980). Investigation of the effect of insulin upon regional glucose metabolism in the rat in vivo. *Endocrinology* **107,** 1827–1837.

103. Friedman, G., Waye, J. D., and Janowitz, H. D. (1963). Persistent dynamic phase of aurothioglucose obesity. *Am. J. Physiol.* **205,** 919–921.

104. Rietveld, W. J., Ten Hoor, F., Kooji, M., and Flory, W. (1979). Maintenance of 24 hours eating rhythmicity during gold thioglucose induced hypothalamic hyperphagia in rats. *Physiol. Behav.* **22,** 549–554.

105. Debons, A. F., Tse, C. S., Zurek, L. D., Abrahamsen, S., and Maayan, L. A. (1986). Adrenalectomy induced anorexia in gold thioglucose-treated obese mice: Metabolic and hormonal changes. *Physiol. Behav.* **38,** 111–117.

106. Soyka, L., Haesler, G., and Crawford, J. (1969). Altered composition and lipolysis of adipose tissue of GTG injected mice. *Am. J. Physiol.* **210,** 1088–1093.

107. Igushi, A., Burleson, P. D., and Szabo, A. J. (1981). Decrease in plasma glucose concentration after microinjection of insulin into VMN. *Am. J. Physiol.* **240,** E95–100.

108. Hatfield, J. S., Millard, W. J., and Smith, C. J. V. (1974). Short-term influence of intraventromedial hypothalamic administration of insulin on feeding in normal and diabetic rats. *Pharm. Biochem. Behav.* **2,** 223–226.

109. Szabo, C., and Szabo, A. J. (1975). Neuropharmacological characterization of insulin sensitive CNS glucoregulators. *Am. J. Physiol. (London)* **229,** 663–668.

110. Woods, S. C., Lotter, E. C., McKay, L. D., and Porte, D., Jr. (1979). Chronic intracerebroventricular infusion of insulin reduces food intake and body weight in baboons. *Nature (London)* **282,** 503–505.

111. Brief, D. J., and Davis, J. D. (1984). Reduction of food intake and body weight by chronic intraventricular insulin infusion. *Brain Res. Bull.* **12,** 571–575.

112. Ramsay, D. S., Richardson, R. D., Kott, J., Lernmark, A., and Woods, S. C. (1989). Intraventricularly transplanted pancreatic islets reduce body weight of rats. *Appetite* **12,** 233. [Abstract.]

113. Plata-Salaman, C., Oomura, Y., and Shimizu, P. (1986). Dependence of food intake on acute and chronic ventricular administration of insulin. *Physiol. Behav.* **34,** 717–734.

114. Nagai, K., and Nakagawa, H. (1982). Time dependent hyperglycemic action of centrally administered 2-deoxyglucose, D-mannitol and D-glucose. *Biomed. Res.* **3,** 175–180.

115. Mori, P., Nagai, K., Hara, M., and Nakagawa, K. (1988). Time dependent effects of insulin in the central nervous system on blood glucose. *Am. J. Physiol.* **249,** R23–30.

116. Havrankova, J., Roth, J., and Brownstein, N. (1978). Insulin receptors are widely distributed in the central nervous system of the rat. *Nature (London)* **272,** 827–829.

117. Van Houten, M., and Posner, B. I. (1981). Insulin receptors in the central nervous system: Localization and characteristics. In D. Andreani, R. de Pirro, R. Lauro, J. M. Olefski, and J. Roth (eds.), *Current views and insulin receptors* (pp. 75–90). New York: Academic Press.

118. Partridge, W. M. (1981). Transport of nutrients and hormones through the blood–brain barrier. *Diabetologia* **20,** 246–254.

119. Corp., E. S., Bohannon, N. J., Wilcox, B. J., Figlewicz, D. P., Woods, S. C., Porte, D. Jr., Dorsa, D. M., and Baskin, D. G. (1986). Insulin specific binding sites are heterogeneously distributed in rat hypothalamus as determined by in vitro quantitative autoradiography. Abstract presented at the IXth International Conference on the Physiology of Food and Fluid Intake, Seattle, 12.

120. Haskell, J. F., Meezan, E., and Pillion, D. J. (1985). Identification of the insulin receptor of cerebral microvessels. *Am. J. Phys.* **248,** E115–125.

121. Melnyck, R. B., and Martin, J. B. (1984). Starvation-induced changes in insulin binding to hypothalamic receptors in the rat. *Acta Endocrinol.* **107,** 78–85.

122. Pacold, S. T., and Blackard, W. G. (1979). Central nervous system insulin receptors in normal and diabetic rats. *Endocrinology* **105,** 1452–1457.

123. Melnyck, R. B., and Martin, J. M. (1984). The failure of chronic experimental hyperinsulinemia to alter insulin binding to hypothalamic receptors in the rat. *Acta Endocrinol.* **107,** 86–90.

124. Wilcox, B., Corps, E., Siglewicz, C., Woods, S., and Porte, D., Jr. (1986). Characteristics of insulin binding in the olfactory bulb of three genotypes of Zucker rats by in vitro quantitative autoradiography. International Conference on the Physiology of Food and Fluid Intakes (ICPFFI), Seattle.

125. Melnyck, R. B., Mrosovski, N., and Martin, J. M. (1983). Spontaneous obesity and weight loss: Insulin binding and lipogenesis in the dormouse. *Am. J. Physiol.* **245,** R396–402.

126. Beck-Nielsen, H., (1975). Diurnal variation in insulin binding to human monocytes. *J. Clin. Endocrinol. Metab.* **47,** 385–390.

127. Larue-Achagiotis, C., and Le Magnen, J. (1984). Insulin infusion during a nocturnal fast suppresses the subsequent daytime intake. *Physiol. Behav.* **33,** 719–722.

128. Kasser, T. R., Deutch, A., and Martin, R. J. (1986). Uptake and utilization of metabolites in specific brain sites relative to feeding status. *Physiol. Behav.* **36,** 1161–1165.

129. Kasser, T. R., Harris, O., and Martin, R. J. (1989). Level of satiety in vitro energy metabolism in brain during hypophagia and hyperphagia and body weight recovery. *Am. J. Physiol.* **257,** R1322–1327.

130. Panksepp, J., and Pilcher, C. W. T. (1973). Evidence for an adipokinetic mechanism in the ventromedial hypothalamus. *Experientia* **29,** (7), 793–794.
131. Fleming, C. G. (1969). Food intake studies in parabiotic rats. *Ann. N. Y. Acad. Sci.* **157,** (Art. 2), 985–1003.
132. Parameswaran, S. V., Steffens, A. B., Hervey, G. R., and De Ruiter, L. (1977). Involvement of a humoral factor in regulation of body weight in parabiotic rats. *Am. J. Physiol.* **132,** R150–157.
133. King, K. (1976). Lipostatic control of body weight: Humoral mediation. *Physiol. Psychol.* **4,** 415–418.

Chapter Ten

Obesity

Many books have been published on obesity that deal mainly with pathological and clinical aspects. Specialized scientific societies exist at both the international and national levels in which physicians and experimenters, physiologists and psychologists, meet for various reasons of wide interest. Spontaneous obesity may be considered a disruption of the normal physiological process, studies in preceding chapters, through which the body fat mass varies only in a range compatible with its function in the body energy balance. The physiologist is interested because the identification of causes of this abnormality will reveal the peculiar role and the fragility of some components of the complicated mechanism involved in normal regulation. Another reason of interest is that obesity may be considered a disease and, as such, is a medical topic.

However, obesity is not considered a disease only because it is a deviation from the normal physiological state. Extreme and permanent leanness in humans is not generally considered an illness; obesity is because it is both a social and physical handicap. Indeed, the social handicap of a moderate obesity depends on body prettiness (particularly of women) in a given time and a given culture. One can recall Greek statues erected as models of feminine beauty who are very chubby and not attractive at all to our modern eyes. At the present time and in various African tribes, girls prior to marriage and boys in other circumstances are ritually submitted to gavage to become fat. But beyond this culturally specific handicap, moderate and massive obesities are indeed physical handicaps involving motor impairment, the prevention of normal activities, sports, etc. Finally, obesity is a disease because it is a cause of other pathologies: diabetes, cardiovascular diseases, etc. Correlations between excess body weight and mortality are convincing.

Finally, another reason for the interest toward obesity in some countries is the striking difference in the spread of the disease between various developed and normally nourished populations. Statistics revealed that compared to European populations, obesity is a specific North American disease. In the United States, it is increasingly becoming a calamity. Because of this, the physiologist is interested by the fact that in some manner the American mode of life and of feeding could be found a cause of this specific American illness.

I. *Animal Obesities*

Generally speaking, obesity is a specific human abnormality. Except for some primates and domestic animals, such as cats and dogs overnourished by their master (often obese him- or herself), animals do not become spontaneously obese. Preceding chapters discussed experimental obesity induced in intact rats by hypothalamic stimulation, insulin treatment, or forced-feeding. The common feature of these induced obesities is in their reversibility, which demonstrates the existence of a regulatory control of the body fat mass. Obesity induced by the ventromedial (VMH) lesion was also described. The lack of reversibility of this obesity indicated that the VMH was involved in the regulatory process. In this chapter, interest will be focused on three other exceptions of spontaneously occurring obesity in extensively studied animal models. Two of these models are called dietary obesity, because their obesity is induced by the nature of the offered food or by its mode of presentation. The third one is genetic obesity, widely studied in a strain of mice (the ob-ob mouse) and rats (the Zucker rat). Whether or not these spontaneous obesities act like human obesity and whether or not they teach something about the regulatory mechanisms must be asked.

A. Cafeteria-Induced Obesity in Rats

As already discussed, hyperphagia and obesity in rats offered a choice of various high-palatability foods are results of the cafeteria regimen. Such a choice stimulates a 50–100% increase of the daily caloric intake. This provoked primary hyperphagia was obviously a cause of the increase of body weight; however, convincing evidence indicated that a dietary-induced thermogenesis by the brown adipose tissue limits the induced obesity. This obesity is reversible unless the development of an adipose tissue hyperplasia prevents this reversibility after some weeks. This finding supported the notion, confirmed in humans, that the adipocyte hyperplasia, contrary to hypertrophia, leads to irreversible obesities.

Two points, already noted, must again be pointed out. Hyperphagia is maintained only if the offered high-palatability foods are changed everyday. If the same foods are maintained, the rat returns rapidly to the caloric intake exhibited on its familiar complete diet and compatible with its weight maintenance. This is a new evidence on the learning of palatability adjusted to the daily caloric requirement. This learned adjustment requiring some delay is prevented by the daily change of the offered foods.

Another instructive fact is that the induced hyperphagia is a meal overeating. Rats eat huge meals, in which they eat successively the four or five offered foods. Thus, hyperphagia is a consequence of the sensory-specific satiation observed early in rats and commonly in humans as a cause of meal overeating. Rats exposed to the cafeteria regimen selected fat- and protein-rich diets (1). Based on this result, the cafeteria-induced hyperphagia was claimed to be identical to hyperphagia and obesity exhibited on a high-fat diet. This suggestion, however, was ruled out. Consuming high-fat diets (and not consuming the cafeteria regimen) induced obesity only in some strains of rats. On the other hand, the effective role of variety and palatability was assessed and confirmed. A choice of foods made differently and highly palatable by addition of noncaloric materials (e.g., artificial sweeteners, greasy mineral oil) also induced hyperphagia (2). A role for the cephalic phase of insulin secretion was also experimented with (3). A blood sampling and insulin determinations during the varied meal showed that the resumed eating from food to food is associated with a succession of peaks of increase in plasma insulin concentration. Elsewhere (Chapter 6, Section IV) the relationships between the preabsorptive release and palatability responses were discussed. These repeated reflexly induced insulin releases by the cafeteria regimen may contribute to renew the eating of the differently flavored foods. Consistent with this notion is the finding that rats exhibiting the greatest cephalic phase are those that are the most hyperphagic on a cafeteria regimen (4). However, this effect can only be contributing factor. Clearly, hyperphagia results from sensory-specific satiation. This is ascertained by the behavior of vagotomized rats, which develop cafeteria-induced hyperphagia despite the elimination of the preabsorptive insulin release (5). It is of interest to note that sympathectomy also does not prevent the hyperphagia. Rats treated by guanetidine as weanlings exhibit palatability-induced overeating in adulthood (6).

Female rats are more hyperphagic on the cafeteria regimen than males are. Exercising by wheel-running, females nevertheless become obese; males do not (7, 8). Young rats exposed to the cafeteria regimen after weanling and again in adulthood develop less obesity in this new exposure than controls do. The result is interpreted as an effect of the development of the thermogenic capacity of the brown adipose tissue by the prior exposure (9). Finally, the weight gain by the cafeteria regimen was shown to be positively correlated to the intensity of the startle reflex provoked by a sound stress and also individually correlated to the overeating by tail-pinching (10). In Chapter 6, Section I both the cafeteria-induced hyperphagia and the tail-pinch-stimulated eating were shown to be opiate-dependent and to lead to a morphine tolerance and a morphinelike dependence.

B. High-Fat Diet-Induced Obesity

Some rats offered a fat-rich diet become obese. This dietary obesity was the subject of a considerable number of works. These works, often unclear in their design and their results, are difficult to clarify.

A high-fat diet, containing, for example, 50–60% fat, is a fat-rich diet, a high-energy diet with a caloric density reaching 6 kcal/g and more, and a low-carbohydrate and/or a low-protein diet. In many works, the absence of any comparison between responses to the high-fat diet and those on a low-fat iso-caloric diet does not permit dissociation between the high-fat and high-energy intake and the low-carbohydrate intake and of their interactions in either the behavioral or the metabolic responses. In addition, a fat-rich diet of high or normal caloric density is a high-palatability diet. The relationships between palatability and postingestive effects of both fat and energy intake and the difference of these relationships between rats becoming obese and those not obese must be examined. Rats of some strains (e.g., Osborne-Mendel or Fisher rats) offered the high-fat, high- or low-caloric diet became obese, whereas rats of other strains (e.g., Wistar) did not. Within the same strain (e.g., Sprague-Dawley), only about half of the rats became obese. A comparison between these rats resistant or not to the induction of obesity will be fruitful.

Rats presented a high-fat, high-energy diet substituted for their chow diet demonstrated an initial increase of their volume intake and, therefore, of their fat and energy intake. This initial hyperphagia was clearly a palatability effect. An identical increase was obtained with a diet adulterated by the addition of a greasy mineral oil, i.e., by a low-fat, low-energy diet (11, 12). However, after 4 or 5 days, the caloric intake on the high-fat, high-energy diet dropped and was reestablished at the control level. Thus, the rats responded to the high-caloric density of the diet as they responded to caloric dilution (11). On the low-fat, low-caloric, greasy diet, rats did the opposite and increased their volume intake to reestablish the previous caloric intake (12).

In the two cases, the adjustment is independent of the fat intake, but only depends on caloric density and energy intake. In this first comparison, however, two questions remain unanswered. Following the adjustment, rats offered the high-fat, high-energy diet continue to have a high-fat intake: Will this high-fat intake without a higher energy intake induce obesity? If this adjustment to the initial palatability effect differs, will the rats become obese or not? Further comparisons will provide partial answers to these questions. Here, regarding the high-palatability role of the high-fat diet, various experiments confirm its limited action. During the first 2 days of initial hyperphagia, the augmentation in the daily intake is achieved by an increase in meal sizes and the eating rate without changes in meal frequency. This meal pattern change is a typical effect of the enhanced palatability of the diet (13). With an open gastric fistula, rats sham-fed more of a liquid high-fat diet than of a low-fat isocaloric diet in 30 min, but the difference diminishes after three or four trials of sham-feeding. Thus, the postingestive effect of the high-fat diet is required for a persistency of hyperphagia. However, the same result seems to indicate that the postingestive action is not required anymore for the disappearance of the initial hyperphagia (14).

After a weight gain on a high-fat diet rats returned to a low-fat diet lost the gained weight. Thus, dietary obesity is reversible, as shown with experimental obesity in intact rats. During the weight loss, rats were hypophagic. Following exposure to the high-fat diet, PFFA, glycerol, and ketone bodies were elevated, indicating a high level of fat oxidation. This persistent fat oxidation on the low-fat diet appears to be a cause of the hypophagia.

Understanding whether or not metabolic perturbation is caused by the concomitant high-fat intake with an excessive fat deposition is important. Rats were fed high-fat or high-carbohydrate diets. The former showed glucose intolerance as a main symptom. Injecting ^{14}C-glucose produced the expiration of $^{14}CO_2$ more slowly (15). After the intragastric glucose load (glucose tolerance test) the slower decline of blood glucose in rats fed a high-fat diet was associated with lower plasma insulin and PFFA concentrations (16). However, the glucose-stimulated insulin release is normal, and evidence exists that the glucose intolerance is due to insulin resistance. The euglycemic–hyperinsulinic clamp confirmed this insulin resistance. At a mean level of plasma insulin, glucose utilization was reduced. The insulin resistance was maximal in muscle and in the brown adipose tissue. At an identical level of resting metabolic rate in rats fed a high-fat or high-carbohydrate diet, the weight gain on the former, in rats nonresistant to obesity, may be due to a reduced thermogenesis by the brown adipose tissue (17). Indeed, the insulin resistance is associated with a low insulin secretory response to glucose tested *in vitro* (18). Thus, in obesity induced by a high-fat diet, not only the hyperinsulinism is absent but the hyposecretion of insulin is associated with insulin resistance contrary to all other cases of obesity. This insulin resistance, therefore, cannot be an effect of hyperinsulinemia, as it suspected in other forms of obesity (18).

The impairment of carbohydrate metabolism by a high-fat intake persists for some time then the rat is shifted to a high-carbohydrate diet. This persistent disturbance of the carbohydrate metabolism leads to a temporary undereating. This was demonstrated early by Lundbaeck and Stevenson (19). Rats offered a weekly alternation of high-fat and high-carbohydrate diets exhibited a reduction of the high-carbohydrate diet intake for the first 3 or 4 days of its respective week. This important observation was extended by Geary et al. (15). Rats were fed either a high-fat or high-carbohydrate isocaloric diet. With both diets, their caloric intake in a chronic feeding was identical, but, of course, the fat intake was higher on the high-fat diet. The weight gain was also identical in the two groups; therefore, the strain was resistant to dietary obesity. The transition from the high-fat to the high-carbohydrate diet was associated with a transient 20% fall of the caloric intake. At this time, rats demonstrated the glucose intolerance and the insulin resistance as a sequel of the high-fat feeding, which led to the reduced caloric intake. This effect of the high-fat feeding is an aspect of the general phenomenon called the glucose–PFFA cycle.

A systematic comparison between two strains or individuals of a same strain, resistant and not resistant to the high-fat-induced obesity, could somewhat clarify the problem. The first comparison was carried out by Schemmel et al. (16). Osborne-Mendel rats fed a diet containing 60% fat for 20 wk deposited 285 kcal body fat per 1000 kcal consumed versus 129 kcal on a low-fat, high-carbohydrate diet. In another strain resisting obesity, the retention by fat deposition was 100 and 128 kcal/1000 kcal consumed on the high-fat and high-carbohydrate diet, respectively. On the two diets, the Osborne-Mendel rats consumed one-third more than the rats of the other strain. In Sprague-Dawley rats offered the high-fat diet, only half of them became obese. Lab chow-fed rats resisting obesity exhibited less intake, less weight gain, and a metabolic efficiency of the food compared to the mean in the strain. Those becoming obese on the high-fat diet exhibited a mean weight gain and metabolic efficiency on the lab chow diet. Rats that become obese alone have a strong hypoinsulinemia on the high-fat diet (20, 21). The same Sprague-Dawley rats offered a high-energy diet (a high-fat or high-sucrose diet) demonstrate a comparable increase of their caloric intake on the two diets. However, only half of them become obese. According to this result (22), the induced obesity would be an effect of the high energy and not of the high-fat intake.

Research indicates that rats who become obese compared to resistant counterparts do so because of a deficiency of the dietary-induced thermogenesis by the brown adipose tissue (23). Compared to Osborne-Mendel rats, those of strains resisting obesity have a more elevated brown adipose tissue guanosine-diphosphate binding and a prandial thermogenesis; however, those rats resisting obesity on the high-fat, high-energy diet become obese on the cafeteria regimen. A deficient sympathetic activation of the brown adipose tissue could participate in the development of obesity in Osborne-Mendel rats, as it also suggested in VMH-lesioned rats (24). *In vivo* the synthesis of norepinephrine from ^{3}H-DOPA by DOPA-hydroxilase is reduced by 50% in obese rats (25).

Another and very different explanation of dietary obesity could derive from the observation (widely documented) that long-chain fatty acid triglycerides induce obesity and middle-chain triglycerides do not. This was tentatively explained by a difference between the two types of fats in the complex digestive process preceding the intestinal absorption. Long-chain triglycerides not hydrolized or reesterified would be transported via the lymphatic route and directly deposited in fat reserves, whereas middle-chain triglycerides (via the portal vein and the liver) would contribute to the pool of oxidizable metabolites. (unexplored as yet). Perhaps the difference between strains and individuals resisting or developing obesity is based on a difference in this digestive processing of fats, but this possibility remains unexplored.

Although some points have been clarified, high-fat diet-induced obesity is still poorly understood. As such, this animal model is difficult to extrapolate to human obesity. Why is the excessive lipogenesis, regardless of its cause, not repressed by the lipostatic mechanism in VMH-lesioned rats? Another unsettled point is the relationship between fat deposition and intake. When the high-fat deposition is associated with a high-energy intake, the high intake is probably due, as it is in many other conditions, to the energy retention as fats and to the compensation for the resulting deficit in the pool of oxidizable metabolites by the enhanced intake.

C. Genetic Obesities

Two models of genetic obesity (the ob-ob mouse and the Zucker rat) have been extensively studied. Hundreds of works have been published and continue to be carried out. Here only a summary of works relevant to human obesity will be given. Most of these works look for similarities to and differences from human obesity, in which the role of genetic factors has been clearly confirmed. The genetic transmission of the obese phenotype obeys the Mandel laws. In the two species, a recessive gene is responsible. ob-ob mice and fa-fa Zucker rats become obese; homozygotes OB-OB and FA-FA and heterozygotes OB-ob and FA-fa do not.

The expression of genes toward obesity is not at the adipocyte level. In ob-ob rats, a transplantation in the renal capsule of adipose tissue from obese to lean and from lean to obese showed, after some months, the differentiation of the adipose tissue sample characteristics of the receptor and not of the donor (26). Thus, studies in tissue culture of the development of preadipocyte in the two phenotytes will be instructive only by testing the role of the medium because only the environment of the cell is involved in its development as a hyper-trophized adipocyte.

Obese ob-ob and fa-fa rats are hyperphagic. However, the hyperphagia referred to the body weight is manifested only from the third to the eleventh weeks. These Zucker obese rats eat 30% more than the nonobese rats. They eat larger and fewer meals. The diurnal cycle is absent (27). The cessation of the hyperphagia corresponds to a maximal hypertrophy of adipocytes and to a later development of adipose tissue hyperplasia (28). Hyperphagia in the two species is not a cause of either the rapid weight gain of obese versus lean rats or of hyperinsulinism, which is the main feature of obese phenotypes. The rapid weight gain and high insulin secretion are detectable as early as the suckling period. Pair-feeding experiments or restriction in mice and rats do not prevent obesity, although it is attenuated (29, 30). In restrained mice, the hyperinsulinism is reduced, but the increase in the carcass lipids is maintained at the level observed in the free-feeding condition.

Thus, hyperinsulinism apparently is the primary factor of the development of obesity in the two strains. The basal insulin level and the glucose-stimulated insulin release are considerably elevated in obese compared to lean rats. In ob-ob rats, the pancreatic content of available insulin is elevated three fold and the insulin secretory capacity six fold compared to OB-OB lean mice (31). *In vitro* the glucagon release of pancreatic islets is reduced, but the circulating glucagon is unchanged (32). In the obese Zucker rats, an oral, not intravenous, glucose load manifested a glucose intolerance despite the high-insulin release. This indi-cates the development of insulin resistance (33). Insulin resistance of muscles more than that of adipocyte is reported (34). This insulin resistance is not due to a reduction of membrane insulin receptors but, rather, to an intracellular alteration of the insulin activity (35).

ob-ob mice both hyperglycemic and glucose-intolerant are typically diabet-ic mice. However, this diabetes, associated with hyperinsulinemia, is a model of the human type 2 diabetes. *In vivo* and *in vitro* studies have confirmed the suggestion that the excessive lipogenesis does not result from a genetically inherited disturbance of the adipose tissue metabolism but, rather, from an effect of hyperinsulinism, which is the inherited agent of the ongoing obesity. Streptozotocin- or alloxan-induced diabetes prevents but does not reverse the established obesity. Vagotomy also does not reverse obesity in obese Zucker rats, contrary to its effect on the VMH lesion-induced obesity. This is the first indication that the mechanism of the genetic obesity is different from that of the VMH syndrome (36). Contrasted behavioral responses confirm this difference. For example, Zucker obese rats as well as ob-ob mice, contrary to VMH rats, respond to a caloric dilution of the diet, although more slowly than their lean counterparts (37, 38). In lean ob-ob mice, a VMH lesion induces hyperphagia, hyperinsulinism, and obesity but does not significantly alter obesity and hyperinsulinism in obese mice. The conclusion of the authors is that the VMH is functionally intact in genetic obesity (39). A LH lesion in the Zucker rat introduces aphagia in lean and obese rats and suppresses differences between the two phenotypes during aphagia. After recovery from aphagia, obese rats gain weight, but less than unlesioned obese controls. A comparable result is reported in ob-ob mice (40).

A large number of convergent results in both Zucker rats and ob-ob mice revealed a possible mechanism of genetic obesity. In the two species, various tests demonstrated a strong impairment of the adrenergic activation of the brown adipose tissue in obese phenotypes. Cold-induced nonshivering thermogenesis and the dietary-induced thermogenesis are absent in obese rats (41, 42). The adrenergic activation and the norepinephrine turnover of the brown adipose tissue are reduced (43). Many indications imply that this loss of the adrenergic activation of brown adipose tissue is of central origin (44). Consistent with this notion is the observation that the increase of plasma norepinephrine in Zucker obese rats is less than in nonobese rats. The rate of norepinephrine synthesis in the heart and pancreas as well as in the brown adipose tissue is lower (45).

Based on present knowledge, the genetically induced defect seems to reside in an impairment of a central and peripheral sympathetic–adrenergic system. This impairment could explain both the loss of the sympathetic inhibition of the pancreas insulin release responsible for hyperinsulinism and the loss of brown adipose tissue thermogenesis, both recognized as the main effectors of the development of obesity.

II. *Overfeeding and Obesity in Humans*

Including human obesity in the content of this book was an untenable challenge. A review, even summarized, about the metabolic, biochemical, neural, and clinical aspects in one chapter was not possible. Many books and proceedings of international meetings exist on the matter. In this volume, only the relationships of obesity and feeding will be outlined.

A. Feeding Behavior of Obese Subjects

How and how much does an obese human eat? Is this obesity due to previous overeating (as is observed) or a secondary effect of a primary metabolic disturbance? Is the mode of eating, in addition to amounts eaten, responsible for the obesity? Contradictory answers to these questions are presented in the literature. Dogmatic controversies often originate from superficial and unsettled concepts.

The existence of hyperphagia in obese humans is questioned. On the bases of interviews and questionnaires of obese patients, many clinicians claim that these patients are not hyperphagic. However, to decide that obese subjects are hyperphagic widely depends on the obesity criteria and on what hyperphagia means. Most affirmations are based on a comparison of the daily intake, either declared or controlled in a group of overweight subjects, with that of the population or a chosen group of normal-weight subjects. The first bias of such a comparison is that the intake is not referred to the body weight. Although less expensive than the lean tissue, the maintenance of 20 or more kg of body fat increases the total oxidative metabolism or metabolic rate. For this reason, the body energy balance and the maintenance of the high body weight require a proportional increase of the energy intake, which is not hyperphagia. If the obese subject eats the same amount of food as normal-weight controls, he or she may be considered hypophagic. Another bias is in the comparison itself and in the references to a group of normal-weight controls. As discussed elsewhere in this volume, the total metabolic rate referred to the weight and, thus, the energy intake required for the body energy balance and weight maintenance are interindividually extremely variable.

Therefore, a comparison between intake of an obese group with the mean intake of a normal-weight control group is misleading. Heavy eaters in the obese group may be suspected to have developed their obesity because of this overeating: In the same normal-weight group, heavy eaters forced to eat like small eaters may lose weight, whereas small eaters forced to eat like heavy eaters may become obese.

Thus, the only valid criterion is the presence or absence of a positive energy balance: excess of energy intake over energy expenditures. However, even on this bases of the body energy balance, a source of misunderstanding is still present. In all comparisons between normal-weight and obese subjects, obese subjects are static obese; i.e., they have established their obesity. A distinction is rarely made between the dynamic and static phases of obesity. In the initial buildup of obesity from a normal body weight, either in infancy or in adulthood, a persistent positive energy balance is necessarily associated with the accumulation of fats. Therefore, the existence of the hyperphagia in this phase cannot be logically denied. The question remains as to whether or not the hyperphagia is an overphagia (as defined above), i.e., whether the overeating is a cause or an effect of the unrepressed accumulation of fats and why this accumulation is not repressed. Irrespective of these considerations, the overeating is not questionable. Furthermore, stable obese patients who maintain a permanent high body weight are rarely differentiated from obese patients who continue to gain body weight, progressing from a moderate to a massive obesity. A persistent positive energy balance also is necessarily present in the latter case. The last bias is the presence or absence of a voluntary and often undeclared diet. In an acute observation of his or her food intake, an obese subject may be suspected to repress intake voluntarily.

Despite these numerous sources of mistake, overeating of obese subjects compared to normal-weight controls is reported in most controlled studies (46, 47).

B. Feeding Pattern of Obese Subjects

Eating too much or not, how do obese subjects eat? Are their modes of eating and responses to internal sensory and external stimulations to eat responsible for their obesity?

Obese subjects are reputed to take fewer daily meals than normal-weight people do. In one particular study (48), 50–60% of observed obese subjects took only two meals daily, the biggest meal taken in the evening or at night. Only 16% of normal-weight females and 6% of males also took two daily meals; 70% took four or more meals daily. The distribution of calories taken within the daily meals also differs in fatty compared to lean subjects. In another careful study (49), 330 children (6–12-yr-olds) were separated into five groups according to their corpulence and to their body mass index (varying between 13 and 22). A control of calories taken at breakfast, lunch, and dinner showed that subjects of fat and obese groups took less at breakfast and more at dinner. The proportion of calories taken at the two main meals was higher in fat than in lean groups. This high intake at evening meals, also indicated in various other studies, is of interest. Previous chapters have mentioned the hypothesis that obesity could result from both deficient lipolysis and fat oxidation at night, not compensating for the diurnal lipogenesis. The high intake before or during the night could be viewed as an anticipatory filling of the gastrointestinal tract taking over the deficient nocturnal mobilization of the internal fuel.

The microstructure of the meal also differs between obese and normal-weight subjects. The evolution of the eating rate from the beginning to the end of a meal was studied early in the very peculiar condition of a liquid food taken from a reservoir called the "feeding machine" (50, 51). A negatively accelerated curve of the cumulative intake, seen by others, was demonstrated by normal-weight subjects: 60% of the meal intake was consumed in the first half of the meal. A linear curve was observed in obese subjects: 50% of the meal intake was consumed during the first half of the meal duration. Recordings of the chewing–swallowing patterns compared between normal-weight and obese subjects confirmed and widely extended this finding. As already mentioned elsewhere, normal-weight subjects were shown to slow their eating rate from the beginning to the end of the meal. The meal was divided into four quarters, in terms of amounts eaten, and the last quarter was compared to the first one. The chewing time and number of masticatory movements on each food unit and the intermouthful pauses augmented in the last quarter by 11.1, 10.3, and 17% respectively. Thus, the eating rate was augmented by the approach of satiety. This was totally absent in obese subjects offered either a high- or low-palatability food. Another difference between lean and obese subjects was in prandial drinking. Contrary to lean subjects, obese subjects drank during their meals as much as they ate (Figs. 10.1 and 10.2).

Eating is stimulated by internal signals, by the palatability of foods, and also by external factors acting as conditioned stimuli. What are the roles played by these stimuli in obese subjects? Many works agree that obese subjects are highly responsive to sensory and external cues and less responsive to internal signals than normal subjects are.

In chewing–swallowing recordings, comparisons among various foods and between single-course and varied meals suggested the role of food palatability. Obese subjects, offered identically sized and differently flavored pieces of food, swallowed the food unit sooner and after less mastication than lean subjects. This was observed in the food rated by subjects as highly palatable in a preference test. Treating this chewing time and number of masticatory movements before swallowing as objective measures of the sensory stimulation to eat, the high sensitivity to this stimulation by obese subjects was confirmed (52, 53). These results were consistent with other works suggesting that obese subjects eat faster

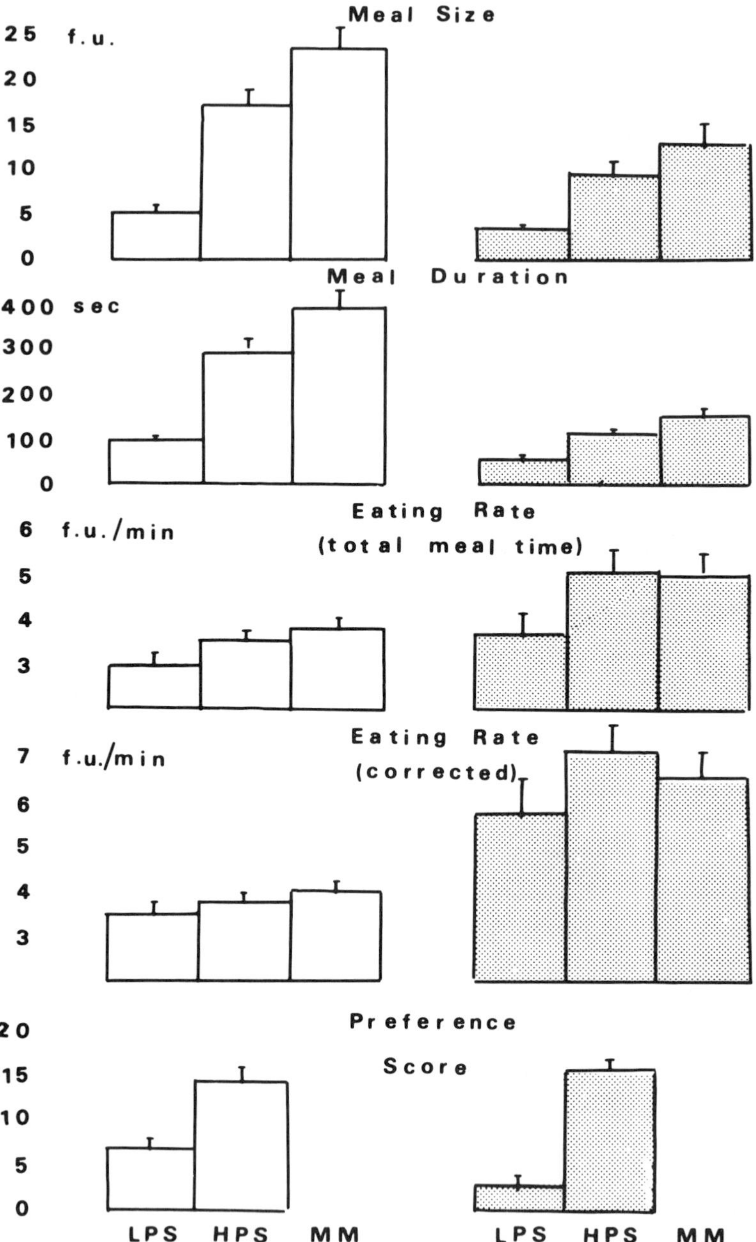

Figure 10.1 Comparison of responses to the palatability of foods of normal weight and obese subjects. f.u., food units; LPS, low palatability single-food meals; HPS, high palatability single food meals; MM, mixed meals.

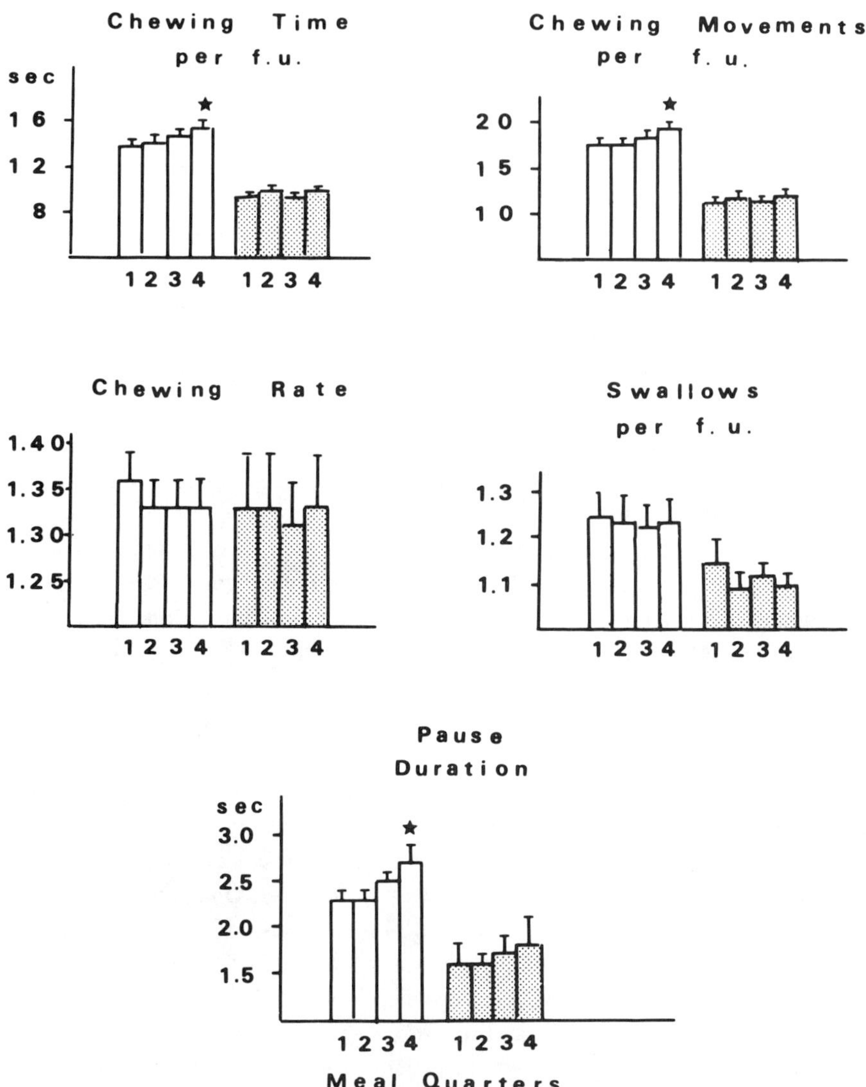

Figure 10.2 Evolution of chewing and swallowing parameters from the beginning to the end of a meal compared in normal weight and obese subjects. f.u., food unit.

than normal-weight subjects (54); however, this high eating rate, irrespective of the food palatability, was not observed by others (55).

Generally speaking, obese subjects are reported to eat more foods they like and to reject more foods they dislike than normal-weight subjects do (56). Their hierarchy of preference differs significantly from those manifested by lean subjects. Pleasantness ratings of 20 different mixtures of sweetened creams were tested in groups of obese and normal-weight subjects (57). The optimally preferred mixture was 20% fats and <10% sucrose in normal-weight subjects and 34% fats and 5% sucrose in obese subjects. The ratio "sucrose : fat preference" was found to be negatively correlated to body weight.

In some obese subjects, this highly stimulating action of the sensory activity of the food looks like food addiction. They eat for pleasure, sensory food reward, and probably contingent endogenous opiate release subserving this food reward (see Chapter 5, Section III). Eating becomes a chemosensory self-stimulation, which overcomes satiating and the satiety processes. Many clinicians reported that patients submitted to a restrictive diet manifested a syndrome resembling an opiate-withdrawal syndrome.

Various signs indicating that obese subjects are relatively unresponsive to internal signals are reported. Alimentary stimuli were presented to normal-weight and obese subjects after either a 400- or 900-kcal meal. In normal-weight subjects, the 900-kcal meal inhibited salivation, but not in obese subjects. In normal-weight subjects, the pleasantness rating of the presented food was reduced by a 541-kcal meal but not in obese subjects (58). Salivation stimulated by the presentation of the food is augmented by a previous food deprivation in normal-weight subjects, but not in obese subjects (59). Obese patients adjust their intake more slowly to a change of the caloric density of their familiar diet than normal-weight subjects (60). Lean and obese subjects were tested after exercise, increasing their daily expenditures by 100 kcal. On control days, obese subjects ate more than lean subjects did: 1800 versus 1100 kcal. On days of exercise, obese subjects ate 180 kcal less and lean subjects 150 kcal more (61).

A role for environmental cues in obese subjects, greater than in normal-weight subjects, was demonstrated in various experiments. A role for the social environment (eating in groups vs. in isolation) was studied (62). Finally, in a series of experiments, Schachter (63) tested this sensitivity of obese subjects to conditioned stimuli to eat such as the time of meals. Obese subjects, asked to take their meal when hungry in the presence of a clock slowed or accelerated by the experimenter, took their meal later and sooner, respectively.

References

1. Prats, E., Monfar, M., Castella, J., Iglesias, R., and Alemany, M. (1989). Energy intake of rats fed a cafeteria diet. *Physiol. Behav.* **45,** 263–272.
2. Louis-Sylvestre, J., Giachetti, I., and Le Magnen, J. (1984). Sensory versus dietary factors in cafeteria-induced overweight. *Physiol. Behav.* **32,** 901–905.
3. Louis-Sylvestre, J., and Le Magnen, J. (1983). Phase céphalique de la sécrétion d'insuline et variété des aliments au cours du repas chez le rat. *Repro. Nutri. Develop.* **23,** 351–356.
4. Berthoud, H., Bereiter, D. A., Trimble, E. R., Siegel, E. G., and Jeanrenaud, B. (1981). Cephalic phase of insulin secretion: Anatomical and physiological characteristics. *Diabetologia* **20,** 393–401.
5. Gold, R. M., Sawchenko, P. E., De Luca, C., Alexander, J., and Eng, R. (1980). Vagal mediation of hypothalamic obesity but not of supermarket obesity. *Am. J. Physiol.* **238,** R447–453.
6. Tordoff, M. G. (1985). Influence of sympathectomy on body weight of rats given chow or supermarket diets. *Physiol. Behav.* **35,** 455–463.
7. Rolls, B. J., Van Duijvenvoorde, P. M., and Rowe, E. A. (1983). Variety in the diet enhances intake in a meal and contributes to the development of obesity in the rat. *Physiol. Behav.* **31,** 21–28.
8. Sclafani, A., and Gorman, A. N. (1977). Effects of age, sex and prior body weight on the development of dietary obesity in adult rats. *Physiol. Behav.* **18,** 1021–1026.
9. Brooks, S. L., Rothwell, N. J., and Stock, M. J. (1981). Increased resistance to obesity following early exposure to the cafeteria diet. *Proc. Nutr. Soc.* **40,** A58.
10. Harrington, M. E., and Coscina, D. V. (1983). Body weight gain and behavioral responsivity as predictors of dietary obesity in rats. *Physiol. Behav.* **30,** 763–770.
11. Strominger, G. L., Brobeck, J., and Cort, F. L. (1953). Regulation of food intake in normal rats and in rats with hypothalamic hyperphagia. *Yale J. Biol. Med.* **20,** 55–74.
12. Hamilton, C. L. (1964). Rat's preference for high fat diets. *J. Comp. Physiol. Psychol.* **58,** 459–460.
13. Richardson, A., Rumsey, R. D. E., and Read, N. W. (1989). The effect of a high fat (HF) diet on food intake and eating behavior in the rat. *Appetite* **12,** 232.
14. Ramirez, I., and Friedman, M. I. (1983). Food intake and blood fuels after oil consumption: Differential effects in normal and diabetic rats. *Physiol. Behav.* **31,** 847–850.
15. Geary, N., Sharrer, E., Freudlsjerger, R., and Raab, W. (1979). Adaptation to high-fat and carbohydrate-induced satiety in the rat. *Am. J. Physiol.* **237,** R139–146.
16. Schemmel, R., Mickelsen, O., and Gill, J. L. (1969). Dietary obesity in rats: Body weight and body fat accretion in seven strains of rats. *J. Nutr.* **100,** 1041–1048.
17. Storlien, L. H., James, D. E., Burleigh, K. M., Chisholm, D. J., and Kraegen, E.

W. (1986). Fat feeding caused widespread insulin resistance, decreased energy expenditure and obesity in rats. *Am. J. Physiol.* **251**, E576–583.

18. Malaisse, W. J., Malaisse-Lagae, F., and Coleman, D. L. (1968). Insulin secretion in experimental obesity. *Metabolism* **17** (9), 802–807.

19. Lundbaeck, K., and Stevenson, J. A. F. (1947). Reduced carbohydrate intake after fat feeding in normal rats and rats with hypothalamic hyperphagia. *Am. J. Physiol.* **151**, 530–537.

20. Levin, B. E., Hogan, S., and Sullivan, A. C. (1989). Initiation and perpetuation of obesity and obesity resistance in rats. *Am. J. Physiol.* **256**, R766–771.

21. Levin, B. E., and Sullivan, A. C. (1987). Glucose-induced norepinephrine levels in obesity resistance. *Am. J. Physiol.* **253**, R475–481.

22. Levin, B. E., Triscari, J., Hogan, S., and Sullivan, A. C. (1987). Resistance to diet-induced obesity: Food intake, pancreatic sympathetic tone and insulin. *Am. J. Physiol.* **252**, R471–478.

23. Fisler, J. S., Lupien, J. R., Wood, R. D., Bray, G. A., and Schemmel, R. A. (1987). Brown fat thermogenesis in a rat model of dietary obesity. *Am. J. Physiol.* **253**, R756–762.

24. Yoshida, T., Fisler, J. S., Fukushima, M., Bray, G. A., and Schemmel, R. A. (1987). Diet, lighting, and food intake affect norepinephrine turnover in dietary obesity. *Am. J. Physiol.* **252**, R402–R408.

25. Levin, B. E., Triscari, J., and Sullivan, A. C. (1983). Altered sympathetic activity during development of diet-induced obesity in rat. *Am. J. Physiol.* **244**, R347–355.

26. Meade, C. J., and Ashwell, M. (1980). Size cell differences in fat cells in New Zealand obese mice: A transplantation study. *Metabolism* **29(9)**, 884.

27. Becker, E. E., and Grinker, J. A. (1977). Meal patterns in the genetically obese Zucker rat. *Physiol. Behav.* **18**, 685–692.

28. Vaselli, J. R., and Maggio, C. A. (1989). Relationship of adipocyte filling to feeding behavior: Measurements in the hyperphagic Zucker fatty rat. *Appetite* **12**, 242.

29. Chlouverakis, C. (1972). Effect of caloric restriction on body weight loss and body fat utilisation in obese hyperglycemic mice. *Metabolism* **21** (1), 10–17.

30. Dubuc, P. U. (1976). Effects of limited food intake on the obese–hyperglycemic syndrome. *Am. J. Physiol.* **230**, 1475–1479.

31. Malaisse, G., Lemonnier, D., Malaisse-Lagae, F., and Mandelbaum, F. M. (1969). Secretion of and sensitivity to insulin in obese rats fed a high-fat diet. *Horm. Met. Res.* **1**, 9–13.

32. Bryce, G., Sullivan, A., and Stein, J. (1977). Insulin and glucagon plasma levels and pancreatic release in genetically obese Zucker rats. *Horm. Metab. Res.* **1**, 366–370.

33. Ionescu, E., Sauter, J. F., and Jeanrenaud, B. (1985). Abnormal glucose tolerance in genetically obese (fa/fa) rats. *Am. J. Physiol.* **248**, E500–506.

34. Czeck, M., Richardson, D. K., and Smith, C. J. (1977). Biochemical basis for fat cell insulin resistance in rodent and men. *Metabolism* **26**, 1027–1028.

35. Le Marchand-Brustel, Y., Jeanrenaud, B., and Freychet, P. (1978). Insulin binding

and effects in isolated soleus muscle of lean and obese mice. *Am. J. Physiol.* **234,** E348–358.

36. Opsahl, C. A., and Powley, T. L. (1974). Failure of vagotomy to reverse obesity in the genetically obese Zucker rat. *Am. J. Physiol.* **226,** 33–38.

37. Bray, G. A., and York, D. A. (1972). Studies on food intake of genetically obese rats. *Am. J. Physiol.* **223,** 176–179.

38. Fuller, Q. L., and Jacoby, G. (1955). Central and sensory control of food intake in genetically obese mice. *Am. J. Physiol.* **183,** 279–289.

39. Chlouverakis, C., and Bernardis, L. L. (1973). Ventromedial hypothalamic lesions in obese hyperglycemic mice (ob ob). *Diabetologia,* **3,** 391–395.

40. Chlouverakis, C., and Bernardis, L. L. (1972). Ventro-lateral hypothalamic lesions in obese hyperglycemic mice. *Diabetologia,* **8,** 179–184.

41. Rothwell, N. J., Saville, M. E., and Stock, M. J. (1982). Metabolic responses to food, atropine and 2-deoxy-D-glucose in Zucker rats. *Proc. Nutr. Soc.* **41,** 37A.

42. Lee, T. F., Wang, L. C., and Russell, J. C. (1987). Enhancement of cold-stimulated thermogenesis in the corpulent rat (LA/Ncp) by aminophylline. *Am. J. Physiol.* **252,** R737–742.

43. Van der Tuig, J. G., Kerner, J., and Romsos, D. (1984). Adrenalectomy increases norepinephrine turnover in brown adipose tissue of obese mice. *Life Sci.* **84,** 1433–1434.

44. Himms-Hagen, J. (1985). Food restriction increases torpor and improves brown adipose tissue thermogenesis in ob/ob mice. *Am. J. Physiol.* **248,** E531–539.

45. Rothwell, N. J., and Stock, M. J. (1983). Acute effects of fat and carbohydrate on metabolic rate in normal, cold-acclimated and lean and obese (fa/fa) Zucker rats. *Metabolism* **32,** 371–376.

46. Lee, C., and Lawler, C. H. (1983). Nutrient intake in relation to body weight in non-obese and obese elderly females. *Nutrit. Res.,* **3,** 149–155.

47. Slattery, J., and Potter, R. (1985). Hyperphagia: a necessary precondition to obesity. *Appetite* **6,** 133–142.

48. Pawan, G. L. S. (1972). Feeding patterns in "resistant" obesity. *Proc. Nutr. Soc.* **31** (3), 90A.

49. Bellisle, F., Rolland-Cachera, M. F., Deheeger, M., and Guilloud-Bataille, M. (1988). Obesity and food intake in children: Evidence for a role of metabolic and/or behavioral daily rhythms. *Appetite* **11,** 111–118.

50. Meyer, J. E., and Pudel, V. (1972). Experimental studies of food intake in obese and normal weight subjects. *J. Psychosom. Res.* **16,** 305–308.

51. Pudel, V., and Meyer, J. E. (1974). Obesity, I—Experimental investigations of appetite and satiety regulation. *Dtsch. Med. Wochenschr.* **99,** 618–628.

52. Bellisle, F., and Le Magnen, J. (1981). The structure of meal in humans: Eating and drinking patterns in lean and obese subjects. *Physiol. Behav.* **27,** 649–658.

53. Spiegel, T. A., Shrager, E. E., and Stellar, E. (1989). Responses of lean and obese subjects to preloads, deprivation, and palatability. *Appetite* **13,** 45–69.

54. Gaul, D. J., Craighead, W. E., and Mahoney, M. J. (1975). Relationship between eating rates and obesity. *J. Consult. Clin. Psychol.* **43,** 123–125.
55. Warner, K. E., and Balagura, S. (1975). Intrameal eating patterns of obese and nonobese humans. *J. Comp. Physiol. Psychol.* **89,** 778–783.
56. Rodin, J. (1975). The effects of obesity and set point on taste responsiveness and ingestion in humans. *J. Comp. Physiol. Psychol.* **89,** 1003–1009.
57. Drewnowski, A., and Greenwood, M. R. C. (1983). Cream and sugar: Human preferences for high-fat foods. *Physiol. Behav.* **30,** 629–633.
58. Wooley, O. W., Wooley, S. C., and Woods, W. A. (1975). Effects of calories on appetite for palatable food in obese and nonobese humans. *J. Comp. Physiol. Psychol.* **89,** 619–625.
59. Wooley, O. W., and Wooley, S. C. (1981). Relationship of salivation in humans to deprivation, inhibition and the encephalization of hunger. *Appetite* **2,** 331–350.
60. Campbell, R. G., Hashim, S. A., and Van Itallie, T. B. (1971). Studies of food-intake regulation in man. *N. Engl. J. Med.* **285,** 1402–1407.
61. Durrant, M., Royston, K., and Bloch, R. (1982). Effect of exercise and energy intake on eating pattern in lean and obese women. *Physiol. Behav.* **29,** 449–457.
62. Edelman, B., Engell, D., Bronstein, A., and Hirsch, E. (1983). Environmental effects on the intake of overweight and normal weight men. *Appetite* **7,** 71–83.
63. Schachter, S. (1968). Obesity and eating. *Science* **161,** 751–756.

Chapter Eleven

Selective Nutrient Appetites

Animal species as well as human select their foods. First of all, they discriminate between food and nonfood. Materials in their environment are selected as "food" inasmuch as they are sources of metabolizable energy. In animals, but rarely in humans, intakes of inedible materials such as earth (geophagia), hairs (trichophagia), bones, etc. are reported. They are classified as perverted appetites, or pica (1). Their significance is not clear, although their relation to some unknown specific deficiencies is suspected. Foods, as sources of metabolizable energy are (as is well known) carbohydrates, fats, and proteins. Throughout the preceding chapters, intakes of these foods were manifested as responses to their energetic or caloric properties and, as such, were involved in body energy balance. However, the question was raised earlier about the interchangeability of the three macronutrients as suppliers of calories. Is their substitution to each other limited or not? Is an exclusive carbohydrate, fat, or protein feeding possible or not? Does a nutritional requirement of carbohydrate, fat, and protein calories exist? Long ago, the nutritional requirement was apparently not solely calories, and specific need for any of the three macronutrients, permanently or occasionally, existed. The question is whether or not these vital needs also are satisfied by selective intakes (often designated "specific appetites"), and what are their particular mechanisms.

I. *Selective Appetites for Macronutrients: Protein Appetite*

A. Specific Regulation (or Not) of Fat and Carbohydrate Intakes

The three macronutrients are interchangeable in the pool of oxidized metabolites; however, we have seen that this is not true for all tissues. In peripheral tissue, fat oxidation may be substituted for carbohydrates. This is achieved through the use of repleted or overrepleted body fat reserves. Fats must also be substituted for carbohydrates in muscular exercise. This is not true in brain tissue, which does not oxidize fats and proteins and, thus, creates a specific need for glucose. However, these specific requirements of fats or carbohydrates over time and among tissues are not necessarily reflected in steady-state conditions by selective intakes, inasmuch as fats are convertible into carbohydrates and, vice versa, some proteins convertible into carbohydrates in intermediary metabolisms. Nevertheless, the possibility of specific regulations of the carbohydrate and fat intakes *per se* corresponding to their respective turnover has been investigated. Such a selective appetite is *a priori* a certainty regarding proteins. Nitrogen and a series of amino acids cannot, of course, be manufactured by the body from fats and carbohydrates. Body reserves of proteins do not exist. Thus, the protein turnover and nitrogen balance must be necessarily covered on a short-term basis by a protein intake separately regulated.

In the steady-state condition, the existence of selective macronutrient appetites was extensively studied by the self-selection method. Rats were offered the permanent choice among carbohydrate, fat, or protein diets separately and simultaneously presented in three food cups. This self-selection method was initiated and elegantly developed by Curt Richter in the late 1930s. In his outstanding work, Richter postulated implicitly that the intake of various diets was necessarily adjusted to the nutritional requirement. Therefore, he considered that the free intake, better than studies of metabolic turnovers, was a means to assess the nutritional requirements and their variations.

Richter *et al.* (2, 3) showed that young rats thus tested received an average of 54% of their calories from carbohydrates, 23% from fats, and 23% from proteins. They separated from this mean a different repartition of intakes in a group of rats that received 82, 17, and 1% of their calories, respectively, from carbohydrates, proteins, and fats. The authors found a justification of their postulate that intakes reflect nutritional requirements in that the self-selecting rats had a more rapid growth than did controls fed a chow diet (the MacCollom diet). On the other hand, they interpreted the difference recorded between the two groups as an indication of different nutritional requirements among rats. In a later reexamination by Rozin (4) of the self-selection of rats, this interindividual difference was found far larger than that found by Richter. In the Rozin study, rats received 51–84% of their calories from carbohydrates, 9–14% from proteins, and 1–59% from fats. These and other data provided evidence for a relative stability of the protein intake among rats (around a general mean of 13%). By contrast, the extreme variations of the repartition of the caloric intake between fats and carbohydrates provide the opposite indication that selective fat and carbohydrate appetites do not exist. This does not prove that, as mentioned earlier, specific needs of the two nutrients among tissues and over time do not exist but does prove that the interchangeability and interconvertibility of these nutrients avoid a selective regulation by food intake.

More evidence on this lack of a regulatory carbohydrate and fat intake is supported by Rozin's experiment. Rats were offered a choice among proteins, starch solutions, and oil. After stabilization of intakes from the three items, the response of rats to a dilution and reconcentration of the three nutrients was tested. Rats augmented and reduced their intake of the protein solution so that they maintained their protein intake regardless of the concentration. They did not on the diluted and reconcentrated carbohydrate and fat sources.

This lack of a specific regulation of the carbohydrate and fat turnovers by food intake was observed in a steady-state condition of these turnovers. The opposite is observed in various conditions in which one of these nutrients becomes a nonmetabolizable material or an accentuated and specific need of one of them is created. We have already mentioned that diabetics, unable to oxidize carbohydrates in their insulin-sensitive tissues, come to reject high-carbohydrate diets and to prefer fats. The same is reported in rats made deficient in the vitamin B complex, which is required in carbohydrate metabolism. Those rats reject sugar solutions and exhibit a considerable appetite for fats (5). Preceding chapters described that rats refed after starvation augment their intake of fats; after exercise and under the effect of exogenous insulin, they augment their preferential intake for carbohydrates. In lactating females, a strong fat and protein appetite is demonstrated (6).

B. Protein Appetite

The existence of a selective protein appetite reinforced by the amino acid requirement has been early recognized and widely investigated. This was at first assessed by the observation that rats (irrespective of conditions of presentations) demonstrated their capacity to maintain a limited and relatively constant proportion of their protein intake adequate to their normal growth. This was observed in the self-selection condition, as mentioned earlier, by the early Richter experiments. This constancy of the selection is also manifested in mixed diets. In one particular experiment (7), rats were offered successively mixed diets containing various concentrations of proteins. They maintained a perfectly constant daily intake on these various diets and a constant ratio of their caloric intake from proteins to the total caloric intake. A comparable result is reported in rats given a solution of the 11 essential amino acids in addition to a protein-free solid diet. They again manifested their capacity in this condition to take the amount needed to avoid the deficiency from the solution and to maintain the growth rate (8). Rats made previously deficient by a protein-free diet immediately prefer a protein-containing diet to the deficient one.

Rats who were presented a permanent choice between two diets containing, respectively, 55 and 0% proteins maintained a constant ratio of their caloric to protein intake. Examination of the lever-pressing pattern showed a dissociation between the diurnal meal pattern of caloric versus protein intake. Fifty-five percent of meals were mixed meals, 25% protein only, and 20% nonprotein meals (9). In another work, this dissociation was confirmed. Rats took fewer proteins during the night at a time of high-caloric intake than during the day (10). As mentioned elsewhere, during the night the intake of fats and proteins is augmented for the last half of the period (11).

In an elegant experiment, Leathwood and Ashley (9) further tested this capacity of rats to adjust their intake of proteins. Weanling or adult rats assigned to four groups were presented a choice of two diets: either 0 and 20%, 0 and 40%, 0 and 60%, or 10 and 40% proteins. In the four groups, rats initially preferred the diet lacking or poor in proteins; then, after 1–10 days they took, regardless the choice, a constant amount of proteins, adequate to ensure normal growth in young rats. After 14 days, the choice of the four groups was suddenly changed. As an effect of this change, weanling rats drastically altered their caloric intake from proteins; Adults reestablished it. Based on the observation of considerable interindividual differences, the authors concluded that a specific protein appetite does exist but that its underlying mechanism is not precise and reliable.

More surprising is the response of rats to diets either poor or rich in proteins. Rats given a low-protein diet (e.g., <3% proteins) rapidly refused this diet and lost weight; therefore, on such a diet, rats do not augment their intake to obtain the necessary amount of proteins and, thus, seem to avoid an overeating of calories; in other words, they seem to sacrifice the regulation of their protein requirement to the maintenance of normophagia. This was confirmed by the fact that such rats submitted to cold or after exercise augmented their caloric, and thereby their protein, intakes up to a level sufficient for a normal-weight gain (12). However, instead of maintaining a regulated caloric intake and a low-protein intake, rats, by refusing the diet, in fact sacrificed both their caloric and protein requirements. Thus, a low or null protein intake induced aphagia. Rats also refused a high-protein diet (e.g., >50% protein). This high-protein diet, also low in carbohydrates and fats, may be a cause of this aversion. The excess of protein *per se* associated with an abnormal level of blood and brain amino acids, which act as a toxic agent, may also be a cause of this rejection of the high protein diet.

Rats also reject amino acid-imbalanced diets. In a choice between an amino acid-balanced diet and the same diet lacking one of the essential amino acids (e.g., histidin, threonin) rats after some time preferred the former and entirely rejected the latter. This aversion for imbalanced diets is very strong. Protein-deficient rats presented a choice between a protein-free diet and a diet in which only histidin is lacking, prefer the protein-free diet despite the aggravation of their deficiency by this choice (13–15).

C. Mechanism of Protein Appetite

Three different concepts have been presented and taken as a working hypothesis in investigating the mechanism of protein appetite. The first one came from Richter (16). This outstanding experimenter speculated that all the self-selections of diets he demonstrated, including that of proteins, were developed by deficiencies of a peripheral taste sensitivity for stimuli of the normally or newly required diet. As already mentioned, such an interpretation did not resist experimentation. The taste threshold to salts of adrenalectomized rats is unchanged (17). Another and more robust theory was that a disturbed blood amino acid pattern produced by an inadequate intake of proteins and by protein deficiency was reflected by parallel disturbances of the brain amino acid content. Brain sites sensitive to concentrations of various amino acids would govern the response like energy deficit governs the caloric appetite. Indeed, neuronal responses to locally applied amino acids in the lateral hypothalamus have been reported (18). However, experiments of lesion did not support the hypothesis. Lesions in various parts of the brain (particularly in the ventromedial hypothalamus or in the lateral hypothalamus) did not change the responses to protein-complete or protein-free diets (19–21). The only exception is the lesion of the prepyriform cortex, which is the primary olfactory central projection. This is consistent with the fact that the olfactory bulbectomy strongly disturbs the regulation of protein intake (22). Although amino acids are claimed to be specific stimuli of taste receptors in rats, sensory stimuli provided by various proteins and involved in the discrimination are mainly olfactory stimuli.

This fact is the link to the third concept of mechanisms, convincingly supported by experimental results. Selective protein appetites are the result of learned flavor preferences and aversions.

Before examining experiments demonstrating the condition, it is of interest to mention observations that suggest an unlearned response to protein sensory cues evoked by protein deficiency. Deutsch *et al.* (23) report that protein-deficient rats presented the choice between the repleting and the protein-free diets manifest their preference for the former in the first minutes of the presentation. This immediate preference varies among proteins and apparently among their sensory properties. It is null with casein and strong with gluten. Further confirmation of these results could support the notion that some protein stimuli are unconditioned stimuli (in the true meaning of these words) of regulatory responses to proteins, like sweetness is the unconditioned stimulus of calorie regulated responses and saltiness is the unconditioned stimulus of regulated responses to salts.

Convincing evidence for a learning process was provided by Simson and Booth (24, 25). Rats are given alternatively two forms of a protein-free diet with different odors added. The consumption of one of them is preceded by a gastric load of a mixture of essential amino acids, and the other one by a gastric load of a saline. Within the first 2 hr of the first presentation, an aversion was manifested for the latter form while a preference was developed for the former. In a final choice, not associated with gastric loads, rats preferred the protein-free diet labeled by the odor previously associated with the amino acid gastric load.

What are the primary postingestive reinforcers of these aversion and preference learnings? Ashley *et al.* (26) Ashley and Anderson (27) showed that the ratio of the blood tryptophan concentration to neutral amino acids was negatively correlated to protein intake. This was confirmed by other researchers. Various other works demonstrated changes of the blood amino acid pattern related to the intake of protein-free or -imbalanced diets. These various disturbances and the associated impairment of intermediary metabolisms are presumably causes of diffuse malaises, which may act as negative reinforcers like in a trivial conditioned taste aversion. The relief of these deficiencies may act in the opposite way to induce the preference for rehabilitating diets.

A number of works have been presented in favor of the imaginative hypothesis that serotoninergic structures in the brain and the entry of tryptophan into the brain as precursor of the serotonin synthesis were critically involved in protein appetites and in the repartition between protein and carbohydrate intakes (26–28). After some years of success, this theory has been totally ruled out (9, 29).

II. *Self-Selection of Vitamins and Minerals*

A. Vitamin Appetites

Since the pioneering work of Harris *et al.* (see Chapter 4, Section III), the development of appetites for vitamin-containing diets in vitamin-deficient rats has been confirmed (30–32). This was assessed in rats made deficient in the various B vitamins: thiamin, riboflavin, and pyridoxin. The preference for the thiamine-containing diet in thiamine-deficient rats appears very rapidly (1 or 2 days) (33, 34). This is not the case in vitamin A-deficient rats. Their preference for a diet containing the precursor of the vitamin carotene appears only after weeks and inasmuch as the rat regains weight (35).

B. Mineral Appetites

In the steady-state condition, rats, like in other animal species as well as humans, exhibit selective appetites for various ions contributing to the hydro-osmotic homeostasis, essentially Na and Cl and for other ions such as Ca and K in relation to their specific physiological turnover. The physiological basis for Na appetite, out of the scope of this volume, will not be detailed here. However, it is of interest to recall that its study was initiated by the dramatic and famous experiment by Richter (36), who showed the immediate enhancement of this Na appetite in adrenalectomized rats. In a choice of various salts, they developed enhanced intakes proportional to the fall of the plasma concentration of various ions provoked by adrenalectomy: maximal for NaCl and less for Ca^+ and K^+ (36). The role for a learning and/or for an unconditioned response was debated (37). The unconditioned response was recently supported by Schulkin (38), who demonstrated that Na-deficient rats exhibit a short-term preference for salty tasting salts lacking Na.

Parathyroidectomized rats augment 3.9-fold their intake of a Ca^+ lactate solution (39). Rats made deficient by a K^+-free diet and offered a choice between NaCl and KCl solutions developed a preference for NaCl, the more salty of the two salts (40). Physiological requirements other than those provoked by deficiency generate the development of Na^+ appetite. This is the case in pregnant females (41). Interestingly, rats, strongly stressed by an intense noise, demonstrate an enhanced intake of NaCl (42).

In other deficiencies of various oligoelements (zinc, lead, iron) created by free diets, selective choices of rehabilitating diets have been demonstrated (43–45).

C. Mechanisms of Vitamin and Mineral Appetites

The initial experiment by Harris *et al.* convincingly demonstrated the learning of specific appetites. The response is mediated by sensory cues associated with the postingestive relief of the deficiency. This response is a typical conditioned taste preference that is a learned palatability (46). Later, an acquired taste aversion to the diet, introducing the deficiency, was shown to be associated to the taste preference for the rehabilitating diet. Thiamine-deficient rats received a choice between their familiar diet and a novel food. Irrespective of the presence of absence of thiamine in one or the other diet, they preferred the novel food contrary to the neophobic response of controls (47). This aversion is manifested even if the deficiency is alleviated by thiamine injection before testing.

The illness or imbalance is clearly a reinforcer of such aversions, and the relief of this illness or imbalance by the postingestive action of the deficient vitamin or mineral is the reinforcer of the selective appetite. Thus, to designate these selective appetites as partial, or specific, hungers is quite inappropriate. Looking for as many brain sites as possible vitamin or mineral deficiencies, sites detecting these deficiencies and governing the specific appetites is, of course, vain.

In humans, except the salt appetite relevant to the hydroosmotic regulation, the emergence of specific appetite for vitamins and minerals is not evident. In developed countries, available foods are so varied that the ordinary foods (e.g., fruits, vegetables) provide the required vitamins and minerals without an apparent specific appetite for some of them. On the other hand, a cognitive learning, recommendations, and prescriptions of some sources of these nutrients contribute to avoid deficiency symptoms. Nevertheless, some of these deficiencies, mainly of calcium and iron, are currently diagnosed. This seems to confirm that the development of an adequate appetite has not prevented these deficiencies; however, historical reports prove the opposite. When Jacques Cartier discovered Canada and landed in an Indian village, his crew was stricken by scurvy. Indians recommended that Jacques Cartier treat his sailors with hawthorn berries. Four centuries later, these berries were found to be one of the vegetables containing the highest concentration of ascorbic acid.

References

1. Green, H. H. (1925). Perverted appetites. *Physiol. Rev.* **5,** 336–348.
2. Richter, C. P., Holt, L. E., and Barelare, B. (1937). The effect of self-selection of diet—food (protein, carbohydrates and fats), minerals, and vitamins—on growth, activity, and reproduction in rats. *Am. J. Physiol.* **119,** 388–389.

3. Richter, C. P., Holt, L. E., and Barelare, B. (1938). Nutritional requirements for normal growth and reproduction in rats studied by the self-selection method. *Am J. Physiol.* **122,** 734–744.

4. Rozin, P. (1968). Are carbohydrate and protein intakes separately regulated? *J. Comp. Physiol. Psychol.* **65,** 23–29.

5. Richter, C. P., Holt, L. E., Barelare, B., and Hawkes, C. D. (1938). Changes in carbohydrate, fat and protein appetites in vitamin E deficiency. *Am. J. Physiol.* **124,** 597–602.

6. Soegel, C., and Collier, G. (1972). Dietary self-selection by pregnant and lactating rats. *Physiol. Behav.* **8,** 151–154.

7. Schoenfeld, T. A., and Hamilton, L. W. (1976). Multiple factors in short-term behavioral control of protein intake in rats. *J. Comp. Physiol. Psychol.* **90,** 1092–1104.

8. Harthed, T. C., and Galagger, B. (1972). Auto-regulation of amino acid intake by albino rats. *J. Comp. Physiol. Psychol.* **85,** 107–111.

9. Leathwood, P. D., and Ashley, D. V. M. (1983). Strategies of protein selection by weanling and adult rats. *Appetite* **4,** 97–112.

10. Li, E. T., and Andernon, G. H. (1982). Meal composition influences subsequent food selection in the young rat. *Physiol. Behav.* **29,** 779–784.

11. Shor-Posner, G., Azar, A. P., Filart, R., Temple, D., and Leibowitz, S. F. (1986). Morphine-stimulated feeding: Analysis of macronutrient selection and PVN lesion. *Pharm. Biochem. Behav.* **24,** 931–40.

12. Meyer, J. H., and Hargus, W. A. (1959). Factors influencing food intake in rats fed low-protein rations. *Am. J. Physiol.* **197,** 1350–1352.

13. Sanahuja, J. C., and Harper, A. E. (1962). Effect of amino acid imbalance on food intake and preference. *Am. J. Physiol.* **202,** 165–168.

14. Sanahuja, J. C., and Harper, A. E. (1963). Effect of dietary amino acid pattern on food intake. *Am. J. Physiol.* **204,** 686–690.

15. Sanahuja, J. C., Rio, M. E., and Lede, M. N. (1965). Decrease in appetite and biochemical changes in amino acid imbalance in the rat. *J. Nutr.* **86** (4), 424–432.

16. Richter, C. (1939). Salt threshold of normal and surrenalectomised rats. *Endocrinol.* **24,** 367–371.

17. Pfaffmann, C., and Bare, J. K. (1950). Gustatory nerve discharges in normal and adrenalectomized rats. *J. Comp. Physiol. Psych.* **43,** 320–324.

18. Wayner, M. J., Ono, T., De Young, A., and Barone, F. C. (1975). Effects of essential amino acids on central neurons. *Pharm. Biochem. Behav.* **3** (Suppl. 1), 85–90.

19. Leung, P. M., and Rogers, Q. R. (1973). Effect of amygdaloid lesions on dietary intake of disproportionate amounts of amino acids. *Physiol. Behav.* **11,** 221–226.

20. Leung, P. M., and Rogers, Q. R. (1970). Effect of amino acid imbalance and efficiency on food intake of rats with hypothalamic lesions. *Nutr. Res. Int.* **1,** 1–10.

21. Leung, P. M., and Rogers, Q. (1971). Importance of pre-pyriform cortex in food intake response of rats to amino acids. *Am. J. Physiol.* **221**, 329–335.

22. Larue, C., and Le Magnen, J. (1971). Modification des séquences alimentaires et de la régulation calorique après ablation des bulbes olfactifs chez le rat. *C.R. Acad. Sci. (Paris)* **272**, 90–93.

23. Deutsch, J. A., Moore, B. O., and Heinrichs, S. C. (1989). Unlearned specific appetite for proteins. *Physiol. Behav.* **46**, 619–624.

24. Simson, P. C., and Booth, D. A. (1973). Olfactory conditioning by association with histidine-free or balanced amino-acid loads in rats. *J. Exp. Psychol.* **25**, 354–359.

25. Simson, P. C., and Booth, D. A. (1974). The rejection of a diet which has been associated with a single administration of an histitine-free amino acid mixture. *Br. J. Nutr.* **31**, 285–296.

26. Ashley, D. V. M., Coscina, D. V., and Andernon, G. H. (1979). Selective decrease in protein intake following brain serotonin depletion. *Life Sci.* **24**, 973–984.

27. Ashley, D. V. M., and Anderson, G. H. (1975). Correlation between tryptophane to neural amino acid ration and protein intake in the self-selecting rat. *J. Nutr.* **105**, 1412–1421.

28. Wurtman, R. J., and Wurtman, J. J. (1977). Nutrition and the Brain. New York: Raven Press.

29. Fernstrom, J. D. (1987). Food induced changes in serotonine synthesis: Is there relationship to appetite for specific macronutrients? *Appetite* **8**, 163–186.

30. Scott, E. M., and Verney, E. L. (1949). Self-selection of the diet: IX. The appetite for thiamine. *J. Nutr.* **37**, 81–97.

31. Scott, E. M., and Verney, E. L. (1947). Self-selection of the diet: VI. The nature of appetite for B vitamins. *J. Nutr.* **34**, 47–80.

32. Rozin, P., Wells, C., and Mayer, J. (1964). Specific hunger for thiamine: Vitamin in water versus vitamin in food. *J. Comp. Physiol. Psychol.* **57**, 78–84.

33. Rozin, P. (1965). Specific appetite for thiamine: Recovery from thiamine deficiency and thiamine preference. *J. Comp. Physiol. Psychol.* **59**, 98–101.

34. Tribe, D. E., and Gordon, J. C. (1953). Choice of diet by rats deficient in numbers of the vitamin B complex. *Brit. J. Nutr.* **7**, 197–201.

35. Hartman, B. D. (1955). Provitamine A selection by vitamine A depleted rat. *J. Genet. Psychol.* **86**, 45–51.

36. Richter, C. P. (1936). Increased salt appetite in adrenalectomized rats. *Am. J. Physiol.* **115**, 155–161.

37. Smith, M. H. (1972). Evidence for a learning component of sodium hunger in rats. *J. Comp. Physiol. Psychol.* **78** (2), 242–247.

38. Shulkin, J. (1982). Behavior of sodium-deficient rats: The search for the salty taste. *J. Comp. Psychol. Physiol.* **96**, 628–634.

39. Richter, C. P., and Eckert, J. F. (1937). Increased calcium appetite of parathyroidectomized rats. *Endocrinology* **21**, 50–54.

40. Zucker, M. I. (1965). Short term salt preference of potassium-deprived rats. *Am. J. Physiol.* **208,** 1071–1074.

41. Woodside, B., and Millelire, L. (1987). Self-selection of Ca during pregnancy and lactation in rats. *Physiol. Behav.* **39,** 291–295.

42. Griffiths, W. J. (1956). Diet selection of rats subjected to stress. *Ann. N.Y. Acad. Sci.* **67,** 1–10.

43. Hesse, G. W., Hesse, K. A. F., and Catalanottto, F. A. (1979). Behavioral characteristics of rats experiencing chronic zinc deficiency. *Physiol. Behav.* **22,** 211–215.

44. Snowdon, C. T. (1977). A nutritional basis for lead pica. *Physiol. Behav.* **18,** 885–894.

45. Woods, S. C., Vasselli, J. R., and Milam, K. M. (1977). Iron appetite and latent learning in rats. *Physiol. Behav.* **19,** 623–626.

46. Smith, M., Pool, R., and Weinberg, H. (1958). Evidence for a learning theory of specific hungers. *J. Comp. Physiol. Psychol.* **51,** 758–763.

47. Rodgers, W. L., and Rozin, P. (1966). Novel food preferences in thiamine-deficient rats. *J. Comp. Physiol. Psychol.* **61,** 1–4.

Chapter Twelve

Development of Feeding

Examining the ontogeny of feeding behavior from birth to adulthood at the end of this volume may be considered paradoxical. However, a comprehensive analysis of the body of works devoted to this developmental aspect is now easy, following the previous deciphering of matured mechanisms in adults.

Mammals are born in a state of brain, and therefore behavioral, immaturity, which varies widely among species. In all species, this premature birth before the end of development prevents the animal from feeding itself. However, the maturation of the central nervous system, which allows it to seek, select, and eat the correct amount of foods, requires that it is fed. Natural selection solved the dilemma by the lactation and suckling of the young. This ingenious solution specific to high vertebrates gave its name to the family of which humans are the top representatives. In this last chapter, again only experimental rats will be compared to humans, providing evidence for the understanding of the human level.

I. *Development of Feeding in the Rat Model*

Following 21 days of gestation, the suckling of pups lasts another 21 days before they can freely eat solid foods, as weanling rats. At birth, the weight of the young, depending on the litter size, averages 10 g. As a weanling, weight averages, in a litter of eight pups, 70 g. In addition to the energy retention for this rapid growth, feeding the young must match a very high level of basal metabolism. Referred to the weight, this metabolic rate is, according to Kleiber (1), 2.5 times that of adults. The daily caloric requirement of sucklings is, thus, considerably greater than it is afterward during development and in adulthood.

We will now examine how this nutritional requirement is satisfied by the suckling behavior of rats and whether a regulatory control of the milk intake is effective or only progressively developed during this period.

A. The Suckling Behavior

Soon after birth, the pup, competing with other pups in the litter, finds a nipple on its mother and remains permanently attached to it. This nipple attachment does not depend on an active behavior of the dam; it also occurs on an anesthetized mother. Pups open their eyes after 2 or 3 days only. Thus, the question was raised as to the nature of sensory stimuli that allow the rat to find the nipple. Experiments by Teicher and Blass (2) revealed the role for an olfactory stimulus or pheromone. Washing the nipple prevents the attachment. It is immediately reestablished by the distillate of the washing solvent; it is also reestablished by brushing the washed nipple with saliva or amniotic fluid of the mother. The saliva from a virgin female is without any effect. The author suggests that the licking of the nipples by the mother the day after parturition could provide the stimulus needed for the attachment. Oxitocyn may play a role in this stimulus production. Injecting the mother with oxitocyn after washing its nipples reestablishes the attachment (3). Other stimuli are excluded. Cooling the nipples or shaving the fur around them does not prevent the attachment (4). An olfactory bulbectomy at birth is lethal; however, 90% of pups bulbectomized at postnatal day 4 survive. They exhibit great difficulties in finding the nipple and attaching to it. Contrary to controls, washing the nipple, however, does not prevent this attachment. Thus, in the absence of olfaction, other unknown stimuli permit the attachment (5).

The olfactory system is demonstrated to be functional at birth. Stimulating pups the first day of life by an odorant (methylacetate) provoked the sniffing. The threshold of dilution to obtain the response decreases after the suckling period, indicating a progressive maturation of the system (6).

An evolution of the suckling from birth to weanling is observed. Until the day 11 the attachment to the nipple is permanent and the suckling is continuous even in the absence of milk ejection. Retrieved from the dam, young have the same latency to attach again either deprived or not. After 11 days, undeprived pups demonstrate an increasing latency to attach again (7). Other results indicate that the suckling is not a fixed reflex. Pups repetitively detached from the dam from the first to the fifth day showed difficulty in attaching again, demonstrating a possible role for experience in suckling development (8). The mother imposed suckling during the day, even for blinded pups. A partial nocturnal suckling appears only at the level of weanling (9). An experiment by Hall *et al.* (10) revealed the detailed daily suckling pattern. The suckling was recorded through a rubber cannula between the nipple and the pup mouth. This technique allowed milk intake control in volume and periodicity. Five to 17-day-old rats drink up to 10% of their body weight in one burst of suckling and stop in a state of considerable gastric fullness and distention. After this enormous meal, they sleep. Later (day 19), their unitary milk intake is reduced to 5% body weight and they remain awake afterward (11, 12).

B. Control of Milk Intake

A highlighting experiment was the following (13). Rats, less than 15 days old, were fed by several lactating dams, each of them treated by oxitocyn. Pups were offered an unlimited availability of milk. In this condition, they took an enormous volume of milk. When, on the contrary, the young were deprived by various means, they took from their dam a volume of milk proportional to the time of deprivation. Thus, a response to deprivation is effective, whereas the satiation–satiety system is not.

Responses to deprivation were widely confirmed (14). At the approach of weanling, postingestive factors become the main controller of the amount and periodicity of intake, as on the solid food after weanling (15).

C. Maturation of Brain Regulatory Mechanisms

The above data imply that brain mechanisms that control food intake in adults as a function of nutritional requirements mature late, are activated at the end of the suckling period, and presumably permit this end (16). Before weanling, rats do not respond to 2-deoxy-D-glucose (17, 18) or insulin (19). However, lateral hypothalamic lesions suggest that the system of feeding initiation is at work early. Before the tenth day, the LH lesion is lethal. Rats die totally anorexic. They recover (like adults) only when the lesion is preformed later. The recovered rats then exhibit the same sequel as adults, particularly with the loss of the prandial periodicity (20, 21).

Other brain functions in sucklings are unexplored. As mentioned earlier, in weanlings the ventromedial hypothalamic lesion-induced hyperinsulinemia was later followed by hyperphagia and abnormal body weight gain, an intriguing result that is still unreplicated. Repetitively injecting newborn rats with insulin makes them obese in adulthood (22).

D. From Weanling to Adulthood

In free access to solid foods, a progressive development of the meal pattern is observed. Until 4 wk, the postprandial correlation is absent and, contrary to adults, a preprandial correlation is exhibited. The prandial periodicity is slowly established by augmentation of meal sizes and reduction of meal frequency (23). According to Bernstein (24), the diurnal cyclicity is also progressively established. Initially, the night : day ratio is very low. The prevailing nocturnal feeding of adults is reached only at 6 wk of age. This progressive establishment of the adultlike meal pattern has been viewed as comparable to the successive stages of the recovery of feeding after the LH lesion. Slowing the development by restriction or thyroidectomy augments this similarity. The conclusion drawn by the authors is that a maturation of the LH is needed for the emergence of the normal feeding pattern after weanling (25).

E. Acquisition of Food Preferences

As mentioned elsewhere, food preferences or aversions may be conditioned *in utero* by injecting odorant or tastes into the amniotic fluid during pregnancy. During lactation, rats acquired a preference for the diet eaten by the mother as manifested in a choice after weanling (26). The eating of a garlic-flavored diet by the mother induces a high palatability for the same diet in the young (27). In addition to this role of the mother's diet, the presence of adults at the place of feeding is a determinant factor of the acquisition of preferences. The young eats the food that its mother eats, and this behavior is seen as parental education (28) (Fig. 12.1). A role for the sensory postingestive association identical to that demonstrated in adults is clear. Young rats fed a starch solution acquired a preference for the flavor associated with this solution. The acquisition is slower with a protein solution; it is null with a solution of fats (29). However, the weanling rats could separately and adequately select proteins (30). More surprising is the following finding (31). Mothers were fed during pregnancy and lactation by a diet containing either 10, 20, or 40% proteins. Young rats at weanling were offered a choice between two diets containing 10 and 19% protein, respectively. Their intake of protein relative to the energy intake from this choice was found to be correlated to the concentration of casein content of the mother's diet during pregnancy and/or during lactation.

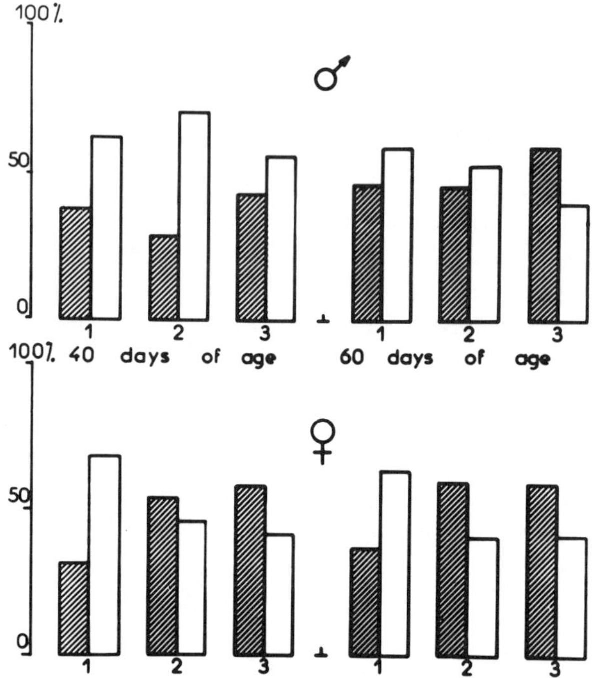

Figure 12.1 Injecting an odorant to mother rats during lactation induces preference of youngs for a food flavored by the odorant. ▨, solid diet flavored by Citrol; ☐, natural diet.

II. *Feeding in Children*

Like other matters in preceding chapters, the development of human feeding behavior from birth to adulthood could be the topic of a full volume. Moreover, many works dealing with this developmental aspect have been, when appropriate, already described. Thus, this last chapter on children's feeding, far from exhaustive, will give only some essential facts and ideas.

A. Newborns

Human infants double their birth weight at 4 mo and triple it at 12 mo. In a 1-yr-old infant, the daily energy requirement is estimated to be 1000 kcal/day. This includes about 50 kcal of energy retention for growth. The rest, referred to body weight, indicates a metabolic rate about three times its value in adults.

In breast-fed and nonbreast-fed newborns, this energy requirement is manifested by the immediate emergence of the suckling behavior and of the feeding demand by periodic and noisy crying. The intensity of these cries while waiting to eat suggests that this high energy requirement of the baby is associated with a very intense and painful hunger.

During the first weeks and months, the feeding of the infant is not free in its frequency nor in the amount of milk drunk. Both are prescribed by the mother and to the mother by pediatricians. In old experiments, the spontaneous periodicity of the food demand was tested by asking the mother to respond to all demands of the baby. In one particular study, the breast-fed baby manifested no demand at all on the first day. From day 3 to day 6 he asked and accepted the breast 11 times/day.

This frequency was progressively reduced thereafter and stabilized at 5–6 times/day after 10 wk. The intervals became constant at this time (about 90 min) during the day and were by far, longer at night (32, 33). For some months, until the time at which the baby begins to accept or to refuse some semiliquid offered foods, the newborn has no choice. He or she only accepts and drinks the milk of his or her mother or its replacement. This acceptability of the milk is manifested the first day of life. Thus, the flavor (taste and smell) of the milk is an unlearned, inherited stimulus to eat. Surprisingly a preference for the milk of the mother compared to the milk of a nursing mother has not been tested. A preference for mother's milk may be due to *in utero* conditioning experimented in animals, or to a self/non-self discrimination by odor sensitivities also experimented in animal species.

In addition to this innate palatability of the mother's milk, a suckling stimulation effect of sweetness was recognized early (34–36). Adding sugar to the milk formula increases the rate of suckling. The use of a rubber nipple for sensing and allowing the recording of the tongue, cheeks, and jaw movements showed that the baby responds readily as a function of concentration of the added sugar. Therefore, he or she exhibits a qualitative and quantitative taste discrimination as soon as the first day of life (37).

The aversive response to bitterness is also manifested in the first days of life. Responses to saltiness are clearly manifested after only 2 mo (35, 36). Changes in the respiratory rhythm under stimulation by various odorant mixtures assess this functional activity of the olfactory system in newborns (37–39). Steiner (39) initiated the original technique confirming this olfactory activity at birth and, in addition, the exhibition of appetitive or aversive responses to these stimuli. The videotape of newborns stimulated by various odorants revealed orofacial expressions, which clearly differentiated these two types of responses (Figs. 12.2 and 12.3). However, no research has demonstrated that a particular

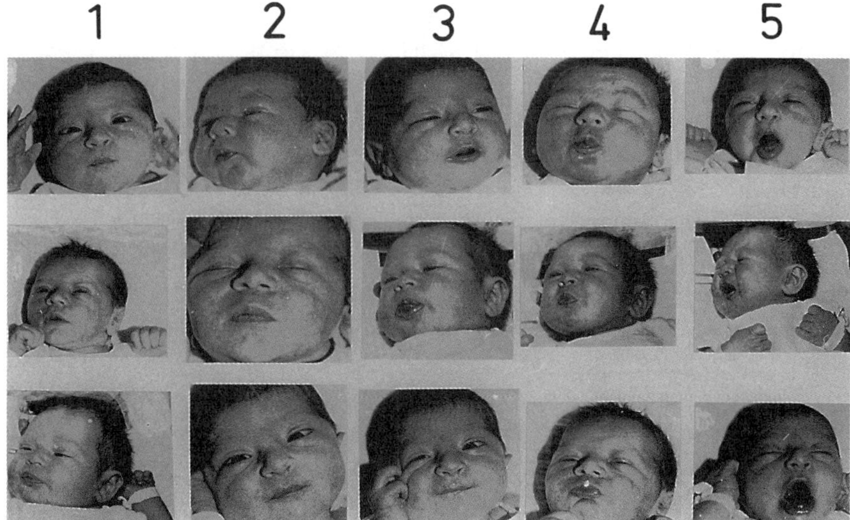

Figure 12.2 Differential facial expressive features to gustatory stimuli in termborn neonates prior to the first exposure to food intake, at the age of a few hours after birth. 1, without stimulation; 2, in response to stimulation with distilled water; 3, in response to stimulation with sucrose solution; 4, in response to stimulation with citric acid; 5, in response to stimulation with quinine HC1. [From J. M. Steiner (1977). Innate discriminative human facial expressions. J. M. Weiffenbach (Ed.) In *Taste and Development*, pp. 173–189. U.S. DHEW-N111, Bethesda.]

olfactory stimulant acted as an unconditioned stimulus to suck, like, for example, milk flavored with sweetness.

As already mentioned, these initial inherited preferences or aversions, manifested during the suckling period, are later profoundly modified by experience. Comparisons of sweet preference and aversion in various groups from 2–20 yr of age show the highest preference in children compared to adults. Although quantitatively measured, this is not, of course, a discovery (40–42). This preference progressively decreases until adulthood, less rapidly in girls than in boys. These preferences for taste and smell of liquid and solid foods are mostly differentiated according to the food. Some sweetened foods and beverages remain highly appetitive. Sweetening other foods or beverages makes them aversive. The same is true for saltiness.

Comparable evolution has been observed in responses to food-related olfactory stimuli. In an old and still valid study, three groups of 30–40 children from 1 to 5, 8 to 9, and 12 to 14 yr of age were compared to adults. Percentages of appetitive responses were as follows: for a fruity odor, 56, 66, 69 and 95%,

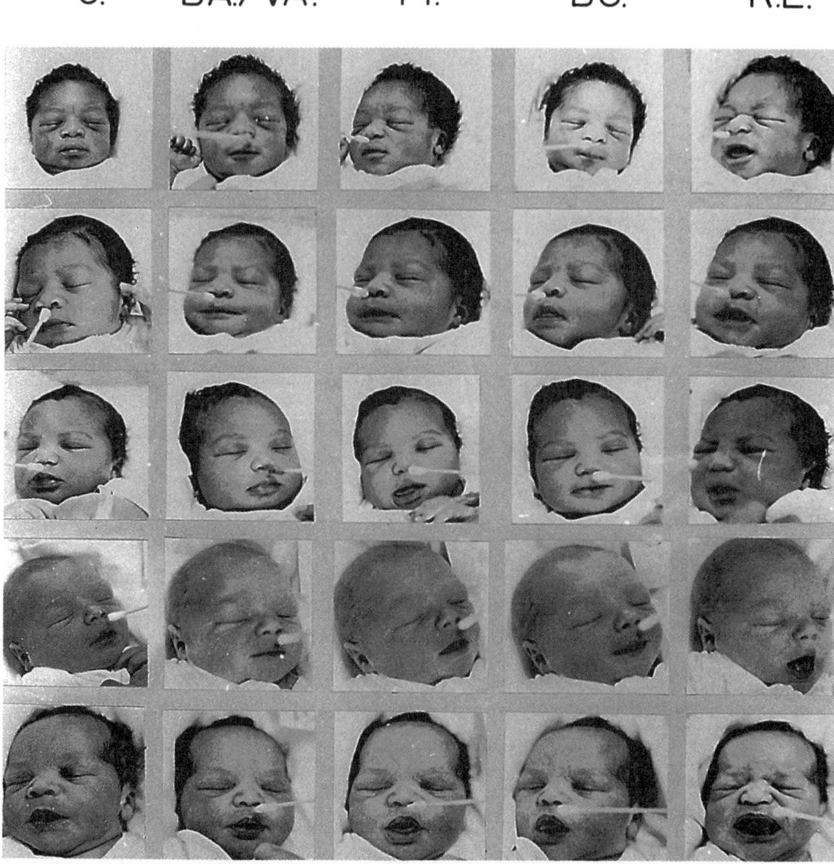

Figure 12.3 Neonate response to food-related odors. c, control, infant is presented with an "odorless" applicator; BA/VA, the stimulus is an artificial banana or vanilla aroma; F1, response to a fish-type flavor (artificial shrimp flavor); BU, an artificial butter flavor, sampling "milky" odors; RE, the smell of rotten eggs. [From J. E. Steiner (1979). Human facial expressions in response to taste and smell stimulation. H. W. Reese and L. P. Lipsitt (Eds.) In *Advances in Child Development*, Vol. XIII, pp. 257–293. Academic Press, New York.]

for a mercaptan odor (fecal odor), 33, 36, 0, and 2%; and for a floral odor, 45, 34, 23, and 78% (43). On the basis of these comparisons and others, the author notes the singularity of the third group of children around puberty and the accentuation of both positive and negative responses in adults compared to those of children. In a more recent study, 300 children from 8 to 9 and from 15 to 16-yr old were compared in their likes and dislikes for 10 alimentary odors. An evolution, only slightly significant, was observed for some of them without any apparent effect of puberty. The most significant evolution in the direction of preference was for coffee (44).

B. Inheritability or Acquisition of Food Preference

The question was raised about the inheritability of preferences or aversions for the complex flavor of various solid foods in children. Various studies on homozygote and heterozygote twins have totally excluded this genetic origin in the spectrum of individual preferences and aversions for fruit, vegetables, meat, fish, etc, and their various preparations (45, 46). Resemblances in preferences and aversions among children of the same family and between children and their parents have been widely investigated. These works were designed to confirm the absence or limited role of genetic factors and the prevailing social versus nutritional factor in the acquisition of responses to foods. An exemplary study by Pliner and Levin Pelchat (47) was carried out on 84 Canadian families with 120 food items. Resemblances between brothers and sisters, between them and their parents, and between children and parents of other families were statistically treated. The resemblance among children of the same family, irrespective of sex, was stronger than that between children and their parents. The role of imitation and of frequent experience with a particular food in each family is pointed out by the authors. Similarities among children of different families indicate the strong influence of sociocultural factors (47).

The role of education by parents in this sociocultural transmission of food habits is self-evident. Children learn to like and dislike those foods their parents present to them as good or bad to eat. This is particularly self-evident in food aversion exhibited by children for foods never tasted and acquired from their parents (48). The role of both neophobia (49) and familiarization by repeated exposures has been researched (50). The preference for fruit juice tasted 20, 10, 5, and 0 times by children was scored proportionally to the number of tastings. The role of imitation was also tested (51).

In such an evolution of human preferences and aversions, the social pressure and, more generally, the environment play the role of external reinforcer eventually substituted for the postingestive nutritional reinforcement. The extreme variety of this environmental context throughout childhood may explain the origin of individual cravings or aversions and their often sudden changes associated with unknown events of the individual feeding history. As mentioned elsewhere, experiments demonstrated the conditioning of the stimulation to eat in children by arbitrary external sensory cues (52).

However, like rats and adults, children learn to eat the appropriate amount of food to ensure their growth. They also learn to like foods appropriate for their nutritional requirements by the postingestive reinforcement. As described in preceding chapters, this learning process was experimentally proven in the adjustment to changes of the caloric density of food (53). Like in adults, the initial palatability of that food and its associated pleasantness as well as the satiating capacity of the food are acquired by children in a subconscious learning until adulthood and beyond.

References

1. Kleiber, M. (1961). *The fire of life: An introduction to animal energetics*. New York: Wiley, 428 pp.
2. Teicher, M. H., and Blass, E. M. (1977). First suckling response of the newborn albino rat: The roles of olfaction and amniotic fluid. *Science* **198,** 635–637.
3. Singh, P. J., and Hofer, M. A. (1978). Oxytocin reinstates maternal olfactory cues for nipple orientation and attachment. *Physiol. Behav.* **20,** 385–389.
4. Blass, E. M., Teicher, M. H., Cramer, C. P., Bruno, J. P., and Hall, W. G. (1977). Olfactory, thermal and tactile controls of suckling in preauditory and previsual rats. *J. Comp. Physiol. Psychol.* **91,** 1248–1260.
5. Teicher, M. H., Flaum, L. E., Williams, M., Eckhert, S. J., and Lumia, A. R. (1978). Survival growth and suckling behavior in neonatally bulbectomized rats. *Physiol. Behav.* **21,** 553–562.
6. Alberts, J. R., and May, B. (1980). Ontogeny of olfaction: The development of the rat's sensitivity to urine and amyl acetate. *Physiol. Behav.* **24,** 965–970.
7. Hall, W. G. Cramer, C. P., and Blass, E. M. (1975). Developmental changes in suckling of rat pups. *Nature (London)* **258,** 318–320.
8. Stoloff, M. L., Kenny, E. M., Blass, E. M., and Hall, W. G. (1980). The role of experience in suckling maintenance in albino rats. *J. Comp. Physiol. Psychol.* **94,** 847–856.
9. Levin, R., and Stern, J. M. (1975). Maternal influence on ontogeny of suckling and feeding rhythms in the rat. *J. Comp. Physiol. Psychol.* **89,** 711–721.
10. Hall, W. G., Cramer, C. P., and Blass, E. M. (1977). Ontogeny of suckling in rats: Transitions toward adult ingestion. *J. Comp. Physiol. Psychol.* **91,** 1141–1155.
11. Hall, W. G., and Rosenblatt, J. S. (1977). Suckling behavior and the intake control in the developing rat pup. *J. Comp. Physiol. Psychol.* **91,** 1232–1247.
12. Hall, W. G. (1979). The ontogeny of feeding in rats: I. Ingestive and behavioral responses to oral infusions. *J. Comp. Physiol. Psychol.* **93,** 977–1000.

13. Cramer, C. P., and Blass, E. M. (1983). Mechanisms of control of milk intake in suckling rats. *Am. J. Physiol.* **245,** R154–160.

14. Cramer, C. P., and Blass, E. (1985). Nutritive and non-nutritive determinants of milk intake of suckling rats. *Behav. Neurosci.* **99,** 578–581.

15. Hall, W. G., and Rosenblatt, J. S. (1978). Development of nutritional control of food intake in suckling rat pups. *Behav. Biol.* **24,** 423–427.

16. Satinoff, E., and Stanley, W. C. (1963). Effect of stomach loading on sucking behavior in neonatal puppies. *J. Comp. Physiol. Psychol.* **56,** 66–68.

17. Houpt, K., and Houpt, T. R. (1975). Effect of gastric loads and food deprivation on subsequent food intake in suckling rats. *J. Comp. Physiol. Psychol.* **88,** 764–772.

18. Gisel, E. G., and Henning, S. J. (1980). Appearance of glucoprivic control of feeding behavior in developing rats. *Physiol. Behav.* **24,** 313–318.

19. Lytle, L. D., Moorcroft, W. H., and Campbell, B. A. (1971). Ontogeny of amphetamine anorexia and insulin hyperphagia in the rat. *J. Comp. Physiol. Psychol.* **77,** 388–393.

20. Almli, C. R. (1978). The ontogeny of feeding and drinking in rat: Effect of early brain damage. *Neurosci. Behav. Rev.* **2,** 281–300.

21. De Castro, J. M., and Balagura, S. (1975). Ontogeny of meal patterning in rat and its recapitulation during recovery from hypothalamic lesion. *J.C.P.P.* **89,** 799–802.

22. Dorner, G., and Gotz, F. (1972). Hyperglycemia and overweight of adult male rats treated as newborn animals with insulin. *Acta Biol. et Medica Germanica* **29,** 467–470.

23. De Castro, J. M., and Balagura, S. (1976). A preprandial intake pattern in weanling rats ingesting a high fat diet. *Physiol. Behav.* **17,** 404–405.

24. Bernstein, I. L. (1976). Ontogeny of meal patterns of the rat. *J. Comp. Physiol. Psychol.* **90,** 1126–1132.

25. Cheng, M. F., Rozin, P., and Teitelbaum, P. (1971). Starvation retards development of food and water regulations. *J. Comp. Physiol. Psychol.* **76,** 206–218.

26. Galef, B. G., and Clark, M. M. (1972). Mother's milk and adult presence: Two factors determining initial dietary selection by weanling rats. *J. Comp. Physiol. Psychol.* **78,** 220–225.

27. Bronstein, P. M., Levine, M. J., and Marcus, M. (1975). A rat's first bite: The nongenetic, cross-generational transfer of information. *J. Comp. Physiol. Psychol.* **89,** 295–298.

28. Galef, B. G. (1985). Ontogeny of diet selection and the control of ingestion. *Appetite* **9,** 331.

29. Booth, D. A., Stoloff, R., and Nichols, J. (1974). Dietary flavor acceptance in infant rats established by association with the effects of nutrient composition. *Physiol. Psychol.* **2** (3A), 313–319.

30. Mursten, B., Meach, M., and Anderson, G. H. (1975). Protein intake regulation in the weanling rat: Self-selection of proteins and energy. *J. Nutri.* **104,** 573–582.

31. Leprohon, C. E., and Anderson, G. H. (1980). Maternal diet affects feeding behavior of self-selecting weanling rats. *Physiol. Behav.* **24**, 553–559.
32. Symparian, G., and MacLendon, S. (1945). Further record of the self demand schedule in infant feeding. *J. Pediatr.* **27**, 109–114.
33. Trainham, G., Symparian, G. J., and Kzait, R. M. (1945). A case history of twins fed a self-demand regimen. *J. Pediatr.* **27**, 97–109.
34. Desor, J., Maller, O., and Turner, R. (1973). Taste in acceptance of sugars by human infants. *J. Comp. Physiol. Psychol.* **84**, 496–501.
35. Desor, J. A., Greene, L. S., and Maller, O. (1975). Preferences for sweet and salty in 9- to 15-year-old and adult humans. *Science* **190**, 686–687.
36. Desor, J. A., Maller, O., and Andrews, K. (1975). Ingestive responses of human newborns to salty, sour, and bitter stimuli. *J. Comp. Physiol. Psychol.* **89**, 966–970.
37. Nowlis, G. H., and Kessen, W. (1976). Human newborns differentiate differing concentrations of sucrose and glucose. *Science* **191**, 865–866.
38. Engen, T. (1965). Psychological analysis of the odor intensity of homologous alcohols. *J. Exp. Psychol.* **70**, 611–616.
39. Steiner, J. E. (1973). The human gustofacial response. J. F. Bosma (ed.), In *Oral sensation and perception.* (pp. 254–278). Bethesda: U.S. Department of Health, Education, and Welfare.
40. Beauchamp, G. K., and Cowart, B. J. (1985). Congenital and experiencial factors in the development of human flavor preferences. *Appetite* **6**, 357–372.
41. Beauchamp, G. K., and Moran, M. (1984). Acceptance of sweet and salty tastes in 2-year-old children. *Appetite* **5**, 291–305.
42. Desor, J. A., and Beauchamp, G. K. (1987). Longitudinal changes in sweet preferences in humans. *Physiol. Behav.* **39**, 639–641.
43. Forster. (1950). The Development of olfactory preferences. Perfumer and Essential Oil Record, **41**, 244–247.
44. Laing, D. G., and Clark, P. J. (1983). Puberty and olfactory preferences of males. *Physiol. Behav.* **30**, 591–598.
45. Rozin, P., and Millman, L. (1987). Family environment, not heredity accounts for family resemblance in food preferences and attitudes; a twin study. *Appetite* **8**, 125–134.
46. Greene, L. S., Desor, J. A., and Maller, O. (1975). Heredity and experience: Their relative importance in the development of taste preference in man. *J. Comp. Physiol. Psychol.* **89**, 279–284.
47. Pliner, P., and Levin Pelchat, M. (1986). Similarities in food preferences between children and their siblings and parents. *Appetite* **7**, 333–342.
48. Fallon, A. E., Rozin, P., and Pliner, P. (1984). The child's conception of foods: The development of food rejections with special reference to disgust, and contamination sensitivity. *Child Dev.* **55**, 566–575.
49. Birch, L. L., McPhee, L., Shoba, B. C., Pirok, E., and Steinberg, L. (1987). What kind of exposure reduces children's food neophobia? *Appetite* **9**, 171–178.

50. Pliner, P. (1982). The effects of mere exposure on liking for edible substances. *Appetite* **3**, 282–290.
51. Birch, L. L. (1980). Effects of peer models' food choices and eating behaviors on preschoolers' food preferences. *Child Dev.* **51**, 489–496.
52. Birch, L. L., McPhee, L., Sullivan, S., and Johnson, S. (1989). Conditioned meal initiation in young children. *Appetite* **13**, 105–113.
53. Birch, L. L., and Deysher, M. (1985). Conditioned and unconditioned caloric compensation: Evidence for self-regulation of food intake in young children. *Learn. Motiv.* **16**, 341–355.

Subject Index